with best wishes

Colin Jones.

Aug '57.

Earth reinforcement and soil structures

Colin J. F. P. Jones, BSc, MSc, PhD, CEng, FICE

Professor of Geotechnical Engineering
University of Newcastle upon Tyne, UK

 Thomas Telford

Published by Thomas Telford Publishing, Thomas Telford Services Ltd, 1 Heron Quay, London E14 4JD, and co-published in the United States of America with the ASCE Press, American Society of Civil Engineers, 345 East 47th Street, New York, New York 10017–2398.

First published 1985 by Butterworth & Co (Publishers Ltd)
Revised reprint 1988
Thomas Telford edition published 1996

Distributors for Thomas Telford books are
USA: American Society of Civil Engineers, Publications Sales Department, 345 East 47th Street, New York, NY 10017–2398
Japan: Maruzen Co. Ltd, Book Department, 3–10 Nihonbashi 2-chome, Chuo-ku, Tokyo 103
Australia: D.A. Books and Journals, 648 Whitehorse Road, Mitcham 3132, Victoria

A catalogue record for this book is available from the British Library

Classification
Availability: Unrestricted
Content: Original analysis
Status: Author's invited opinion
User: Engineers, academics and students

UK ISBN: 0 7277 2525 4
USA ISBN: 0 7844 0194 2

Typeset in Great Britain by Alden, Oxford, Didcot and Northampton.
Printed and bound in Great Britain by Redwood Books, Trowbridge, Wiltshire.

Preface

J. A. Gaffney, CBE, DSc, FEng
Past President of the Institution of Civil Engineers

Few subjects in recent years have raised the general interest and imagination of the Civil Engineering profession as the concept of reinforcing soil. The basic simplicity of the principles and the economic benefits which may be gained are very attractive to the designer constrained by economic problems, while the possibility of producing alternative and innovative structural concepts gives scope to the engineer's imagination.

In some areas, developments in the use of earth reinforcement and soil structures has been dramatic; elsewhere, use has been modest; nowhere has the subject been ignored. Although academic treatment has been intense, the driving force and major developments have come from the original practitioners, government research bodies and the material suppliers, who have skilfully developed and marketed the benefits of earth reinforcement.

Acceptance and general application and use of any technique require comprehensive specifications, workable technical standards and reference examples. In the field of earth reinforcement, these technical specifications have been provided and the subject is recognized as an important and rapidly expanding field. The rate of growth has resulted inevitably in a demand for information on the subject. This textbook is aimed at bridging this information vacuum.

The book provides a general treatment of the subject of reinforced soil; it is not exhaustive and is aimed at the practising engineer and the post graduate student. Although the book covers the theoretical elements in some depth, the main emphasis is with the practical aspects of the subject in that the subjects of analysis, economics, construction details, materials and durability are considered in greater depth than is usual with textbooks.

Acknowledgements

This book includes material which was published originally by the Department of Transport and is reproduced with the permission of The Controller of Her Majesty's Stationery Office. The author wishes to make it clear that he takes sole responsibility for any use he has made of, and opinions expressed with respect to, this material.

The author wishes to acknowledge the assistance and help received from the members of the former West Yorkshire Metropolitan County Council and colleagues in the University of Newcastle in preparing this textbook.

Particular thanks are due to Mrs Christine Earle Storey and Mrs Ann Bridges for preparing the text.

The author and publishers would like to thank the undermentioned firms and organisations for giving their permission for the use of material and illustrations reproduced in this book.

American Society of Civil Engineers
Association Amicale des Ingénieurs Anciens Elèves de l'Ecole Nationale des Ponts et Chaussées
R. D. Bassett, University College, University of London
R. Bonaparte, GeoSyntec, USA
E. R. L. Cole, Lancashire County Council, County Surveyor's Department
M. R. Dyer, Geotechnical Consulting Group
T. W. Finley, Senior Lecturer in Civil Engineering, Glasgow University
Professor M. Fukuoka, Science University of Tokyo
Professor Dr-Ing G. Gudehus
The Controller, Her Majesty's Stationery Office
T. S. Ingold, Consultant
R. A. Jewell, GeoSyntec, Belgium
Dr R. A. King, Deputy Manager, CAPCIS/UMIST
Professor i.r. P. C. Kreijger
C. Lawson, Ten Cate
B. V. Lee, BSC Sheffield Laboratories
P. Mallinder, R. & D. Laboratories, Pilkington Bros. PLC
Professor A. McGown, Strathclyde University
Netlon Limited
Okasan Kogyo Company Limited
Professor F. Schlosser, Ecole Nationale des Ponts et Chaussées
Thorburn Associates
TRL, Crowthorne, Berks

Notation

a		factored bearing capacity under toe of structure
a		ratio parameter
A_{des}		spectral acceleration
a_m	m/s^2	maximum horizontal acceleration
a_r	m^2	cross-sectional area of any unit of reinforcement
A_R	m^2	total surface area of a reinforcement element
A_s	m^2	area of a critical plane acted on by a single element of reinforcement
a_{si}	m^2/m	structural cross-sectional area parallel to the face of the structure of the ith layer of reinforcing elements, per metre 'run' of structure. In the case of grid reinforcement, the structural cross-sectional area of the longitudinal elements, per metre 'run' of the structure
b	m	width of the loading strip contact area at right angles to the facing
		width of footing or foundation, or span of void
B	m	width of an element of reinforcement, or anchor
B^1	m	half effective width of embankment
C		conceptual shear zone
c	kN/m^2	cohesion of soil
c'	kN/m^2	cohesion of the fill in the structure under effective stress conditions
c'_r	kN/m^2	adhesion between the fill and the reinforcing elements under effective stress conditions
CR		reduction in ultimate strength of reinforcement for creep rupture
c_u		undrained shear strength
C_u		coefficient of uniformity
D	m	design diameter of a void
d	m	diameter of longitudinal reinforcing element in a grid or anchor
		perpendicular distance between the centre line of the strip load contact area and the rear of the structure
D	m	depth to base of foundation
D_F		dynamic force
D_m	m	embedment depth of reinforced soil walls and abutments
D_p		dynamic earth pressure
d_s/D_s		maximum allowable differential deformation occurring at the surface of an embankment or pavement
e	m	eccentricity of an applied force
		eccentricity of vertical strip load with respect to the centre line of the contact area of the load on top of a structure

Symbol	Units	Definition
E	kN/m^2	elastic modulus
E_p	mV	potential of platinum electrode
E_r	mV	redox potential
E_r	kN/m^2	elastic modulus of soil reinforcement
F_A	kN	allowable pullout resistance
F_F	kN	friction resistance of a grid reinforcement
F_i	kN/m	horizontal shear applied to the strip contact area of width b on top of the structure per metre 'run'
F_p	kN	pullout resistance of transverse element of a grid
F_R	kN	anchor resistance of a grid reinforcement
FS		factor of safety
F_t	kN	pullout resistance of an anchor
F_T	kN	total pullout resistance of a grid reinforcement
g	m/s^2	acceleration due to gravity
g		gravitational constant
G_s		specific gravity of soils
h	m	height of wedge of reinforced soil depth of nominal section or element of soil in a structure
H	m	overall height of reinforced structure or fill
H_c	m	critical height of structure (energy method)
h_{bi}		height of section under consideration above the base
h_i	m	height of the reinforced soil above the ith layer of reinforcement elements
h_0	m	critical depth
h_t	m	height above toe of a structure
I		empirical stiffness factor
I'		design stiffness factor
I_c	A	current flow between electrodes
I_D		relative density
I_R		empirical relative density index
I_T	m^3	second moment of area of a traverse reinforcing element
K_a		coefficient
K_c		critical horizontal acceleration
K_{des}		coefficient of lateral earth pressure used in design
K_m		coefficient of seismic design
K_0		coefficient of earth pressure at rest
K_p		coefficient of passive earth pressure
l		strip length
L	m	length of element of reinforcement or embedded length length at right angles in plan to the face of the structure of the bottom layer of reinforcing elements
L_A	m	length of anchor element
L_b	m	reinforced bond length
L_e	m	effective length of reinforcement
L_i	m	length at right angles to the face of the structure of the ith layer of reinforcing elements
L_{ip}	m	length of that part of the ith layer of reinforcing elements beyond the potential failure plane

LL		liquid limit
L_m	m	base length of a soil nailing structure
L_n	m	nail length
L_r	m	reinforcement length
L_S	m	elastic extension of reinforcement
L_{SP}	m	length of slip plane
M	kg	effective mass of a reinforced soil structure
M_{eff}		effectiveness of a soil structure
M_i	kg	mass of the active zone of the reinforced fill at a depth of h_i below the top of a reinforced soil structure
M_i	kNm/m	bending moment about the centre of the plan section of the structure at the ith layer of reinforcing elements, arising from the external loading acting on the structure per metre 'run'
M_0	kNm/m	overturning moment per metre 'run' of structure
MS		percentage of mobilized shear strength of soil
n		number of effective layers of reinforcing elements
N		number of the first layer of reinforcement to cross the theoretical failure line
		number of reinforcements per area considered
N_γ		Terzaghi bearing capacity coefficient for strip footing
N_c		Terzaghi bearing capacity coefficients for strip footing
N_q		Terzaghi bearing capacity coefficient for strip footing
N_w		number of transverse members
p		dimensionless bond area parameter
P	kN/m	backfill thrust on reinforced soil block per metre 'run'
P	kN/m	horizontal factored disturbing force per metre 'run'
P_a	ohm-cm	measure of resistivity of soil
P_A	kN	resistance of anchor to pull-out
P_{AE}	kN/m	dynamic horizontal thrust on reinforced soil block resulting from a seismic event
P_{at}	kN/m^2	basic permissible axial tensile stress in reinforcing elements
P_B	kN	bearing face on front of anchor
P_f	kN	pullout resistance generated by friction on top of bottom of granular fill within a triangular anchor
pH		value of acidity of an aqueous solution
P_{HS}	kN/m	total horizontal thrust on a reinforced soil block during a seismic event
PI		plasticity index
P_i	m/m	total horizontal width of top and bottom faces of the ith layer of reinforcements per metre 'run' of structure. In case of grid reinforcement the width of the ith layer of grid per metre 'run' of structure
P_{Ii}	kN/m	horizontal inertial face at the ith layer of reinforcing elements
P_{IR}	kN/m	horizontal uncritical face resulting from a seismic event
P_L		horizontal propping force

P_q	kN/m	resultant of active earth pressure on reinforced soil block per metre run due to a uniform surcharge
P_R	kN	reinforcement force
P_s	kN	pullout resistance generated on anchor shaft in cohesionless fill
P_u	kN/m	ultimate pullout resistance per metre 'run'
q	kN/m^2	average contact pressure of footing on reinforced subsoil
q_0	kN/m^2	average contact pressure of footing on unreinforced subsoil
q_r	kN/m^2	factored bearing pressure acting at the base of a structure
q_r		bearing capacity ratio
q_{ult}	kN/m^2	ultimate bearing capacity
R	ohm	resistance
R_A	ohm	anode resistance
R_c	ohm	cathode resistance
R_F	kN/m	resistance to sliding per metre 'run' of wall
R_0	kN m/m	resistance to overturning of structure per metre 'run'
R_v	kN	resultant of all factored vertical loads
S		shear strength of soil
S_{av}		average shear strength in an embankment
S_{a1}		first spectral acceleration
S_{a2}		second spectral acceleration
S_E	m	effective spacing of reinforcement or soil nails
S_h	m	horizontal spacing of reinforcement or soil nails
S_i	kN/m	vertical loading, applied to a strip contact area of width b on top of a structure, per metre 'run'
S_{rm}	kN	shear resistance of geocell or geogrid mattress
S_v	m	vertical spacing of reinforcement or soil nails
t	m	thickness of an element of strip reinforcement, or thickness of steel facing or anchor plate
T	kN/m	total tensile force to be resisted by the layers of reinforcement which anchor a wedge of reinforced soil, per metre 'run' (wedge analysis) period of oscillation
T'	kN/m^2	tensile stress in any unit of reinforcement
T_1		first fundamental period of the reinforced soil structure
T_2		second fundamental period of the reinforced soil structure
T_{ad}	kN	tensile adhesion force of reinforcement
T_{ci}	kN/m	reduction in tensile force due to cohesive fill
T_{CR}	kN/m	peak tensile creep rupture strength at the appropriate temperature
T_D	kN/m	design strength of reinforcement
T_{DS}	kN/m	design strength of the reinforcement in seismic conditions
T_f	kN/m	frictional resistance per unit length
T_F	kN/m	tensile force developed from the horizontal shear applied to the top of the structure to be resisted by the reinforcement anchoring the wedge of reinforced soil (wedge analysis)

T_{fi}	kN/m	tensile force developed from the horizontal shear applied to the top of the structure to be resisted by the ith layer of reinforcing elements, per metre 'run'
T_{hi}	kN/m	tensile force developed from the length of the reinforced soil above the ith layer of elements, per metre 'run'
T_i	kN/m	total maximum tensile force resisted by the ith layer of reinforcement, per metre 'run'
T_{is}	kN/m	maximum ultimate limit state tensile face resisted by the ith layer of reinforcing elements during a seismic event
T_{max}	kN/m	maximum tensile force in the bottom layer of reinforcing elements or elements under consideration
T_{mi}	kN/m	tensile force developed from the bending moment (M_i) caused by external loading, per metre 'run'
T_{pi}	kN/m	tensile force due to self-weight, surcharge and bending moment resulting from an external load
T_{rs}	kN/m	tensile load in a tension membrane, per metre run
T_s	kN/m	total resistance of the reinforcing elements anchoring a potential sliding wedge during a seismic event
T_{si}	kN/m	tensile force developed from the external loading (S_i) on top of the structure, per metre 'run'
T_u	kN/m	ultimate tensile strength of reinforcement
T_{wi}	kN/m	tensile force developed from the uniformly distributed surcharge (w_s) on top of the structure per metre 'run'
u	m	depth of reinforcement beneath footing
U_{ext}		external works done by earth pressure (energy method)
v		volumetric strain
V	m	vertical distance between two successive layers of reinforcement
w	kN/m	loading on subsoil due to weight of structure per metre 'run'
W	kN/m	vertical factored resistance force
W	kN/m	total weight of soil structure per metre 'run'
W_1	kN/m	weight of soil contained within Coulomb failure wedge
w_s	kN/m^2	uniformly distributed surcharge on top of a structure
x		variable
$\bar{x}$		shear displacement
X_m		seismic displacement
y		variable
$\bar{y}$		vertical displacement
Z	m	effective depth of residual lateral pressure; depth of the ith layer of reinforcement beneath a footing
α		maximum ground acceleration coefficient
α'		interaction coefficient relating soil/reinforcement bond angle with $\tan \phi'_{des}$
α'		coefficient expressing μ as a proportion of $\tan \phi'$
α'_1		coefficient relating the soil/reinforcement bond angle to $\tan \phi'_{cv1}$ on one side of the reinforcement
α'_2		coefficient relating the soil/reinforcement bond angle to $\tan \phi'_{cv2}$ on the other side of the reinforcement

α'_{sr}		adhesion coefficient relating soil cohesion to soil/reinforcement bond
α''		a function
α_m		maximum structure accleration coefficient at the centre of a reinforced soil mass
α, β		zero extension directions
β'	degree	inclination of a potential failure plane to the vertical plane
β''		a function
γ		shear strain
γ	kN/m^3	unit weight of the fill in a structure
$\bar{\gamma}$		rate of shear strain
γ_{bc}		partial factor in respect of foundation bearing capacity
γ_d		optimum dry density
γ_{ff}		partial factor for loads
γ_{es}		partial load factor for soil unit weight
γ_m		partial material factor
γ_{ml}		partial material factor related to the intrinsic properties of the material
γ_{ml1}		partial material factor related to the consistency of manufacture of the reinforcement and how strength may be affected by this and possible inaccuracy in assessment
γ_{ml2}		partial material factor related to the extrapolation of test data dealing with base strength
γ_{m2}		partial material factor concerned with the construction and environmental effects
γ_{m21}		partial material factor related to the susceptibility of the reinforcement to drainage during installation in the soil
γ_{m22}		partial material factor related to the environment in which the reinforcement is installed
γ_{ms}		partial load factor for soil materials
γ_n		partial factor related to the economic ramifications of failure
γ_p		partial factor for soil/reinforcement interaction (pullout)
γ_q		partial load factor for external live loads
γ_s		partial factor for soil/reinforcement interaction (sliding)
γ_{ss}		partial factor in respect of horizontal sliding on a soil/soil interface
γ_w	kN/m^3	unit weight of water
δ_θ		strain in soil in direction θ
δ_h		lateral strain of soil under an applied load
δ_r		strain in reinforcement
δ_v		axial compression under an applied load
δ_x		strain of reinforcement of unit length $d(x)$
ΔH	m	zone of action of an individual layer of reinforcement
Δv	volts	measured potential difference
ϵ		initial strain in reinforcement
ϵ		linear strain
ϵ_H		horizontal strain

$\dot{\epsilon}_\theta$		normal strain rate
ϵ_1		major principal strain
$\dot{\epsilon}_1$		major principal strain rate
ϵ_3		minor principal strain
$\dot{\epsilon}_3$		minor principal strain rate
ϵ_i		initial strain in geosynthetic reinforcement
ϵ_{120}		strain in geosynthetic reinforcement after 120 years
ϵ_{max}		maximum allowable strain in reinforcement
ϵ_r		lateral strain of soil in the direction of the reinforcement
η		position of incidence of Coulomb wedge with facing (tie-back analysis)
θ_d	degrees	angle of draw of a fill material, approximately equal to the peak friction angle (ϕ')
θ_{max}		rotation of the reinforced soil structure
λ		coefficient dependent on whether the reinforcement support is to function as a one way or two day load shedding system
μ		coefficient of friction between the fill and reinforcing elements
μ^*		apparent coefficient of friction
μ_f		coefficient of friction between the retained soil and the subsoil
μ_{ws}		coefficient of friction between uniformly distributed surcharge and top of structure
ν		volumetric strain
ν		volumetric strain rate
ν		Poisson's ratio
σ	kN/m^2	normal stress
σ'_ν	kN/m^2	applied vertical load
σ_ν	kN/m^2	vertical stress on an element of soil
σ_h	kN/m^2	lateral stress on an element of soil
σ_N	kN/m^2	stress normal to the reinforcement
σ_1		major principal effective stress
σ_3		minor principal effective stress
τ	kN/m^2	shear stress
ϕ	degrees	Mohr–Coulomb angle of friction
ϕ'	degrees	angle of internal friction of soil under effective stress conditions
ϕ'_{cv}	degrees	minimum angle of internal friction developed at large strains
ϕ'_p	degrees	peak angle of friction under effective stress conditions
ϕ_{des}	degrees	angle of internal friction used in design
ϕ_{peak}	degrees	peak angle of shear resistance under effective stress conditions
ψ	degrees	angle of dilation of soil
ψ_{max}	degrees	maximum angle of dilation

Contents

1

Introduction

The basic principles involved in reinforced soil are simple to grasp and have been used by man for centuries. The basic attributes of soil reinforcement which are of particular advantage in civil engineering are reductions in costs and ease of construction, coupled with a basic simplicity which provides an attraction to engineers. Recognition of, and interest in the subject have gained impetus because of the technical and commercial success that has been demonstrated by the practitioners. The concept of reinforcing soil has also attracted the attention of the academic world, for although the concept is easily grasped the theoretical aspects involved are numerous. As a result, much research and development work has been undertaken in universities and laboratories, and soil reinforcing is now recognized as a separate subject in its own right in the geotechnical field.

Reinforced soils are fundamentally different from conventional earth retaining systems in that they utilize a different mechanism of support. The classification scheme for earth retention systems in Table 1.1 provides a summary of current earth retention methods organized according to the two principal categories of *externally* or *internally* stabilized systems. An *externally* stabilized system uses an external structural wall, against which stabilizing forces are mobilized. An *internally* stabilized system involves reinforcements installed within and extending beyond the potential failure mass. Within this system, shear transfer to mobilize the tensile capacity of closely spaced reinforcing elements has removed the need for a structural wall and has substituted a composite system of reinforcing elements and soil as the primary structural entity. A facing is required on an internally stabilized system, but its role is to prevent local ravelling and deterioration rather than to provide primary structural support.

Nearly all traditional retaining walls may be regarded as externally stabilized systems, Fig. 1.1. Internally stabilized systems are identified by reinforced soils with predominantly horizontally layered elements, such as metallic strips or polymeric grids, and soil nailing, in which metallic bars or dowels are installed during in-situ construction. The key aspect of an internally stabilized system is its incremental form of construction. In effect, the soil mass is partitioned so that each partition receives support from a locally inserted reinforcing element. This process is the opposite to what occurs in a conventional backfilled wall where pressures are integrated to produce an overall force resisted by the

1

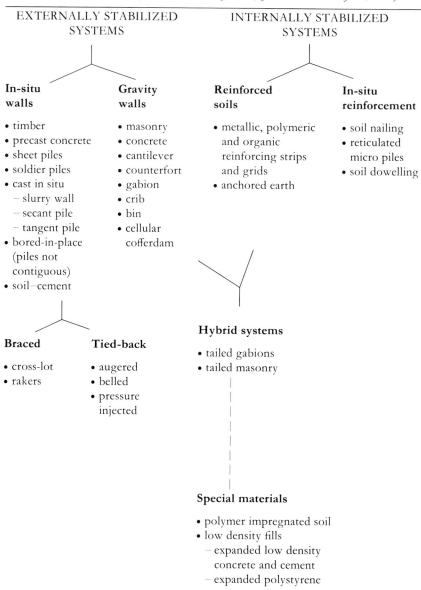

Table 1.1. Classification scheme for earth retention systems (after O'Rourke and Jones, 1990)

EXTERNALLY STABILIZED SYSTEMS

INTERNALLY STABILIZED SYSTEMS

In-situ walls

- timber
- precast concrete
- sheet piles
- soldier piles
- cast in situ
  - slurry wall
  - secant pile
  - tangent pile
- bored-in-place (piles not contiguous)
- soil–cement

Gravity walls

- masonry
- concrete
- cantilever
- counterfort
- gabion
- crib
- bin
- cellular cofferdam

Reinforced soils

- metallic, polymeric and organic reinforcing strips and grids
- anchored earth

In-situ reinforcement

- soil nailing
- reticulated micro piles
- soil dowelling

Braced

- cross-lot
- rakers

Tied-back

- augered
- belled
- pressure injected

Hybrid systems

- tailed gabions
- tailed masonry

Special materials

- polymer impregnated soil
- low density fills
  - expanded low density concrete and cement
  - expanded polystyrene

structure. The overall earth pressure in reinforced soil, for example, is actually differentiated by the multiple layers of reinforcement. In soil nailing, multiple levels of reinforcements interconnect the soil mass so that each potential failure surface is crossed by sufficient reinforcing elements to maintain stability. Hybrid structures combine elements of both internally and externally support soil, they include tailed gabions as in Fig. 1.2 and the improvement of gravity structures as in Fig. 1.3.

It is argued that the rate of development of the modern concept of earth reinforcement could have been greater. However, the critical elements in

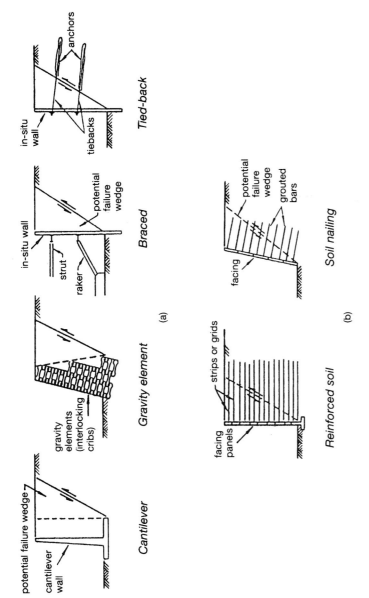

Fig. 1.1. Examples of: (a) externally stabilized and (b) internally stabilized soil retention system

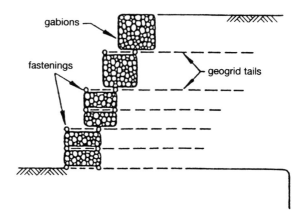

Fig. 1.2. Tailed gabion with geogrids

construction are not necessarily advanced theoretical concepts, but the development of design standards, specifications and technique, without which economical efficient structures cannot be produced. The pioneers of the modern earth reinforcing systems recognized this and have been consistent in establishing reliability and quality control methods. Recognized specifications and standards are available, and the technique has developed to the point where it has reached general acceptance within the engineering profession and in some countries is now considered to be the conventional form of construction for a wide range of structures and applications.

The objectives of this book are to assist in answering some of the questions raised by engineers working in design and construction, and to provide a wider treatment of the subject for the academic and research worker than is usually given in theoretical studies associated with the state of the art. The arrangement of the text has been chosen so as to guide the reader into the subject in a structured approach, and although each chapter can be referred to separately, the order of presentation is chosen to follow in sequence.

A brief history of soil structures and earth reinforcement is provided in Chapter 2 in order to place the subject in context with the work of the pioneers

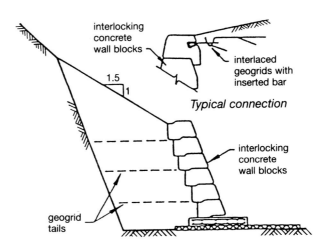

Fig. 1.3. Concrete block wall with geogrid tails

and past generations and civilizations. Examples of modern applications are provided in Chapter 3, which illustrates the breadth of use. These examples are not exhaustive and the implication in presenting this brief catalogue is that the use of reinforced soil and soil nailing is expected to develop in many other applications, particularly as hybrid systems are developed.

Theoretical developments associated with reinforced soil have been extensive in recent years. However, a complete understanding of every aspect and use of the subject has not yet been developed, although an understanding of the fundamental principles was established by the early developers and users and these have since been confirmed or further expanded. The text attempts to provide a balanced overview of the theoretical concepts involved, based on the needs of the practising engineer.

The interaction between soil and reinforcement is critical to the subject and the material properties which influence the interaction are covered in Chapter 5, together with a consideration of the other material properties used in soil structures.

Design and analysis are considered separately from theory in recognition of the systematic procedures and disciplines which are required in developing an actual structure. The importance of idealization and conceptual design in predetermining the decision to use soil reinforcement is explained before the consideration of individual analytical techniques. As with theory, there have been numerous advances in analytical procedures including the concept of *limit modes* used to identify different ultimate and serviceability conditions. The text covers the fundamentals of the main analytical systems, and gives details of the design procedures which may be used in a range of structural problems. Rigid procedures for design are avoided as, in accordance with any design problem, engineering judgement is needed when soil reinforcing techniques are being used.

Successful production of reinforced soil is as dependent on good constructional techniques as on the correct interpretation of theory. The development of workable and effective constructional systems has been a key to the success of the subject without which reinforced soil would have remained an interesting academic toy. Chapter 7 provides a set of general constructional parameters while Chapter 8 provides information of constructional details which have been used successfully. It should be emphasized that the suitability of these details in other structures where conditions, either contractual, material or financial may be different, must not be assumed. As with all design, engineering judgement is essential.

Probably the most contentious element of the text is Chapter 9 which covers costs and economics. It is recognized that the market and financial conditions are volatile; accordingly, the economics of any proposed structure or use of earth reinforcement must be related to the prevailing circumstances and conditions. Similarly, the acceptance or rejection of the concept of any ecology audit depends on an interpretation of economic philosophy.

Durability of earth reinforcement and soil structures is critical and Chapter 10 describes the mechanism of corrosion and the factors which influence the life of any structure. In accordance with conventional design requirements, the measurement of corrosion factors is covered and the durability of potential reinforcing materials is discussed. Particular emphasis is given to polymeric reinforcement, reflecting the growth in the use of these materials.

The final part of the text provides a set of four worked examples. The first is a straightforward design of a retaining wall which serves to illustrate the difference between the theoretical concepts discussed in Chapter 4 and the design and analysis and construction principles covered in Chapters 6 and 7. The second example considers the design of a bridge abutment and, although it uses the same analytical model as the first example, an abutment produces a design problem dominated by external forces. The third example illustrates an entirely different concept of earth reinforcement in the strengthening of the foundations of an embankment. The fourth example shows how reinforced soil can be used to provide support over areas of weak soil or voids, and the final example covers the additional considerations needed in seismic design.

1.1 Reference

O'ROURKE T. D. and JONES C. J. F. P. (1990). An overview of earth retention systems: 1970–1990. *ASCE Speciality Conf. on Earth Retaining Structures*. Cornell, June, 22–51.

2

History

2.1 Ancient structures

The concept of earth reinforcement is not new, the basic principles are demonstrated abundantly in nature by animals and birds and the action of tree roots. The fundamentals of the technique are described in the Bible (Exodus 5, v. 6–9), covering the reinforcement of clay or bricks with reeds or straw for the construction of dwellings. Constructions using these techniques are known to have existed in the 5th and 4th millennia BC.

The earliest remaining examples of soil reinforcement are the ziggurat of the ancient city of Dur-Kurigatzu, now known as Agar-Quf, and the Great Wall of China. The Agar-Quf ziggurat, which stands five kilometres north of Baghdad was constructed of clay bricks varying in thickness between 130–400 mm, reinforced with woven mats of reed laid horizontally on a layer of sand and gravel at vertical spacings varying between 0·5 and 2·0 m. Reeds were also used to form plaited ropes approximately 100 mm in diameter which pass through the structure and act as reinforcement (Bagir, 1944). The Agar-Quf structure is now 45 m tall, originally it is believed to have been over 80 m high; it is thought to be over 3000 years old. Other ziggurats are known to have been built, among them being the structure at Ur which was completed circa 2025 BC and the Sanctuary of Marduk at Babylon, sometimes known as the Tower of Babel, which was completed circa 550 BC (Copplestone, 1963). The Great Wall of China, parts of which were completed circa 200 BC, contains examples of reinforced soil, in this case use was made of mixtures of clay and gravel reinforced with tamarisk branches (Dept. of Transport, 1977).

The Romans also are known to have used earth reinforcing techniques, and reed-reinforced earth levees were constructed along the Tiber. A recent discovery in London of a first-century Roman Army project of a wharf for the Port of Londinium, has shown that past construction techniques are markedly similar to present day methods. The timber wharf, parts of which have been preserved in the Thames mud for 1900 years, is believed to have been 1·5 km in length. The 2 m high structure was formed from oak baulks measuring up to 9 m in length, having a vertical face held in place by timber reinforcing elements embedded in the backfill, Fig. 2.1 (Bassett, 1981).

In parallel with the Romans, the Gauls also made use of an earth reinforcement technique in the construction of fortifications, the technique being to form alternate layers of logs and earth fill (Duncan, 1855).

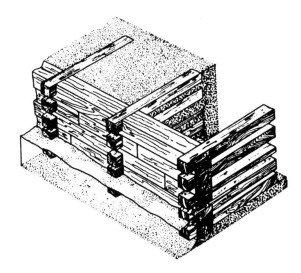

Fig. 2.1. Roman wharf

Reinforcing techniques for military earthworks appeared common up to the last century, although there is little reference in published texts. A notable contribution was made in 1822 when Col. Pasley introduced a form of reinforced soil for military construction in the British Army (Pasley, 1822). He conducted a comprehensive series of trials and showed that a significant reduction could be made in the lateral pressures acting on retaining walls if the backfill was reinforced by horizontal layers of brushwood, wooden planks or canvas; similar observations were made with modern reinforced earth backfills over 150 years later (Saran *et al.*, 1975).

In the past, most use for reinforced soil structures appears to have been in the control of rivers through training works and dykes. Early examples of dyke systems using reed reinforcement and clay fill are known to have existed along the Tigris and Euphrates, well before the adoption of the technique by the Romans. The use of faggoting techniques by the Dutch and the reclamation of the Fens in England are well recorded, as is the construction of the Mississippi levees (Haas and Weller, 1952). The basic technique is illustrated in BS Code of Practice CP No. 2.

The reinforcement of dam structures was introduced at the beginning of the twentieth century by Reed (1904) who advocated the use of railway lines to reinforce rockfill in the downstream face of dams in California. A similar technique, but using grids made up of three-quarter-inch diameter steel bars, was used as late as 1962 in Papua (Fraser, 1962). Other applications of the latter system have been reported in South Africa, Mexico and Australia. Recently, the construction of reinforced earth dams has again been found to be economical.

A significant development to the modern concept of reinforced soil structures was made in the United States in 1925 by Munster (1925). He produced an earth retaining wall using an array of wooden reinforcing members and a light facing. Munster minimized the problem associated with the settling of the backfill by using sliding attachments, between the reinforcing members and the facing. Although the materials and details suggested by

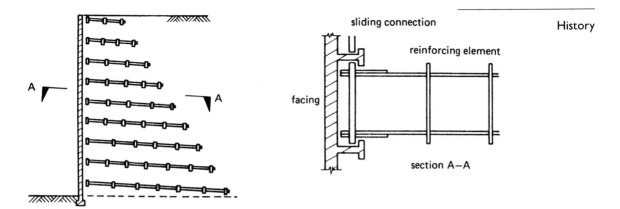

Fig. 2.2. Munster earth retaining structure

Munster would not find favour in modern construction, the techniques inherent in this system are valid and form the core of one of the construction techniques used today, Fig. 2.2.

In the 1930s, French developments came to the fore; first Coyne (1927) introduced the *mur à échelle* (ladder wall), in which the retaining wall consists of a mass of granular filling unified by a row of tie members each having a small end anchor, together with a thin cladding membrane. Settlement of the fill was catered for by the use of flexible tie members, one form of which was a galvanized flat iron strip. Coyne also recognized that the surface cladding needed to be designed for settlement of the fill and advocated the use of flexible gaskets between facing slabs, elsewhere he used a form of overlapping slabs which could move relative to one another, Fig. 2.3. Although Coyne's structures mostly used an anchor block at the end of the tensile reinforcing member, in 1945 he recognized that provided the fill possessed good frictional properties, the ties themselves could provide the necessary bond with the fill without the use of end anchors.

Coyne can be identified with the modern approach to earth reinforcing techniques, not only did he consider the mechanisms but also recognized the problems associated with the technique, such as the need for durability of the reinforcing elements. Recognition of the basic mechanism and what influences performance can be seen in Lallemand's development of the reinforcing

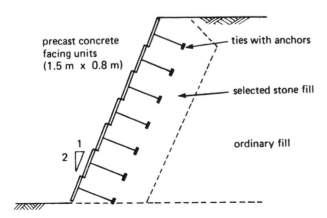

Fig. 2.3. Coyne retaining wall at Brest

9

elements in which a number of rigid claws were arranged along the length of the reinforcement to increase adhesion with the soil (Lallemand, 1959).

2.2 Modern structures

The modern concept of earth reinforcement and soil structures was proposed by Casagrande who idealized the problems in the form of a weak soil reinforced by high-strength membranes laid horizontally in layers (Westergaard, 1938). The modern form of earth reinforcement was introduced by Vidal in the 1960s. Vidal's concept was for a composite material formed from flat reinforcing strips laid horizontally in a frictional soil, Fig. 2.4, the interaction between the soil and the reinforcing members being solely by friction generated by gravity. This material he described as 'Reinforced Earth', a term that has become generic in many countries, being used to describe all forms of earth reinforcement or soil structures. In some countries, including the United States and Canada, the term is a trademark. The first major retaining walls using the Vidal concept were built near Menton in the South of France in 1968, although Vidal had built structures earlier, starting in 1964.

The first structures used a pliant surface cladding made up from horizontally laid U-shaped sheet metal channel members. In 1970 an alternative cladding using a cruciform reinforced concrete member was introduced; concrete-faced structures are now used widely, Fig. 2.5. The first use of Vidal's form of earth reinforcement in the United States was to correct a landslide in California in 1972, while the first reinforced earth structure in the UK was completed in 1973. In the same year another form of construction, the York method, having similarities with the earlier Munster technique, was introduced in the UK, having been developed on behalf of the Department of Transport, Fig. 2.6. The York method has been the subject of continuous development for a period of 15 years and has evolved as a construction philosophy rather than a single technique. Central to the philosophy is that it uses common construction materials wherever possible and can be adapted to use any form of reinforcement or anchor.

The introduction of the Vidal structures led to rapid development. Much fundamental work was sponsored by various national bodies, notably at the Laboratoire des Ponts et Chaussées (LCPC) in France (Schlosser, 1978), by the

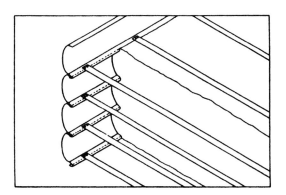

Fig. 2.4. Vidal wall

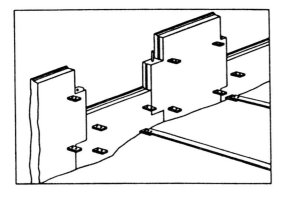

Fig. 2.5. Concrete cruciform faced wall

United States Department of Transportation (Walkinshaw, 1975) and by the United Kingdom Department of Transport (Murray, 1977). This work led to the introduction of improved forms of reinforcement and to a better understanding of the fundamental concepts involved. Fabrics were introduced, although these materials have largely been confined to geotechnical applications other than soil reinforcing. In 1974 the California Department of Transportation introduced the use of mesh or grid as the reinforcing element in retaining walls, which has led to further developments (Forsyth, 1978).

Material development is interrelated with soil structure developments. Whereas the early structures were formed using organic materials such as timber, straw or reed for reinforcement, Pasley recognized the potential of more advanced forms of reinforcement, particularly in his use of canvas as a reinforcing membrane. Canvas could only have been expected to have a limited life before deterioration and Pasley's structures would not have been expected to last for long periods; in the nineteenth century organic reinforcements still remained superior.

It was not until the necessary technical advances had taken place that artificial or engineering materials could be used for reinforcement. Coyne in the first half of the twentieth century was notably conscious of the problems of corrosion, an attitude which is also reflected by Vidal and others. Some structures are not susceptible to corrosion or deterioration of the reinforcement as they have a short life. An example can be found in the mining industry where, as early as 1935, steel wire netting was being used to reinforce roof packs in the Yorkshire coalfield in England (Brass, 1935). The reinforcement

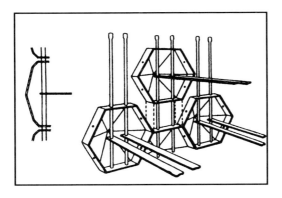

Fig. 2.6. York method (after Jones, 1978)

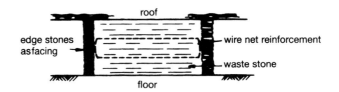

*Fig. 2.7. Wire net reinforced
roof pack in Yorkshire coalfield*

was laid in horizontal layers, dividing the pack into thinner slices, the frictional effect between the wire netting and the waste stone fill being relied upon for stability, Fig. 2.7.

The use of textiles for reinforcement could not be contemplated until the development of synthetic polymer-based materials. Synthetic fabrics were known prior to 1940 but it was not until the late 1960s and early 1970s that the advances in synthetic fabric and geotextile developments led to the construction of reinforced soil structures. Fabric reinforced retaining walls have proved to be economical but are somewhat utilitarian in appearance, and the larger use of geotextile fabrics has proved to be in the areas of separation, filtration and drainage.

Geotextile materials can be divided into two categories: conventional geotextiles and specials. Conventional geotextiles are products of the textile industry and include nonwovens, woven, knitted and stretch bonded textiles. Special geotextiles, usually referred to as geosynthetics, are not usually produced in a textile process. Two major forms have evolved: geogrids and geocomposites. Geogrids have been used in civil engineering since the early 1960s, one of the first major applications being the use of high-density polyethylene grids in the construction of railway embankments in order to reinforce volcanic ash fill, and to enable higher levels of compaction to be attained (Yamamoto, 1966; Watanabe and Iwasaki, 1978). Around the same time, grid reinforcement was used to reclaim land for Nyeta Airport, Tokyo, and to improve the bearing capacity of weak subsoil (Yamanouchi, 1967). Following the examples of the California Highway Authority and the former West Yorkshire Metropolitan County in the UK, high-strength geogrid reinforcement is now used for concrete faced structures.

Geocomposites consist generally of high strength fibres set within a polymer matrix. One of the main uses for these very high strength materials has been as reinforcement of embankment structures over voids or as tension membranes. The development of geosynthetic reinforcements is continuing, a recent innovation being the introduction of 'electro kinetic geosynthetics' with advanced properties combining the functions of drainage, reinforcement and the concept of electro-osmosis.

In the 1980s a special type of reinforcement in the form of an anchor was evolved simultaneously in Europe, Japan and the USA. The multi-anchor system was developed by Fukuoka (1980) for the Japanese Ministry of Construction. The anchor is in the form of a rectangular steel plate, Fig. 2.8. The NEW retaining wall system, developed in Austria, is based on an elevated concrete facing and polymeric ties in the form of a closed loop (Fig. 2.9) (Brandl and Dalmatiner 1986). In the USA and the UK, anchors formed from waste automobile tyres illustrated both the economic and the environmental

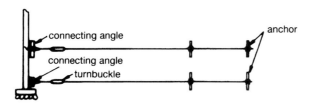

anchor

connecting angle

connecting angle
turnbuckle

*Fig. 2.8. Multi-anchor wall
(after Okasan Kogyo, 1985)*

benefits of reinforced soil. Steel anchors formed from a single piece of rebar were developed by the Transport and Road Research Laboratory in the UK (Murray and Irwin, 1981). The first polymeric anchor was developed in 1992 (Jones and Hassan, 1992).

In 1981, the development of soil structures advanced into a new area of application when synthetic grid materials were employed in the repair of cutting failures on the M1 and M4 motorways in England (Murray *et al.*, 1982). The stabilization of cuttings by earth reinforcing formed in situ, using techniques similar to those employed in ground anchor techniques, had previously been introduced in Germany and the USA. These 'soil-nailing' or 'lateral earth support systems', together with the repair techniques developed on the M4 motorway, epitomize the present stage of earth reinforcing in which the technique is accepted as a conventional design option available for use in the design of geotechnical structures.

Soil reinforcement acting as tension membranes supporting roads, buildings, embankments over voids or acting as construction aids in cases of extremely soft soil (super soft soil) were introduced in the 1980s. Yano *et al.* (1982) describe the problems associated with coastal areas including the bays of Tokyo and Osaka, where soft marine clay has been deposited over wide areas. This material has little or no bearing capacity but can be in a potentially prime location. Soil reinforcement in the form of grids is used to form a primary construction stage providing support for conventional ground improvement techniques.

The use of tensile reinforcing elements to support structures over natural or man-made voids has evolved to the point where the technique is described in the new British Standard on Reinforced Soil (BS 1995).

A multitude of hybrid systems and techniques are now available, one of the

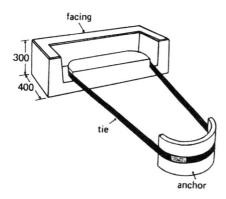

facing

300

400

tie

anchor

*Fig. 2.9. NEW retaining wall
system*

13

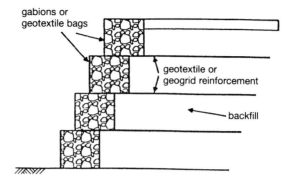

Fig. 2.10. Tailed gabion

most successful of which has proved to be the tailed gabion introduced by Jones and Templeman (1979), thus the stability of conventional gabion structures can be enhanced by the addition of reinforcement, Fig. 2.10. New systems and developments continue to evolve, and even the advantages of pre-stressed reinforced soil have been demonstrated (Barvashov *et al.*, 1979).

2.3 References

BAGIR T. (1944). *Iraq Journal*, pp. 5–6, British Museum.

BARVASHOV V. A., BUDANOV V. G., FOMIN A. N., PERKOV U. R. and PUSHKIN V. I. (1977). Deformations of soil foundation reinforced with prestressed synthetic fabrics. *C.R. Coll. Int. Sols Textiles*, Paris.

BASSETT N. (1981). Prefabrication Roman style. *New Civil Engineer*, August.

BEATON J. L., FORSYTH R. A. and CHANG J. C. (1974). *Design and Field Behaviour of the Reinforced Earth Embankment–Road 39*, Report of California Dept. of Transportation, Division of Highways.

BRASS T. F. S. (1935). Practical experiments in roof control. *Trans. Midlands Inst. Mining Engnrs*, 1934–35.

BRANDL H. and DALMATINER J. (1986). *Struzmaners System 'NEW' und andere konstruktionen nack dem Baden-Aukervabund-Prinzip.* Fed. Min. of Construction and Engineering Road Research, Vienna, **280**.

BRITISH STANDARDS INSTITUTION (1995). *Code of practice for strengthened/reinforced soils and other fills.* BSI, London, 156.

CHANG J. C. (1974). *Earthworks Reinforcement Techniques.* California Dept. of Transportation, Transport Lab. Res. Rept., October.

CHANG J. C., DURR D. L. and FORSYTH R. A. (1974). *Earthworks Reinforcement Techniques.* California Dept. of Transportation, Transport Lab. Res. Rept., February.

CHANG J. C., FORSYTH R. A. and BEATON J. L. (1974). *Performance of a Reinforced Earth Fill*, Transport Research Record No. 510.

CHANG J. C., FORSYTH R. A. and SMITH, T. (1972). Reinforced earth highway embankment—Road 39. *Highways Focus*, **4**, no. 1.

COPPLESTONE T. (1963). *World Architecture.* Hamlyn, Feltham.

COYNE M. A. (1927). Murs de soutenement et murs de quai à échelle. *Le Génie Civil*, Tome XCI, October.

COYNE M. A. (1929). French Patent Specification No. 656692.

COYNE M. A. (1945). Murs de soutenement et murs de quai à échelle. *Le Genie Civil*, May.

DEPARTMENT OF TRANSPORT (1977). *Reinforced Earth Retaining Walls for Embankment including Abutments*. Tech. Mem. BE (Interim), Dept. of Environment, Highways Directorate.

DUNCAN W. (1855). *Caesar*, Harper Brothers, New York.

ENGINEERING CODES OF PRACTICE, JOINT COMMITTEE (1951). *Revetments*. Civil Engineering Code of Practice No. 2 (1951). Earth Retaining Structures, The Institution of Civil Engineers.

FORSYTH R. A. (1978). Alternative earth reinforcements. *ASCE Symp. Earth Reinforcement*, Pittsburgh.

FRASER J. B. (1962). A steel-faced rockfill dam for Papua. *Trans. Inst. of Engrs. Australia*, **CE4.**

FUKUOKA M. (1980). Static and dynamic earth pressure on retaining walls. *Proc. 3rd Australian–New Zealand Conf. on Geomechanics*, Wellington, **3**, 3–37, 3–46.

HAAS R. H. and WELLER H. E. (1952). Bank stabilization by revetments and dykes. *Trans. ASCE*, col. 118, No. 2564.

JONES C. J. F. P. and TEMPLEMAN J. (1979). *Soil structures using high tensile plastic grids (tailed gabions)*. UK Patent No. 794/627.

JONES C. J. F. P. (1978). The York method of Reinforced Earth construction. *ASCE Symposium on Earth Reinforcement*, Pittsburgh.

JONES C. J. F. P. and HASSAN C. A. (1992). Reinforced soil formed using synthetic polymeric anchors. *Geotropika '92, Inst. Teknologia, Malaysia*, 1–8.

LALLEMAND M. F. (1959). French Patent Specification No. 1173383.

MUNSTER A. (1925). United States Patent Specification No. 1762343.

MURRAY R. T. (1977). Research at TRRL to develop design criteria for Reinforced Earth. *Symp. Reinforced earth and other composite soil techniques*. Heriot–Watt University, TRRL Sup. 457.

MURRAY R. T. and IRWIN M. J. (1981) A preliminary study of TRRL anchored earth. TRRL Sup. 674.

MURRAY R. T., WRIGHTMAN J. and BURT A. (1982). *Use of fabric reinforcement for reinstating in situ slopes*. TRRL Sup. 75.

OKASAN KOGYO COMPANY (1985). Private communication, Tokyo.

PASLEY C. W. (1822). *Experiments on Revetments Vol. 2*, Murray, London.

REED F. H. (1904). United States Patent No. 77699.

SARAN S., TALEVAR D. V. and PRAKESH S. (1979). Earth pressure distribution on Retaining Wall with Reinforced Earth Backfill. *Int. Conf. Soil Reinforcement*, Paris, **1**.

SCHLOSSER F. (1978). Experience on Reinforced Earth in France. *Symp. Reinforced earth and other composite soil techniques*, Heriot–Watt University, TRRL Sup. 457.

SCHLOSSER F. and VIDAL H. (1969). La terre armée. *Bull. de Liaison Lab Rout. Ponts et Chaussées*, no. 41, November.

VIDAL H. (1966). La terre armée. *Annales de L'Institut Technique du Bâtiment et des Travaux Publics*, Série Matériaux 30, Supplement no. 223–4, July–August.

WALKINSHAW J. L. (1975). *Reinforced Earth Construction*. Dept. of Transportation FHWA Region 15. Demonstration Project No. 18.

WATANABE S. and IWASAKI K. (1978). Reinforcement of railway embankments in Japan. *Proc. ASCE Spring Convention*, Pittsburgh.

WESTERGAARD H. M. (1938). A problem of elasticity suggested by a problem in soil mechanics. Soft material reinforced by numerous strong horizontal sheets. *Mechanics of Solids*, Timoshenko 60th Ann. Vol., 268–277, Macmillan, New York.

YAMAMOTO K. (1966). *Strengthening of Embankment slopes with Nets*, Annual Report, Morioka Construction Bureau, Jnr, 66–68.

YAMANOUCHI T. (1967). Structural effect of restrained layer on subgrade of low bearing capacity in flexible pavements. *Proc. 2nd Int. Conf. Structural Design of Asphalt Pavements*, Ann Arbor, 381–9.

YAMANOUCHI T. (1970). Experimental study on the improvement of the bearing capacity of soft clay ground by laying a resinous net. *Proc. Symp. Foundations Interbedded Sands*, Perth, 102–8.

YAMANOUCHI T. (1975). Resinous net applications in earth works. *Proc. Conf. Stabilization and Compaction*, University of New South Wales, Sydney.

YANO K., WATARI Y. and YAMANOUCHI T. (1982). Earthworks over soft clay using rope-knotted fabrics. *Proc. Symp. on Recent Developments in Ground Techniques*, Bangkok, 225–237.

3

Application areas

3.1 Introduction

This chapter forms a catalogue of some of the application areas for the use of earth reinforcement and illustrates where soil structures of various forms have been found to provide economic and technical benefits. Each case is an illustration of the concept of earth reinforcements but should not be taken as being the only effective or rational solution to any problem. The diagrams used show typical structural shapes and approximate dimensions and scale. In practice each application should be considered separately. Applications and techniques may often be combined and the introduction of new construction materials enable other applications to be considered. The variety and range of the areas of application for these techniques is unlimited.

3.2 Application: Bridgeworks

3.2.1 **Bridge abutment,** Fig. 3.1

Materials. Reinforced or prestressed concrete facing, strip, grid or anchor reinforcement, frictional or cohesive frictional fill.

Comments. Economical; may be used on poor subsoils; speed of erection high, able to accommodate compressive ground strains, conventional articulation of the bridge.

3.2.2 **Bridge abutment with piled bank seat,** Fig. 3.2

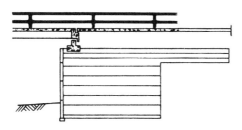

Fig. 3.1. Bridge abutment (after Goughnour and Di Maggio, 1979)

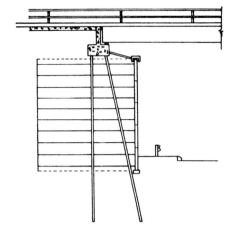

Fig. 3.2. Piled bank seat, reinforced soil abutment

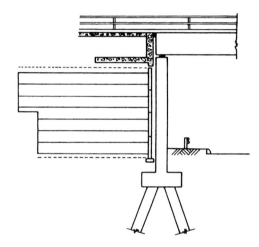

Fig. 3.3. Deck supported on pier, free-standing reinforced soil abutment

Materials. As 3.2.1.

Comments. Economical; reduced settlement of deck support.

3.2.3 Bridge abutment and support to bank seat, Fig. 3.3

Materials. As 3.2.1.

Comments. Economical; reduced settlement of deck.

3.2.4 Sloping bridge abutment, Fig. 3.4

Materials. Masonry or precast concrete paving for facing; geotextile, geogrid, strip or anchor reinforcement; frictional or cohesive frictional fill.

Comments. Abutment becomes an extension of the embankment with very strong abutment/embankment interaction. No bearings required for small structures. Fabric reinforcement suitable for small structures.

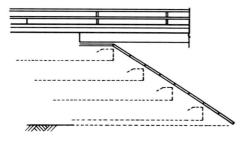

Fig. 3.4. Sloping bridge abutment

3.2.5 Reinforced embankment in place of viaduct, Fig. 3.5

Materials. As 3.2.1.

Comments. Very economical and may be used on poor subsoil; speed of erection high. Opposite faces may be tied together.

3.2.6 Geosynthetic reinforced railway bridge abutments, Fig. 3.6

Materials. Polyester geogrid reinforcement, sandbag temporary facing, full height mass concrete facing; cohesionless or cohesive fill.

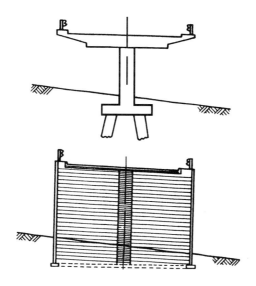

Fig. 3.5. Reinforced embankment in place of viaduct

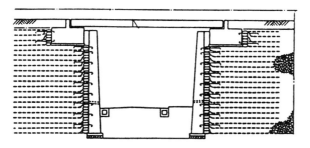

Fig. 3.6. Geosynthetic reinforced bridge abutment

19

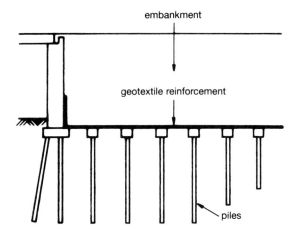

Fig. 3.7. Bridge approach piling
(after Reid and Buchanan,
1984)

Comments. Used in Japan to widen railway embankments for additional tracks. Shown to be stable during major earthquake.

3.2.7 Bridge approach piling, Fig. 3.7

Materials. High strength polymeric reinforcement and specially designed pile caps.

Comments. Used to reduce the settlement of approach embankments for highways or railways when piled abutments are required. Provides a major reduction in cost over alternative methods.

3.3 Application: Dams

3.3.1 Earth fill dam, Fig. 3.8

Materials. Reinforced concrete facing and anchor blocks. Concrete protected ties.

Comments. Special precautions are required with dam structures. Ladder wall dams can accommodate considerable settlements as in the 21 m Conguelac flood control dam in Southern France. An alternative configuration for a Coyne ladder structure is shown in Fig. 3.9.

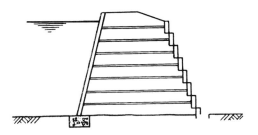

*Fig. 3.8. Coyne ladder wall dam
(after Chabal et al., 1983)*

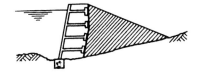

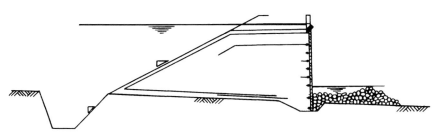

Fig. 3.9. Coyne ladder

Fig. 3.10. Reinforced earth dam (after Cassard et al., 1979)

3.3.2 Reinforced earth dam, Fig. 3.10

Materials. Reinforced concrete facing, metal strip, grid or anchor reinforcement and selected frictional fill.

Comments. Special precautions are required with dam structures.

3.3.3 Reinforced soil structure used to raise the height of an existing dam, Fig. 3.11

Materials. Reinforced concrete facing, resin epoxy coated strip reinforcement and selected frictional fill.

Comments. Special precautions are required with dam structures.

3.4 Application: Embankments
3.4.1 Reinforced embankment, Fig. 3.12

Materials. Geotextile or geogrid reinforcement and indigenous fill.

Comments. Reinforcing embankments may be undertaken for a variety of conditions, including steepening the side slopes, to permit the use of marginal fill, or to strengthen embankments.

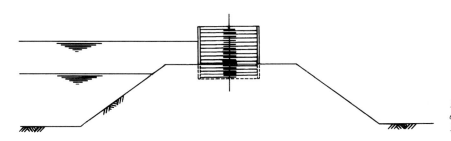

Fig. 3.11. Raising height of existing dam (after Engineering News Record, 1983)

21

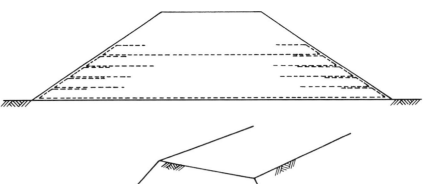

Fig. 3.12. Embankment reinforced to produce stability

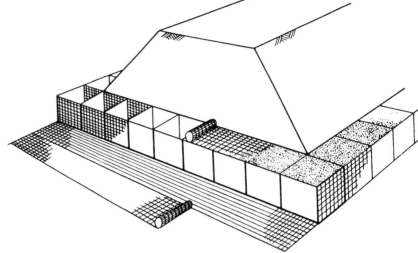

Fig. 3.13. Geocell foundation for embankment

3.4.2 Geocell mattress used to increase embankment stability, Fig. 3.13

Materials. Geogrid material for mattress filled with selected material.

3.4.3 Geogrid vertical web foundation used to produce embankment stability, Fig. 3.14

Materials. Geogrid reinforcement and selected fill.

3.4.4 Tied embankment, Fig. 3.15

Materials. Geotextile or geogrid reinforcement and indigenous fill.

Comments. Tied embankments rely upon the strength of the reinforcing element.

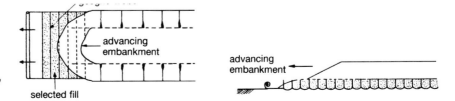

Fig. 3.14. Geogrid vertical web foundation

Fig. 3.15. Tied embankment

3.5 Application: Foundations

3.5.1 Geogrid web or column foundations for embankments on weak subsoil, Fig. 3.16(a) and (b)

Materials. Geogrid reinforcement.

Comments. Reinforcement webs and columns installed in situ.

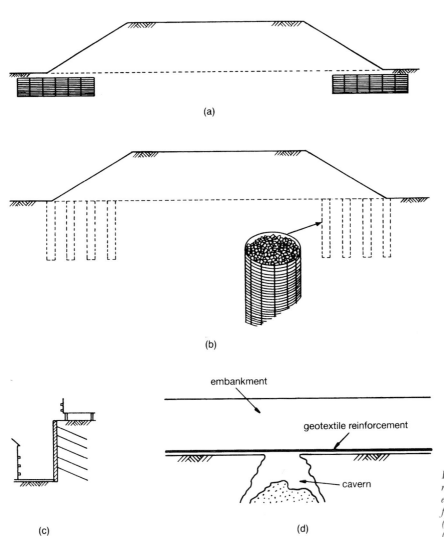

(a)

(b)

embankment

geotextile reinforcement

cavern

(c)

(d)

Fig. 3.16. (a) Geogrid reinforcement of subsoil beneath embankment; (b) stone columns formed from geogrid tubes; (c) excavation in urban area; (d) tension membrane over void

23

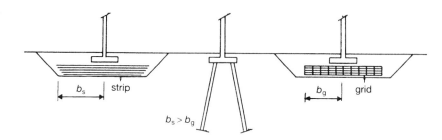

Fig. 3.17. Reinforced foundation

3.5.2 Excavations in urban conditions, Fig. 3.16(c)

Materials. Spray concrete facing and steel nail reinforcement.

Comments. Used in place of anchor systems.

3.5.3 Tension membrane spanning voids or potential voids, Fig. 3.16(d)

Materials. High strength grid reinforcement and selected fill, no facings are required.

Comments. Used to guard against sudden collapse or to retain serviceability following formation of void.

3.5.4 Reinforced footings beneath structures, Fig. 3.17

Materials. Strip or grid reinforcement and frictional fill or cohesive frictional fill.

Comments. Reinforcement used to ensure stability and reduce settlement.

3.5.5 Reinforced foundations beneath storage tanks, Figs 3.18, 3.19

Materials. Grid reinforcement and granular fill.

Comments. Reinforced foundations are used to reduce total and differential settlement.

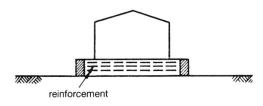

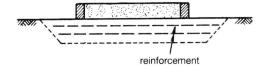

Fig. 3.18. Reinforced foundation (after Chinese Report, 1979)

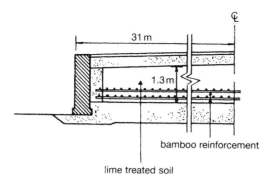

Fig. 3.19. Storage tank foundation in China (after Kim et al., 1982)

3.6 Application: Highways

3.6.1 Reinforced embankments supporting carriageways, Fig. 3.20

Materials. Reinforced concrete or steel facings, with strip, grid or anchor reinforcement and frictional or cohesive frictional fill.

Comments. Earth reinforcement permits design idealization not possible with other forms of construction.

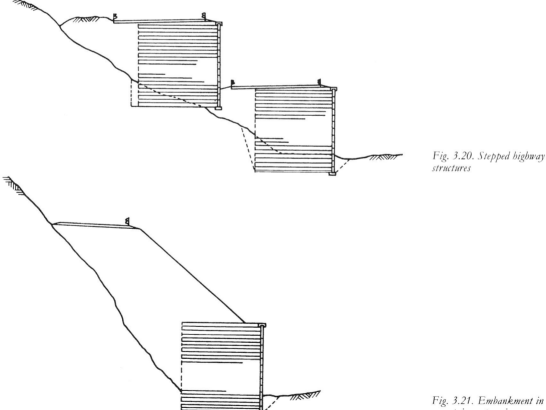

Fig. 3.20. Stepped highway structures

Fig. 3.21. Embankment in mountainous terrain

(a)

(b)

Fig. 3.22. (a) Stepped embankment or retaining wall; (b) gabion faced reinforced soil structure

3.6.2 Reinforced soil support to embankments in mountainous regions, Fig. 3.21

Materials. As 3.6.1.

Comments. As 3.6.1.

3.6.3 Reinforced embankments supporting highways, Fig. 3.22(a) and (b)

Materials. Reinforced concrete or steel facings. Strip, grid or anchor reinforcement.

Comments. As 3.6.1.

3.6.4 Repair of embankment failures, Figs 3.23, 3.24, 3.25

Materials. Waste car tyres and selected fill; the tyres may be tied together with steel hairpins or geotextile tape.

Comments. Economic repair technique.

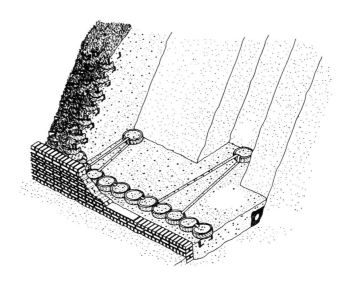

Fig. 3.23. Tyre/Geotextile composite

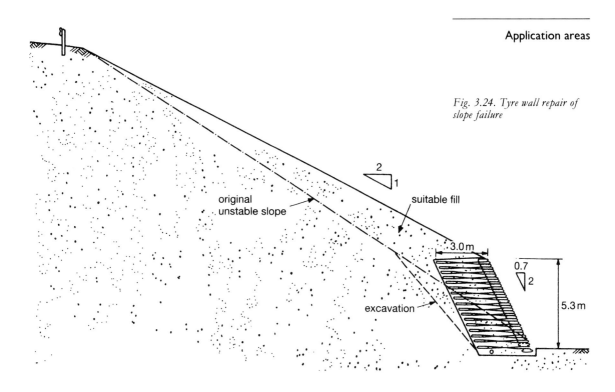

Fig. 3.24. Tyre wall repair of slope failure

3.6.5 Formation of reinforced cutting using soil nailing or the lateral earth support system, Fig. 3.26

Materials. Steel soil nails and spray concrete facing.

Comments. Can only be undertaken if in-situ soil suitable for soil nailing or lateral earth support system, i.e. cohesionless soil conditions.

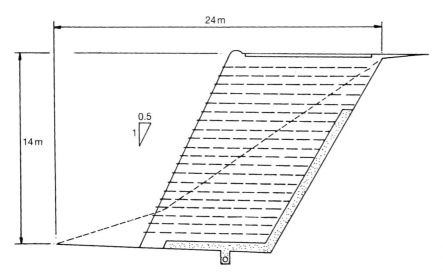

Fig. 3.25. Tyre reinforced embankment slope (Santa Cruz, California)

27

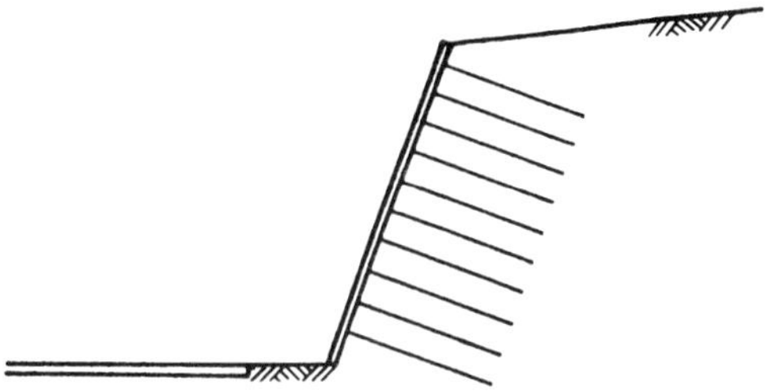

Fig. 3.26. Cutting formed using soil nailing

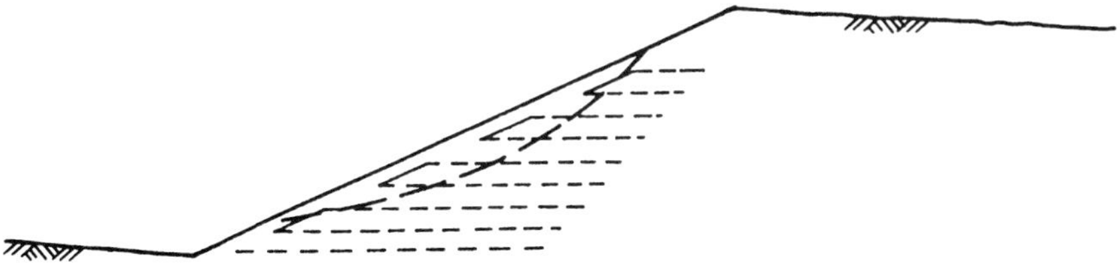

Fig. 3.27. Reinforced soil repair of cutting failure

3.6.6 Reinforced soil repair of cutting failure, Fig. 3.27

Materials. Geogrid reinforcement used with the failed material.

Comments. Eliminates the need to import suitable fill.

3.6.7 Formation of a reinforced soil block to permit widening of a cutting, Fig. 3.28

Materials. Mixture of fine polymeric threads with sand and water.

Comments. An industrialized method to provide an artificial root mass.

3.7 Application: Housing

3.7.1 Reinforced soil used to form terraced housing on sloping sites, Fig. 3.29

Materials. Reinforced or prestressed concrete facing units and strip reinforcement with selected granular fill.

Comments. Application suitable for warm/temperate climates.

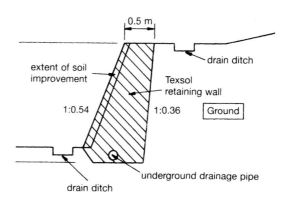

Fig. 3.28. Texsol retaining wall (after Leflaivre et al., 1983)

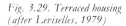

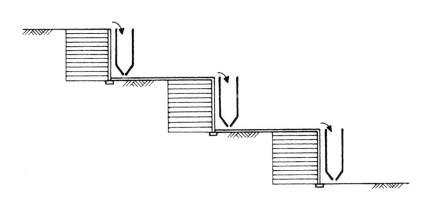

Fig. 3.29. Terraced housing
(after Leviselles, 1979)

Fig. 3.30. Rock crushing plant

Fig. 3.31. Mineral hopper

3.8 Application: Industrial

3.8.1 Rock crushing plant, Fig. 3.30

Materials. Steel or reinforced concrete facing units together with strip or grid reinforcement and selected fill.

Comments. Some of the largest reinforced soil structures have been constructed for industrial use.

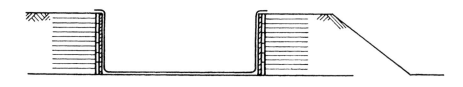

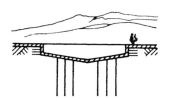

Fig. 3.32. (a) Settlement basin;
(b) settlement lagoon

Fig. 3.33. Containment dyke

3.8.2 **Mineral storage bunkers,** Fig. 3.31

Materials. Sloping reinforced concrete facing units, together with strip, grid or anchor reinforcement and granular selected fill.

3.8.3 **Settlement tanks and lagoons,** Fig. 3.32(a) and (b)

Materials. Reinforced or prestressed concrete facings, and strip, grid or anchor reinforcements, with selected fill.

3.8.4 **Containment dykes,** Fig. 3.33

Materials. As 3.8.3.

3.8.5 **Roof support packs in underground mining,** Fig. 3.34

Materials. Wire grid reinforcement and minestone waste fill.

Comments. Used in place of traditional timber packs.

3.9 Application: Military
3.9.1 **Army bunkers, traverses and blast shelters**

Materials. Polymeric reinforcement formed as grids or sheets.

Comments. The use of non-metallic reinforcements and lightweight materials may be advantageous.

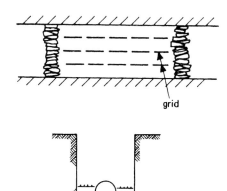

Fig. 3.34. Reinforced pack for roof supporting underground mining

Fig. 3.35. Rigid inclusions around buried pipe structures (after Magyarne Jordau et al., 1979)

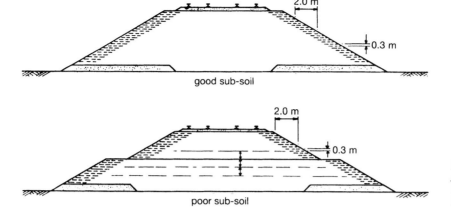

Fig. 3.36. Railway embankment standard, Japanese National Railway (after Uezawa and Kornine, 1975)

3.10 Application: Pipeworks

3.10.1 To provide side support to buried pipe structures, Fig. 3.35

Materials. Grid or special shaped inclusions acting together with the usual pipe backfilling material.

Comments. Produces improved bedding conditions around pipes and increases lateral earth support.

3.11 Application: Railways

3.11.1 Reinforcement of railway embankments to provide stability over poor subsoil and to protect embankments from washout caused by typhoon rains, Fig. 3.36

Materials. Geogrid and strip reinforcement and indigenous fill.

Comments. Low strength thermoplastic netting has been found to be effective as a reinforcing material.

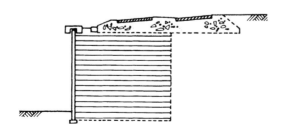

Fig. 3.37. Railway support

31

3.11.2 Railway embankment, Fig. 3.37

Materials. Prestressed or reinforced concrete or steel facings, together with grid or strip reinforcements and selected fill. Waste material may be used with geogrid reinforcement.

Comments. Railway loading on the edge of structures has been accommodated without difficulty or cost penalties.

3.12 Application: Root pile systems

3.12.1 Foundation supports and repair systems, Fig. 3.38

Materials. Minipiles formed of reinforced concrete.

Comments. The reinforcement in these applications are usually acting as compression reinforcement (see Chapter 4, Theory).

3.13 Application: Sports structures

3.13.1 Ski jumping slopes

Materials. Steel or geogrid gabions or reinforced concrete facings, with strip or grid reinforcement.

Comments. Sites may have access problems and gabion facings may be easier to transport than precast elements.

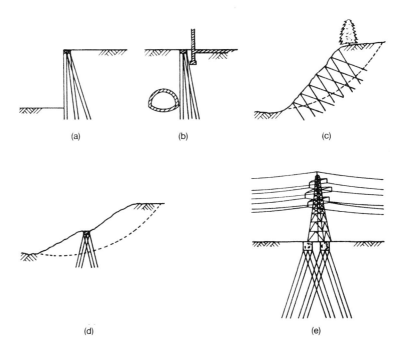

(a) (b) (c)

(d) (e)

Fig. 3.38. Root piles (after Lizzi, 1983)

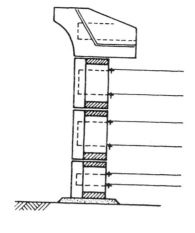

Fig. 3.39. Sea wall (after Gagnon, 1979)

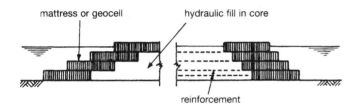

(a) *Construction with reinforced mattresses or geocells*

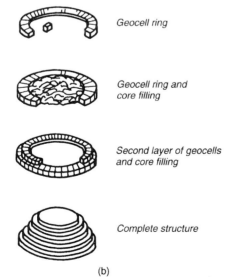

Geocell ring

Geocell ring and core filling

Second layer of geocells and core filling

Complete structure

(b)

Fig. 3.40. Underwater island construction (after Jewell and Wishert, 1983)

3.14 Application: Quays and sea walls and waterway structures

3.14.1 Sea wall, Fig. 3.39

Materials. Special reinforced concrete facing elements, together with strip or grid reinforcement and selected granular fill.

Comments. Special precautions are required to guard against washout.

3.14.2 Islands constructed underwater, Fig. 3.40

Materials. Geocell reinforcement with hydraulically placed fill.

3.14.3 Wall adjacent to river or forming the sides of a canal or quay, Fig. 3.41

Materials. Reinforced or prestressed concrete or timber facings together with grid or strip reinforcement and selected fill.

Comments. The earliest reinforced soil quays built by the Romans used timber for the facing and timber baulks as the reinforcement.

3.15 Application: Underground structures

3.15.1 Vaults, Fig 3.42

Comments. Used as military shelters.

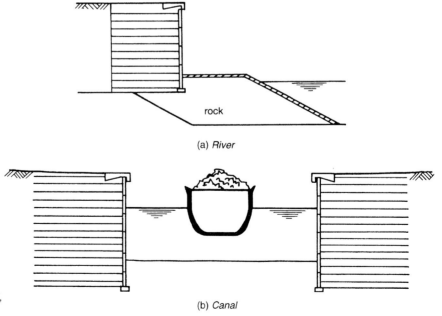

(a) *River*

(b) *Canal*

Fig. 3.41. Reinforced soil structures forming: (a) walk adjacent to a river; (b) sides of a canal (after Patel and Soupal, 1979)

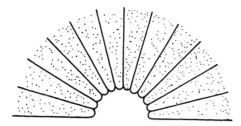

Fig. 3.42. Vault

3.16 Application: Land reclamation

3.16.1 In coastal areas to reclaim land covered by sludge, very soft industrial waste or extremely soft clay, Fig. 3.43

Materials. Polymeric grid and a layer of fine cohesionless soil (sand).

Comments. A number of techniques have been developed. One of the most successful is the primary stage construction technique whereby a working platform is created on the very soft soil, thereby permitting the use of conventional soil improvement techniques to be employed.

3.17 References

CASSARD A., KERN F. and MATHIEU G. (1979). Use of reinforcement techniques in earth dams. *C.R. Coll. Int. Renforcement des Sols*, Paris.

CHABAL J.-P., TARDIU B., CURERBER P. and BATSAND J. (1983). A novel reinforced fill dam. *VIII ESCMFE*, Helsinki.

CHINESE REPORT (1979). *Testing and research of tank foundation in Zheijiang Refinery Factory.* Zheijiang University, Hangzhow, China.

ELIAS V. and MCKITTRICK D. P. (1979). Special uses of Reinforced Earth in the United States. *C.R. Coll. Int. Renforcement des Sols*, Paris.

ENGINEERING NEWS RECORD (1983). Old fill dam gets fast safety lift. January.

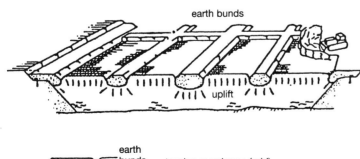

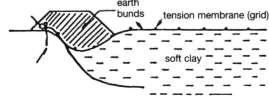

Fig. 3.43. Land reclamation using tension membranes (after Yano et al., 1982)

Application areas

GAGNON G. (1979). Sea wall constructed in Reinforced Earth. *C.R. Coll. Int. Renforcement des Sols*, Paris.

GOUGHNOUR R. D. and DI MAGGIO J. A. (1979). Application of Reinforced Earth in Highways throughout the United States. *C. R. Coll. Int. Renforcement des Sols*, Paris.

HANNA B. E. and McKITTRICK D. P. (1979). Reinforced Earth Retaining Walls. *C.R. Coll. Renforcement des Sols*, Paris.

JEWELL R. A. and WISHERT S. J. (1983). Underwater construction using reinforced hydraulic fill. *VIII ESCMFE*, Helsinki.

KIM Y. S., SHEN C. K. and BRAY S. (1982). *Oil storage tank foundation on soft clay. VIII ECSMFE*, Helsinki.

LEFLAIVRE E., KHAY M. and BLIVET J. C. (1983). Un nouvean material: le Texsol. *Bulletin de Liason des Laboratoire des Ponts et Chaussées*, Paris, No. 125, 105–114.

LEVISELLES J.-F. (1979). Use of the Reinforced Earth technique in the construction of housing. *C.R. Coll. Int. Renforcement des Sols,* Paris.

LIZZI F. (1983). The uticolo di pali radice (Reticulated root piles for the improvement of soil resistance). *VIII ECSMFE*, Helsinki.

MAGYARNE JORDAU M., SCHOULE P. and SZALATKAY I. (1979). Improvement of bedding conditions around pipes by rigid inclusions. *C.R. Coll. Int. Renforcement des Sols,* Paris.

REID W. M. and BUCHANAN N. (1984). Bridge approach support piling. *Proc. Conf. on Piling and Ground Treatment*, Thomas Telford, London, 267–274.

PATEL M. D. and SOUPAL R. C. (1979). Use of Reinforced Earth Walls in Canals. *C.R. Coll. Int. Renforcement des Sols*, Paris.

UEZAWA, H. and KORNINE T. (1975). Reinforcement of embankments using net. *Tetgudo-Doboka*, **17**, no. 5, 21–24 (In Japanese).

YANO K., WATERI Y. and YAMANOUCHI T. (1982). Earthworks on soft clay using rope-netted fabric. *Proc. Symp. on Recent Developments in Ground Techniques*, Bangkok, 225–237.

4

Theory

4.1 Introduction

Birds and animals which use reinforced soil systems do so through instinct; the early applications of the principles of earth reinforcement, such as in the ziggurat at Agar Quf, may have been based upon theoretical studies although an empirical approach seems more likely. The empirical proposals of Pasley in the nineteenth century were based upon the results of a large number of experiments. Pasley's approach is valid when considering narrow fields of applications, however, theory is required to describe basic actions.

In 1924, Coyne (1927 and 1945) introduced the ladder wall in which a series of reinforcing elements usually, but not necessarily, having an anchor were connected to a facing to form a reinforced soil structure, Fig. 4.1.

Coyne used the analogy of a Howe Beam to describe the action of his structures, Fig. 4.2. The Howe Beam differs from the braced girder in that the verticals are tension members as is the bottom flange. The top flange and the diagonals are in compression. By rotating the Howe Beam through 90° so that the beam appears to be erected vertically (i.e. like a ladder), then using Coyne's words, Fig. 4.3 represents

> a beam whose uprights, represented by the anchorages, are in tension and whose compressed diagonals are formed in the fill itself. The compressed member of the beam is the facing AC and its stretched member falls around about the vertical plane BD, passing through the tail of the anchorage. The corresponding extensions (i.e. of the tension members) are neutralized by the weight of the fill. The whole may be considered forming a single block of earth coherent in the whole zone ABCD transversed by the tie rods.

Later, Westergaard (1938) working on a concept suggested by Casagrande

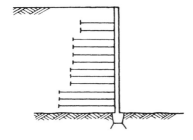

Fig. 4.1. Coyne–ladder wall

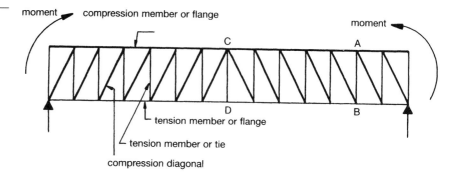

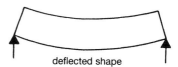

Fig. 4.2. Howe beam

considered a medium made up of soft elastic material reinforced by closely spaced horizontal flexible but unstretchable sheets; this material was known as the Westergaard material and its properties described in terms of the theory of elasticity. Harrison and Gerrard (1972) showed this system to be a limiting case of a cross-anisotropic material.

Reinforced soil is somewhat analogous to reinforced concrete in which the reinforcement is bonded to the soil in the case of reinforced soil, or to the concrete in the case of reinforced concrete. However, direct comparison between the two situations is not completely valid; whereas with reinforced concrete the reinforcement is designed to carry the tensile forces in the structural element, in the case of reinforced soil, particularly with non-cohesive soils, it is likely that a completely compressive stress field will exist. The mode of action of reinforcement in soil is, therefore, not one of carrying developed tensile stresses but of the anisotropic reduction or

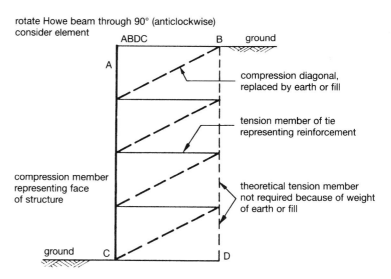

Fig. 4.3. Howe beam analogy of ladder wall

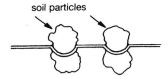

soil particles

Fig. 4.4. Diagrammatic representation of reinforced soil

suppression of one normal strain rate. This suppressive mechanism was described by Vidal (1963, 1966, 1969a and 1969b) and is expressed diagrammatically in Fig. 4.4, which shows individual soil particles tied together, producing a form of pseudo-cohesion.

Consider a semi-infinite mass of cohesionless soil at depth h. Vertical stress,

$$\sigma_v = \gamma h \tag{1}$$

and the at-rest lateral stress,

$$\sigma_H = K_0 \gamma h$$

where $K_0 \approx 1 - \sin \phi$, $\phi =$ angle of friction of the soil.

If the soil expands laterally the lateral stress $(K_0 \sigma_v)$ reduces to the limiting value $(K_a \sigma_v)$ where

$$K_a = \left(\frac{1 - \sin \phi}{1 + \sin \phi} \right) = \tan^2(45° - \phi/2) \tag{2}$$

see Fig. 4.5.

The action and relevance of reinforcement in soil can be illustrated by considering an element of the cohesionless soil, Fig. 4.6. If a vertical load is applied to the soil, the element will strain laterally, δ_h as well as compress axially, δ_v. If reinforcement is added to the soil element in the form of horizontal layers, provided there is adhesion or interaction between the reinforcement and the soil caused by friction or other means, and that the reinforcement is stiff, the soil will be restrained as if acted upon by a lateral force equivalent to the at-rest pressure $(K_0 \sigma_v)$, i.e. the effect of the reinforcement is to restrict anisotropically one normal strain $(\dot{\epsilon}_\theta)$. This is a general condition, valid for any value of vertical stress σ_v, and it can be seen that as σ_v increases so the lateral stress also increases. Reference to Fig. 4.5 shows that the stress circle for the reinforced condition always lies below the rupture curve. Failure can occur only if the reinforcement ruptures or if the adhesion between the reinforcement and the soil fails.

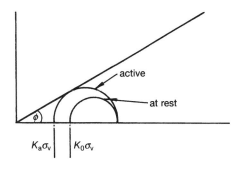

Fig. 4.5. Stress state in soil

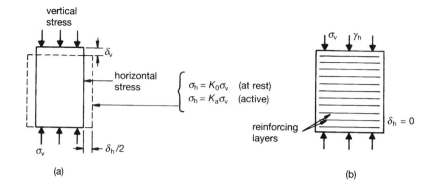

Fig. 4.6. (a) Soil; (b) soil and reinforcement

The force transferred from the unit of soil into the reinforcement is equivalent to the lateral stress $= K_0 \sigma_v$.

Hence, tensile stress in any unit of reinforcement

$$= \frac{K_0 \sigma_v}{a_r}$$

where $a_r =$ cross-sectional area of the reinforcement. Therefore, strain in the reinforcement

$$\delta_r = \frac{K_0 \sigma_v}{a_r E_r} \tag{3}$$

where E_r is the elastic modulus of the reinforcement and the lateral strain of the soil ϵ_r in the direction of the reinforcement.

Therefore, $\epsilon_r = \delta_r = \dfrac{K_0 \sigma_v}{a_r E_r}$

If the effective stiffness $(a_r E_r)$ of the reinforcement is high, then $\epsilon_r \to 0$ and the argument relating to Fig. 4.6 holds. As the effective stiffness decreases, ϵ_r increases, and the earth pressure coefficient $K_0 \to K_a$.

4.2 General theory

4.2.1 Stress–strain relationship of reinforced soil

4.2.1.1 *Vertical structures or walls*

The previous argument relating to the behaviour of reinforced soil holds for vertically faced structures, but because of the anisotropic nature of the action and effect of the reinforcement, does not apply to the general case. Bassett and Last (1978) have considered a more general approach to the concept of earth reinforcement by considering the modification of the strain field of a soil caused by the addition of reinforcement. Fig. 4.7 shows a conventional Mohr circle of stress and the corresponding Mohr circle of strain rate.

The centre of the strain circle represents

$$\frac{\dot{\epsilon}_1 + \dot{\epsilon}_3}{2} = \frac{\dot{\nu}}{2}$$

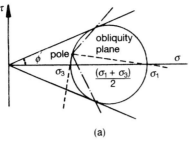

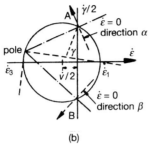

(a)

(b)

Fig. 4.7. (a) Mohr circle of stress; (b) Mohr circle of strain rate

(volumetric strain rate) and the diameter

$$\frac{\dot{\epsilon}_1 - \dot{\epsilon}_3}{2} = \frac{\dot{\gamma}}{2} \max \tag{4}$$

The Pole or Origin of Planes determines the major ($\dot{\epsilon}_1$) and minor ($\dot{\epsilon}_3$) principal strain directions and also gives the α and β planes at A and B across which $\dot{\epsilon} = 0$. The physical conditions delineated by the α and β directions are important, as within the arc segment containing the minor principal strain direction $\dot{\epsilon}_3$ all normal strains will be tensile and hence any reinforcement would be effective. The α and β directions for various points in a strain field can be joined to form zero extension trajectories or characteristics. The α and β extension characteristics also represent the potential slip or rupture planes. Fig. 4.8(a) shows the potential slip planes in a cohesionless backfill of a flexible cantilever rotating about the toe away from the fill, the α and β trajectories indicate the expected form of the strain field, with a constant horizontal direction for the tensile principal strain ($\dot{\epsilon}_3$)

Reinforcement placed within the tensile arc would be effective; inspection indicates that the optimum direction for reinforcement is horizontal in line with the principal tensile strain ($\dot{\epsilon}_3$) Fig. 4.8(b). This direction is used in practice. It can also be concluded that reinforcement placed parallel with the α and β trajectories would be equivalent to placing reinforcement in line with a rupture plane; if the adherence between the reinforcement and the soil was less than the shear strength of the soil alone, then the effect would be to lubricate the rupture plane thereby weakening the soil. Reinforcement placed within the compressive arc, Fig 4.8(b), would have to be capable of resisting compression stresses to be effective.

If the reinforcement is 'stiff' compared with the $\Sigma\dot{\epsilon}_3$ generated within the shearing soil mass and if there is efficient adherence between the reinforcement and the soil mass then the direction of the reinforcement must be aligned to one

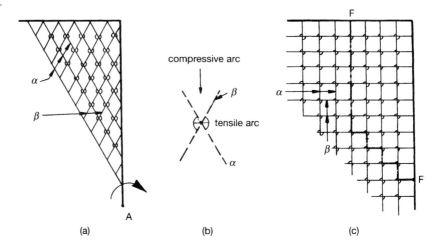

Fig. 4.8. (a) α and β characteristics of reinforced fill produced by wall rotating about A (after Milligan, 1974); (b) location of compressive strain arc; (c) α and β characteristics for reinforced fill. β direction aligned with horizontal reinforcement (after Bassett and Last, 1978)

of the zero extension characteristics. Thus the presence of the reinforcement placed in a horizontal plane causes a rotation of the α and β trajectories, at the same time the dilation rate is suppressed. The potential failure mechanism of the reinforced composite would attempt to align with the amended trajectories, i.e. along line (F–F) Fig. 4.8(c).

4.2.1.2 Sloping embankment

Whereas the tensile principal strain ($\dot{\epsilon}_3$) of a cantilever wall can be assumed to be horizontal, the case of the sloping embankment is not so straightforward. The problem facing the designer is that of determining or predicting the directions of the compressed strain trajectories and the α and β zero extension lines. Failure to do this could result in tensile reinforcement being placed in a position of compressive strain, or along a potential rupture plane. Predictions of the α and β planes can be obtained from centrifuge tests (Bassett and Horner, 1977); from model tests (Roscoe, 1970); by using mathematical models (Sims and Jones, 1979), or from limit equilibrium methods. The task is eased by using the observation that under monotonic loading conditions the axes of principal total stress and incremental strain coincide. Fig. 4.9 shows the idealized zero-extension characteristic fields through and beneath an embankment, together with the directions of the principal compressive stresses. By inspection it can be seen that reinforcement placed horizontally in the majority of the embankment would be advantageous, but horizontal reinforcement restricted to zone C would be potentially dangerous.

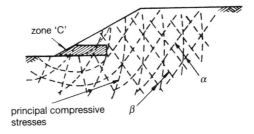

Fig. 4.9. Idealized zero-extension characteristic fields through and beneath an embankment

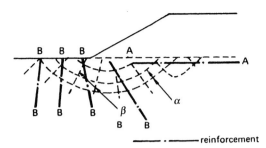

reinforcement

Fig. 4.10. Possible reinforcement

Reinforcement at the base of an embankment can be achieved by two methods. Horizontal reinforcement can be placed at the base in a manner similar to the technique with vertical walls (A–A) Fig. 4.10, which will create a condition of horizontal restraint on the plane of the reinforcement; this method has been discussed by Binquet and Lee (1975). Alternatively, reinforcing tendons can be introduced into the foundation soil beneath the embankment, aligned with the principal tensile strain directions (B–B), Fig. 4.10. This is considered in the next section.

4.2.1.3 Reinforcement below footings

Reinforcement placed within the tensile arc of the strain field causes realignment of the strain field which improves performance in both stiffness and load-carrying capacity. This concept, which has been developed for walls, can be applied to bearing capacity in foundations.

Figure 4.11(a) shows an idealized zero-extension characteristic field for a foundation on a uniformly dilating material, similarly Fig. 4.11(b) shows the case of a collapsing material and Fig. 4.11(c) of an undrained or zero volume change material. Inspection of Fig. 4.11(a),(b),(c), indicates the potential effectiveness of reinforcement beneath the footing in the three conditions; the scope for reinforcing the dilatant material being significantly greater than with the collapsing material. Fig. 4.11(d) shows the influence of horizontal restraint at the base caused by a rough footing or by reinforcement.

In accordance with the previous argument the ideal reinforcing pattern, for the directions of the principal tensile strains indicated in Fig. 4.11, lie along the lines shown in Fig. 4.12(a). The ideal pattern has reinforcement placed horizontally below the footing, which becomes progressively more vertical further from the footing. The form of the reinforcement required in this application, Fig. 4.12(b),(c), would need to be different to that used in walls, small diameter (100 mm) ground anchors being typical of meeting the requirements. Alternatively, a grid could be employed, Fig. 4.12(d).

4.3 Factors affecting the performance and behaviour of reinforced soil

The following factors, listed in Table 4.1, influence the behaviour and performance of reinforced soil. To these should be added the external loading and environmental factors.

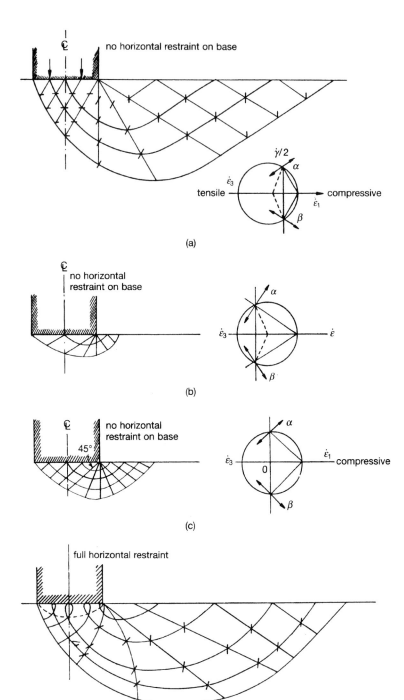

Fig. 4.11. (a) Zero-extension
characteristics for dilating soil
(after Bassett and Last, 1978);
(b) zero-extension characteristics
for a collapsing material;
(c) zero-extension characteristics
for an undrained or zero volume
change material; (d) zero-
extension characteristics for a
dilatant material with horizontal
restraint at the base

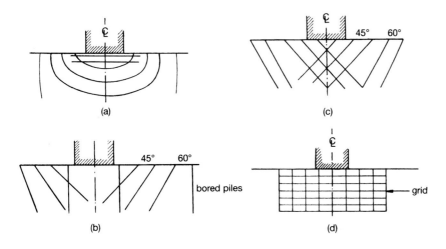

Fig. 4.12. Pattern of reinforcement beneath footing: (a) ideal reinforcement pattern; (b) practical reinforcement pattern (reinforcement placed after structure); (c) practical reinforcement pattern (reinforcement placed before structure); (d) practical reinforcement pattern

Table 4.1. Factors that influence behaviour and performance of reinforced soil

Reinforcement	Reinforcement distribution	Soil	Soil state	Construction
Form (fibre grid, anchor, bar, strip)	Location	Particle size	Density (void ratio)	Geometry of structure
Surface properties	Orientation	Grading	Overburden	Compaction
Dimensions	Spacing	Mineral content	State of stress	Construction system
Strength		Index properties	Degree of saturation	Aesthetics
Stiffness (bending, longitudinal)				Durability (See Chapter 10)

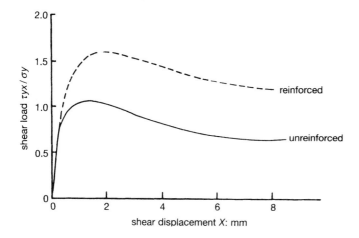

Fig. 4.13. Load–displacement results for shear tests on reinforced and unreinforced dense sand (after Jewell, 1980)

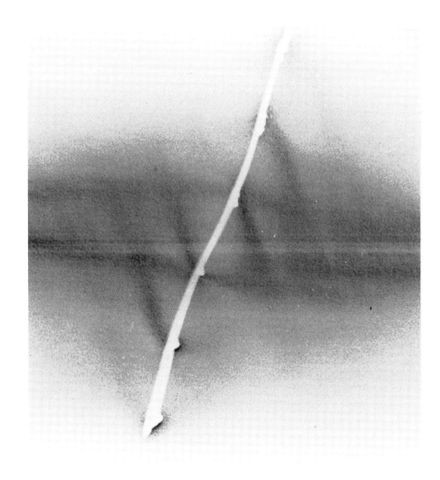

Rupture pattern of sand reinforced by a grid reinforcement in a direct shear apparatus (after Jewell, 1980)

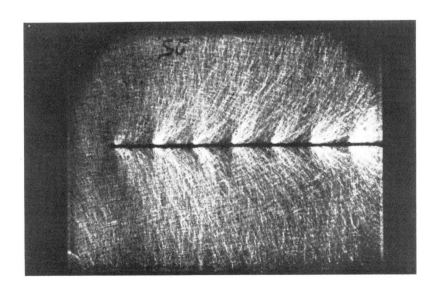

Rupture patterns observed in tests on inter-action of a grid reinforcement in a pullout test on glass (after Dyer, 1985)

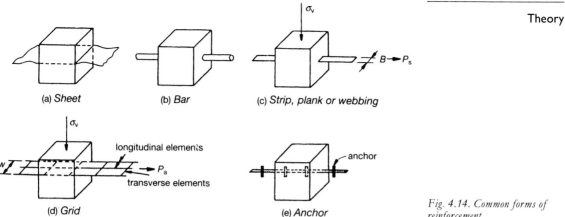

(a) Sheet (b) Bar (c) Strip, plank or webbing

(d) Grid (e) Anchor

Fig. 4.14. Common forms of reinforcement

4.3.1 Reinforcement

Reinforcement when introduced into soil and aligned with the tensile strain arc disrupts the uniform pattern of strain that would develop if the reinforcement did not exist. The reinforcement also inhibits the formation of continuous rupture surfaces through the soil, with the result that the soil exhibits an improved stiffness and shear strength, Fig. 4.13.

From the figure it can be seen that the reinforcement has no initial effect, it is only after the reinforcement has been strained that it has influence. As the soil strains it mobilizes strength to resist the shear loads, soil strain causes strain in the reinforcement, which leads to a further increase in strength in the reinforced soil. Strength is improved until a limiting value is achieved, with further shear displacement the improvement remains constant.

4.3.1.1 Form

In order to improve the performance, the reinforcement must adhere to the soil or be so shaped that deformation of the soil produces strain in the reinforcement. Reinforcement can take many forms depending largely on the material employed. Common forms are sheets, bars, strips, grids and anchors, Fig. 4.14. The forms shown rely on friction to develop bonds between the soil and reinforcement; the grid and the anchor provide a more positive bond by developing an abutment or soil-reinforcement interlock.

Considering the case of a strip length l, width B, the frictional resistance available from the strip can be developed from Figs 4.14 and 4.15.

Value of bond between soil and reinforcement, $\mathrm{d}T_{ad} = T_1' - T_2'$

Normal stress on the strip per element of structure $= \sigma_v$

Normal force acting on the strip $= \sigma_v \mathrm{d}l/B$

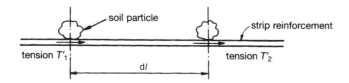

Fig. 4.15. Development of adhesion on a reinforcing strip

Therefore, tensile force generated in the reinforcement, assuming the coefficient of friction between soil and reinforcement is μ, is expressed as

$$\mathrm{d}T_{\mathrm{ad}} = 2^{*}\sigma_{\mathrm{v}}B\,\mathrm{d}/\mu$$

Therefore for no slippage:

$$\mu > \frac{\mathrm{d}T_{\mathrm{ad}}}{2\sigma_{\mathrm{v}}B\,\mathrm{d}l} \tag{5}$$

* (2, as friction developed on both sides)

Stress distribution along reinforcement

In the case of grid reinforcement, the width of the reinforcement is not restricted by the actual material section of the reinforcement but by the dimensions of the transverse elements and the shear strength of the soil. The mechanism of action of a grid in providing resistance to slippage (pullout) is not fully understood. Among the mechanisms proposed is the passive resistance theory (Chang *et al.*, 1977) and the bearing capacity theory (Bishop and Anderson, 1979). The bearing capacity mechanism is a form of passive resistance with a limited failure plane; however, it has been concluded that the passive resistance mechanism may be true for a complete grid but does not hold for individual transverse members. A failure mechanism, therefore, for the grid, has been suggested as being in accordance with Fig. 4.16. This is based upon the Terzaghi–Buisman bearing capacity equation for a strip footing:

$$q_{\mathrm{ult}} = Bc'N_{\mathrm{c}} + \tfrac{1}{2}\gamma\,\mathrm{d}B^{2}N_{\gamma} + \gamma\,\mathrm{d}N_{\mathrm{q}} \tag{6}$$

This can be arranged in terms of F_{p} (pullout resistance provided by the transverse members along where d = diameter, σ_{v} is the overburden pressure and N_{w} is the number of transverse members). For one transverse member:

$$F_{\mathrm{p}}/N_{\mathrm{w}} = dc'N_{\mathrm{c}} + \tfrac{1}{2}\sigma_{\mathrm{v}}d^{2}N_{\gamma} + \sigma_{\mathrm{v}}dN_{\mathrm{q}} \tag{7}$$

In a cohesionless soil where d is small and $c' = 0$ this can be reduced to

$$F_{\mathrm{p}}/N_{\mathrm{w}} = \sigma_{\mathrm{v}}dN_{\mathrm{q}} \tag{8}$$

Some forms of anchor may employ end plates, or be formed from loops,

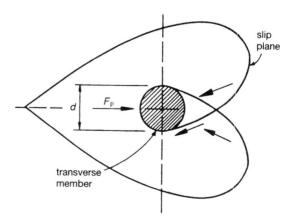

Fig. 4.16. Suggested slip planes for horizontal bearing

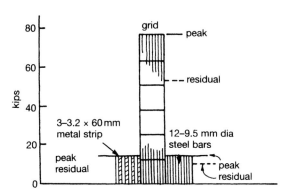

Fig. 4.17. Pullout resistance versus area of reinforcement (after Forsyth, 1978)

zigzags or bends; each will develop different adherence characteristics based on an appropriate theory.

A comparison between the bar/strip form of reinforcement and grip reinforcement, in terms of pullout performance relating to the peak and residual loading, is shown in Fig. 4.17. Although pullout resistance does not illustrate the actions in a soil, it is an important parameter and does provide an indication of reinforcement efficiency.

Grids have been shown to be an effective form of reinforcement for both cohesionless and cohesive soils. Grid reinforcement can be used to increase the shear strength of cohesive soil under both short-term and long-term loading conditions.

4.3.1.2 Surface properties

For sheets, bars and strips, equation (5) indicates that the coefficient of friction between the reinforcement and soil is a critical property, the higher the friction the more efficient the reinforcement. Thus an ideally rough bar, strip or sheet is significantly better than a reinforcement with a smooth surface. An ideally rough surface can be produced by glueing a layer of sand to the reinforcement thereby ensuring a soil-to-soil interface. Alternatively, the surface can be made rough by deforming it, using grooves, ribs or embossing a pattern. Roughened surfaces will tend towards the ideally rough condition depending on the depth and spacing of the deformity and the grading and particle size of the soil. The effect of roughening the surface of the strip can be seen in Fig. 4.18.

The surface properties of grid reinforcement have little or no effect on pull-out resistance, provided that soil particles penetrate through the grid between the transverse members of the grid. The frictional adherence between the longitudinal members of a grid and the soil is influenced by the surface properties and the coefficient of friction between the longitudinal members and the soil. The influence of the horizontal bearing capacity of the transverse elements, as expressed by equation (7), is of an order greater.

Although the pullout test is widely differentiated between the effectiveness of various reinforcing elements, the test itself does not accurately reflect the action of reinforcement in soil. This is illustrated in Fig. 4.19, which shows a comparison between the measured pattern of displacements in sand in a

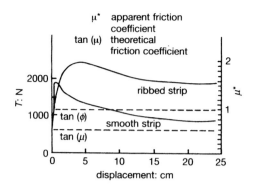

Fig. 4.18. Pullout test in reinforced earth walls—influence of the nature of the strip surface (after Schlosser and Elias, 1978)

direction parallel with a reinforcement grid in a shear apparatus and in a pullout test (Jewell, 1980a).

4.3.1.3 Dimensions

The dimensions of the reinforcement must be compatible with the conditions. The theoretical dimensions of any reinforcement are likely to be modified to conform with the requirements of logistics and durability. In addition the form, strength, stiffness and spacing will all influence the dimensions chosen.

4.3.1.4 Strength

Reinforcement strength is synonymous with robustness; logic demands that any reinforcement should be robust (see sections 4.3.5.4 Durability and 4.3.1.5 Stiffness). Any sudden loss of strength could have catastrophic effects since the improvement in shear strength is directly dependent upon the magnitude of

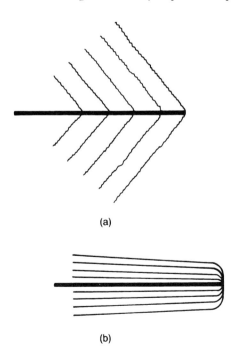

Fig. 4.19. Displacement patterns developed in shear apparatus and pullout test (after Jewell, 1980a): (a) shear apparatus; (b) pullout test

the maximum force generated in the reinforcement, Fig. 4.13. Sudden loss of strength due to failure would have the effect of suddenly reducing the shear strength of the reinforced soil to the shear strength of the soil shown at an equivalent displacement.

The use of a factor of safety against this mode of failure is necessary, the preferred failure mechanism being one of loss of adherence between the soil and the reinforcement, in which case a redistribution of shear stress is possible without total loss of the structure.

4.3.1.5 Stiffness

Bending stiffness (EI/y), the product of the elastic modulus and the second moment of area, has not been shown to have any significant effect on the performance of reinforced soils except in the case of reinforcements used as a tension membrane over super soft soil. In this application, bending stiffness is of major importance (Jones and Zakaria, 1994).

Longitudinal stiffness (Ea_τ), the product of elastic modulus and the effective cross-sectional area, has a marked effect on performance. Longitudinal stiffness of the reinforcement governs the deformity (or strains) which occur in the reinforced soil. The effect of placing reinforcement in soil in the direction of tensile strain is to restrict deformation and a force generated proportional to the resultant strain is developed in the reinforcement. An equilibrium condition is reached dependent upon the longitudinal stiffness of the reinforcement and the load–displacement characteristics of the soil.

The stress–strain characteristics of reinforcement are usually linear (e.g. steel strip reinforcement), this is not the case with soil, as Fig. 4.20 shows. In this case the soil softens once full shear strength has been mobilized.

From Fig. 4.20 it can be seen that as long as the maximum strains that develop in the reinforced soil at any point are less than those required to mobilize the soil, a stable condition exists. Using this hypothesis, a maximum value for the allowable tensile strain of the soil alone can be established. Assuming that the tensile strain in the reinforcement is the same as the strain in the soil (assuming no slip), then the maximum force in the reinforcement can be determined. If this maximum force is less than the pullout force of the reinforcement an equilibrium condition will exist. However, if the limiting force (i.e. the pullout or anchor force) is less than the force generated in the reinforcement, the longitudinal stiffness does not determine the failure shear strength of the reinforcement soil, and the limiting force cannot be used in any stability analysis.

The above argument holds for the case of maximum strains which are less than the strains at peak shear strength; the strain at peak shear strength may be influenced by soil density, stress history and stress level, in addition the effects of any rotation in the principal axis should be considered. However, experience indicates that in many soils failure in shear is fairly insensitive to density and stress level and that the magnitude of tensile strain will, in most cases, be greater than those present for both soils exhibiting dilation and expansion as well as for collapsing soils. Thus if maximum force in the reinforcement can be generated by a strain of less than 3 per cent, the above condition holds.

If the reinforcement is extensible the maximum force in the reinforcement

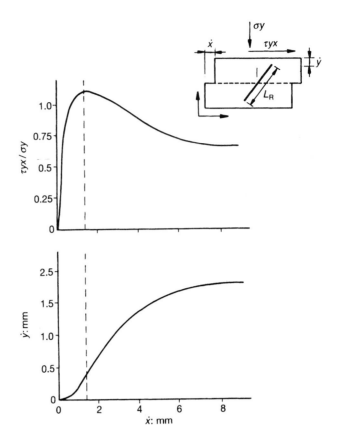

Fig. 4.20. Typical load–displacement results of a shear test on a dense sand (after Jewell, 1981)

will not be generated before the soil has passed the point of maximum shear strength. In these cases the maximum force in the reinforcement is controlled by the deformation in the soil. At strains beyond the peak strength of the soil alone, although the soil may be losing strength, the reinforcement would be gaining strength. Thus the reinforced soil may exhibit a peak load-carrying capacity relating to the peak shear strength of the soil, or it may exhibit an enhanced strength at strains beyond the point of peak shear strength of the soil alone. In either case the strength of the reinforced soil is greater than the soil alone, Fig. 4.21.

The use of extensible reinforcement in a reinforced soil structure can provide additional stability. Laboratory studies using metallic reinforcements have shown that the failure of model reinforced soil walls can be sudden if the failure mechanism is the result of the rupture of metallic reinforcement. With metallic reinforcement, rupture of one reinforcement leads to rapid load shedding and potential overstress of adjacent reinforcements leading directly to further rupture which leads, in turn, to more load shedding and hence structural instability. The use of extensible reinforcement, able to creep, may be immune to this mechanism with stress redistribution being accomplished without reinforcement rupture (Jaber, 1989). An electrical analogy can be used to describe the different potential failure methods of inextensible and extensible reinforcements. The rupture and subsequent rapid failure with

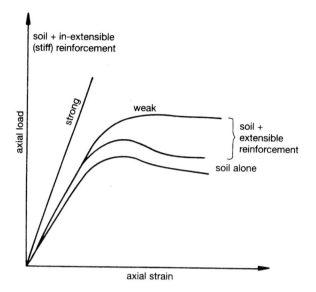

Fig. 4.21. Load–axial strain relationship for soil and soil reinforced with stiff (inextensible) and extensible (low stiffness) materials (after Andrawes and McGown, 1978)

metallic reinforcement can be identified as being a series system failure where failure of one element leads directly and immediately to failure of the whole. Extensible reinforcement can be identified as being equivalent to a parallel system, in which total failure occurs only when all the reinforcements fail simultaneously, Fig. 4.22. If any reinforcement is overstressed, creep will occur, leading to limited load shedding but not reinforcement rupture.

It can be concluded that the most stable reinforced soil structural form uses a rigid facing and is reinforced with a geosynthetic reinforcement.

4.3.2 Reinforcement distribution

4.3.2.1 Location

In order to establish which is the logical area for the reinforcement, potential

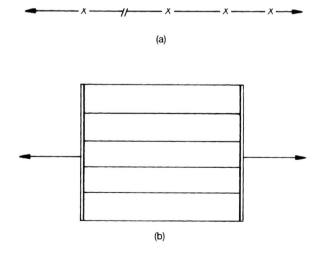

Fig. 4.22. (a) Series failure (inextensible reinforcement); (b) parallel failure (extensible reinforcement)

failure mechanisms and planes have to be established together with the associated strain fields. For optimum effect, reinforcement is positioned within the critical strain fields in the locations of greatest tensile strains.

4.3.2.2 Orientation

The general theory of the behaviour of reinforcement in soil presented earlier emphasizes the importance of the reinforcement being placed along the principal tensile strain directions developed in the soil alone, under the same stress conditions, Fig. 4.8. Changing the orientation of the reinforcement will reduce its effectiveness, and if orientated in the direction of the principal compressive strains, the action of reinforcement changes from that of tensile strain reinforcement to compressive strain reinforcement. If the reinforcement is orientated along the zero extension directions, an overall reduction in the strength of the reinforced soil may result, Figs 4.23 and 4.24.

In most reinforced soil structures the reinforcement is laid horizontally; in vertically faced structures this often results in the reinforcement being orientated in a near optimum plane although some work suggests that the optimum plane occurs with the reinforcement angled downwards at 10–15° from the horizontal (Smith and Birgisson, 1979). In other structural systems the choice of a horizontal plane for the reinforcement does not produce an efficient solution and may even reduce stability.

4.3.2.3 Spacing

In laboratory tests, Smith (1977) and Jewell (1980a and 1980b) have established that the increase in strength of a reinforced soil is not always directly proportional to the number of reinforcing elements in the system (all other things being constant). The spacing between separate reinforcing elements affects the performance of individual reinforcing members. Below a certain

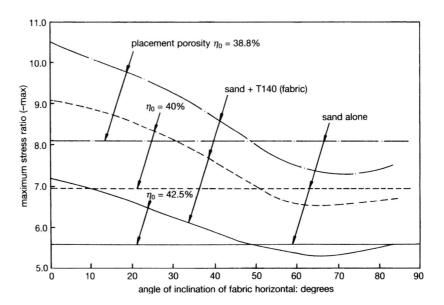

Fig. 4.23. The effect on the peak stress ratio of including T140 fabric in Leighton Buzzard sand for different initial sand porosities at a confining stress of 70 kN/m² (after Andrawes and McGown, 1978)

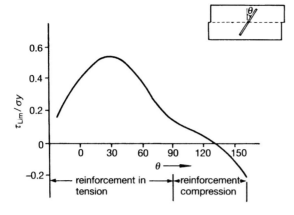

Fig. 4.24. *Variation in the maximum increase in shear strength of sand with the orientation of a single plane of stiff, tough reinforcement (after Jewell, 1980)*

spacing interference occurs, with the consequence that as the spacing reduces the increase in shear strength of the reinforced soil provided by each reinforcement is reduced. The interference between reinforcements in sand depends on the ratio of the spacing S_h, to the effective length of the reinforcement L_e, when L_e is defined as the length of reinforcement extending away from the critical plane in the soil, Fig. 4.25. The critical ratio is $S_h/L_e \geq 1$; below unity the influence of each reinforcement on the shear strength of the sand is progressively diminished.

Translating laboratory tests to practical conditions is difficult, but the implications of Fig. 4.25 suggest that given a choice of reinforcement, the larger reinforcement placed at wider spaces is more efficient than a greater number of smaller reinforcements placed closer together.

4.3.3 Soil

The soil used in a reinforced structure depends upon conditions and circumstances; in some instances the reinforcement function may be to improve a weak soil or waste material. Elsewhere, as in a bridge abutment, the soil may consist of a well-graded granular material compacted to a high density and exhibiting volumetric expansion during shear. The soil properties and the soil state will have a marked influence on behaviour when reinforced.

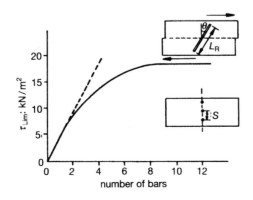

Fig. 4.25. *Test results from shear apparatus showing the influence of the spacing between rough reinforcing bars (after Jewell, 1980a)*

4.3.3.1 Particle size

The ideal partical size for reinforced soil is a well-drained, well-graded granular material, providing every opportunity for long-term durability, stability during construction and having good physiochemical properties (see Chapter 10, Durability). In the normal stress range associated with reinforced soil structures, well-graded granular soils behave elastically, and post-construction movements associated with internal yielding will not normally occur.

Fine-grained soils are normally poorly drained and effective stress transfer between reinforcement and soil may not be immediate, resulting in a slowed construction rate. Fine-grained soils often exhibit elastic-plastic or plastic behaviour, thereby increasing the chance of post-construction movements (see Chapter 5, Materials).

In many countries, such as parts of Japan, the availability of good cohesionless fill is limited and fine grained soils are used in reinforced soil structures. In these cases the use of drainage to create negative pore water pressures can be used to provide the necessary stability. This leads to the concept of combining the functions of reinforcement and drainage. Research into the use of geosynthetics with dual functions of drainage and reinforcement indicates that the increase in shear strength of the fine soil/geosynthetic composite is due in equal measure to the drainage and the reinforcement. Great care has to be taken when combining the materials that provide the reinforcement and drainage function together, as the result can be to produce structures with inbuilt planes of weakness. In order to develop the full potential of the reinforcement and drainage, a properly constructed *composite* geosynthetic is required (Heshmati, 1993). (See also Chapter 7, Construction).

4.3.3.2 Grading

A well-graded soil can be compacted to the required density and provides the most advantageous conditions to optimize the soil-reinforcement properties. Poorly graded soils may lead to the conditions associated with fine-grained soils. Uniform soils are undesirable and may lead to problems of stability (see Chapter 5, Materials).

4.3.3.3 Mineral content

Soils having a beneficial or benign composition with regard to the durability of the reinforcing elements are desirable. It is known that some clay minerals, such as illite, accelerate metal corrosion (see Chapter 10, Durability).

4.3.3.4 Index properties

See Chapter 5, Materials.

4.3.4 Soil state

4.3.4.1 Density

The density of a soil has an effect on the stress–strain relationship in soils, accordingly, relative density will influence some aspects of reinforced soil.

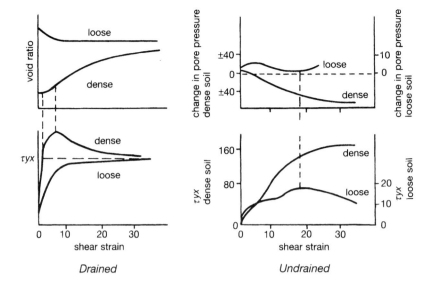

Fig. 4.26. Idealized stress–strain relationship of a granular soil

Fig. 4.26 shows the effect of density on the idealized stress–strain relationship of a granular soil.

The dense soil in the drained condition dilates during shear, whereas the loose soil has a lower deviator stress, no peak value and exhibits volumetric reduction. In the undrained state no volume changes occur; Bassett and Last (1978) has likened this state to conditions in a reinforced structure. The negative pore pressures developed during shear in a dense soil can be used to estimate the apparent increase in overburden stress. The effect is to increase the normal stresses acting on the reinforcement and enhance the apparent coefficient of friction between the soil and the reinforcement.

The effect of a dilating soil on the normal stress of a reinforcing element can be significant, however, the increase in stress may reduce rapidly with increasing shear strain, Fig. 4.27.

Both of the main analytical methods use the peak angle of friction, ϕ'_p, in design. Bolton (1986) has shown that the peak shearing resistance of granular

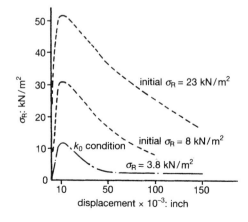

Fig. 4.27. Development of normal stress on the surface of a metal cylinder as it is displaced in sand (Note. Metal cylinder of equivalent size to full-size reinforcing strip) (after Bassett and Last, 1978)

57

fill, which takes account of soil density and mean stress, can be defined in terms of an empirical relative density index, I_R, for slopes and retaining structures up to 10 m in height.

$$I_R = 5I_D - 1 \tag{9}$$

where $I_D = (e_{max} - e)/(e_{max} - e_{min})$ is the definition of relative density. The peak friction angle, ϕ'_p, depends on the relative density index, I_R, such that

$$\phi'_p = \phi'_{cs} + 5I_R \tag{10}$$

Jewell (1990) has shown the link between compaction, specified as a percentage of the Proctor optimum dry density, and relative density, I_D, to be

$$e_{min} = \frac{Gs\,\gamma_w}{\gamma_d} - \quad \text{and } e_{field} = \frac{Gs\,\gamma_w}{0.95\,\gamma_d} - 1 \tag{11}$$

for compaction to 95% Proctor optimum, where G_s is the specific gravity of solids, γ_w is the weight of water, and γ_d is the dry density.

For routine fill where $\phi'_{cs} = 32°$ the relative density index for equation (9) would be $I_R = 2.7$, and the peak angle of friction from equation (2) would be $\phi'_p = 45°$. If compaction of the fill achieved only 90% of the Proctor optimum value, $I_R = 1$, the effect on the peak friction angle would be to produce $\phi'_p = 38°$ compared with $\phi'_p = 45°$. The effect with respect to the lateral earth pressure parameters used in the analytical models is to raise K_a (from 0·2 to 0·4) and K_0 (from 0·3 to 0·4), producing a significant increase in required tension in the reinforcement.

The relation between mobilizing shearing resistance and consolidation is described by the stress dilatancy equation, derived by Rowe (1962)

$$\phi'_p = \mathrm{fn}(\phi'_{cs}, \psi) \tag{12}$$

This may be simplified to the relationship, produced by Bolton (1986)

$$\phi'_p = \phi'_{cs} + 0.8\psi \tag{13}$$

The relation between maximum dilation, ψ_{max}, and relative density index, I_R, may be obtained from equations (1) and (5), and $\psi_{max} = 6.25\,I_R$ for plane strain. As with the angle of friction, the influence of poor compaction is to reduce the maximum dilation, ψ_{max}. If the fill is compacted to 90% of Proctor optimum the reduction of peak dilation ψ_{max} would be from 17° to 6°. One form of reinforcement, noticeably high adherence strip, relies on the restrained dilatancy effect to develop the apparent friction, μ^*, used in design; therefore, any reduction in dilation characteristics of the soil has a direct bearing on this assumption. The influence of dilation on the soil/reinforcement friction or adhesion of sheet material or geogrids may not be critical.

4.3.4.2 Overburden

Figure 4.28 shows the influence of overburden pressure on the pullout resistance of strip reinforcement. The apparent coefficient of friction decreases with increased overburden pressure. This is consistent with the general observation that the peak angle of shearing stress of a granular soil decreases with increase in normal stress.

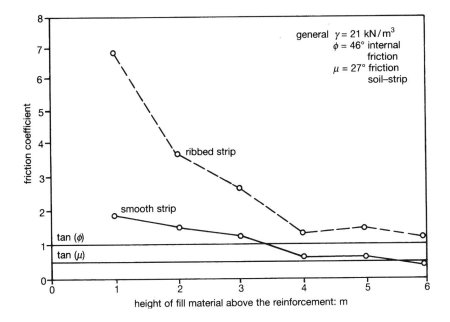

Fig. 4.28. Influence of the overburden pressure in pullout test (after Schlosser and Elias, 1978)

4.3.4.3 State of stress

As the stress within a cohesionless soil increases, so the critical void ratio decreases and the relative strain in the soil ϵ, compared to the strain in the reinforcement, reduces and the effective lateral stress tends to the active condition. Thus the state of stress within a reinforced structure will be different with increasing height and with different quantities and types of reinforcement. At the top of a vertically faced reinforced soil structure the stress state will tend to the at-rest condition, K_0; lower down, the active condition K_a will prevail, Fig. 4.29. This has been found in practice, and conforms to the normal state of stress behind conventional retaining walls (Sims and Jones, 1974 and Jones and Sims, 1975).

4.3.4.4 Degree of saturation

Well-graded cohesionless soils do not produce problems associated with saturation. Fine-grained materials, including cohesive soils, are usually poorly drained and effective stress transfer may not be immediate. The result

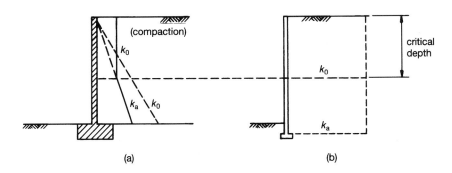

Fig. 4.29. Earth pressure distribution in reinforced soil structure: (a) conventional retaining wall; (b) reinforced soil wall

can be a temporary decrease in shear strength leading to a reduced construction rate, needed to ensure stability.

4.3.5 Construction

4.3.5.1 Geometry of structure

The nature of some soil structure problems demand structures having special geometries. In addition, a change in geometry such as steepening an embankment, will normally alter the strain field profiles within the structure. A change in geometry may increase or decrease the need for, or the effectiveness of, reinforcement.

4.3.5.2 Compaction

The use of modern compaction plant generates significant residual lateral pressures which suggest that at-rest (K_0) pressures predominate in many compacted fills. This condition has been confirmed in the case of earth pressures acting against retaining walls and bridge abutments, as well as in reinforced soil structures.

Compaction is controlled by shear strain; as shear strain is susceptible to the weight of the compaction plant, the larger plant will produce greater degrees of compaction. The degree of compaction is also dependent upon the number of applications or passes of the plant or roller; therefore, for uniformity between structures a method specification is to be preferred. The intensity of the lateral soil pressure produced by compaction is dependent upon the presence of fixed restraints such as abutments, wing walls, or double-sided structures and the direction of the roller. The compaction of soil in a reinforced soil structure is usually accomplished by using a roller running parallel to the face of the structure. As a result the residual lateral pressures parallel to the face of the wall are likely to be higher than those normal to the face.

The action of reinforcing members in soil during compaction will be to resist the shear strain in the fill caused by the plant. Tensile stresses will develop in the reinforcement proportional to the residual lateral pressure acting normal to the face of the wall. The presence of the reinforcement in the soil raises the

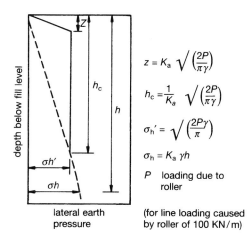

$$z = K_a \sqrt{\left(\frac{2P}{\pi \gamma}\right)}$$

$$h_c = \frac{1}{K_a} \sqrt{\left(\frac{2P}{\pi \gamma}\right)}$$

$$\sigma_h' = \sqrt{\left(\frac{2P\gamma}{\pi}\right)}$$

$$\sigma_h = K_a \gamma h$$

P loading due to roller

lateral earth pressure

(for line loading caused by roller of 100 KN/m)

Fig. 4.30. Compaction pressure (after Ingold, 1979)

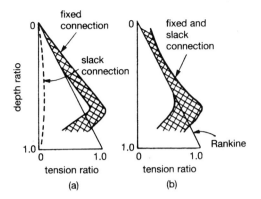

Fig. 4.31. Effect of slackness in reinforcement face connection (after Naylor, 1978): (a) at face; (b) at peak.

threshold for the residual lateral pressure which can be generated in the fill. The intensity of the lateral pressure generated by compaction can be derived from Fig. 4.30.

4.3.5.3 Construction systems

Specific techniques have been developed to aid the construction of reinforced earth and soil structures. Their application predetermines the use of certain materials, reinforcement forms, soil densities and construction geometries, and, therefore, influences the behaviour of the reinforced structure (see Chapter 7, Construction).

The mode of erection or construction of reinforced soil structures is such that internal movements after construction is complete are not normally possible. The result is that the internal stresses built in during construction cannot be relieved. A consequence, in some forms of construction using vertically faced structures, is that the connection of the facing to the reinforcement has to be designed to cater for the highest tensile stress in the reinforcement. If the facing could move laterally a small distance after construction by having some degree of slackness in the reinforcement/face connection, then the lateral pressures acting on the face would be reduced to the (K_a) active condition and the design criteria for the reinforcements would be reduced. This hypothesis has been confirmed by Naylor (1978) using mathematical modelling techniques, Figs 4.31, 4.32 and in laboratory studies by McGown et al. (1987). Adjustable connections between the reinforcement and the facing have been developed by Jones (1979).

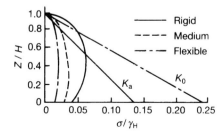
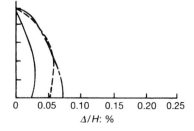

Fig. 4.32. Variation in lateral facing pressures and displacements for different facing stiffness (after McGown et al., 1987)

4.3.5.4 Durability

See Chapter 10.

4.4 Design theories

4.4.1 General

The modern approach to the design of reinforced soil is to use the limit state concept. This can be refined further into the consideration of *limit modes*. Any analytical model can be used within this framework.

A limit equilibrium approach can be used to provide a general analysis of reinforced soil. It is assumed that there is a critical plane through the reinforced soil and that this plane bisects the reinforcement; defining the plan area on the critical plane for each reinforcing member as A_S and the force in each member at the critical plane as P_R, the overall stress resultants in the soil may be calculated, Fig. 4.33.

Taking the reinforcement force P_R positive in tension, it is clear that the reinforcement has increased the normal stress and reduced the shear stress experienced by the soil. Thus, for reinforced cohesionless soils or drained cohesive soils, the increase in shear strength for the soil τ is given by:

$$\tau = \frac{P_R}{A_S}(\cos\theta\tan\phi' + \sin\theta) \tag{14}$$

where ϕ' is the angle of friction for the soil and θ is the orientation of the reinforcement with the critical plane.

For reinforcement with a high longitudinal stiffness in cohesionless soil:

$$P_R = \sigma_N A_R \tan\phi' \tag{15}$$

where σ_N is the stress normal to the reinforcement and A_R is the effective surface area of the reinforcement equal, in the case of a bar or a strip, to the product of the perimeter and the effective length of bar or strip. Or, in the case of a grid, to the product of the plan width and the effective length of the grid.

In the case of cohesive soil, the relevant value of P_R is that which exists at failure of the reinforced soil. The magnitude at failure depends upon the rate of shearing or the degree of drainage. The value for undrained conditions is not

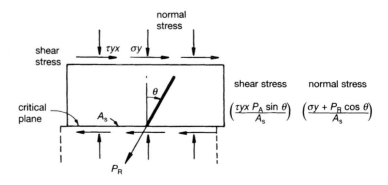

Fig. 4.33. Stress resultants on the critical plane of reinforced soil in direct shear

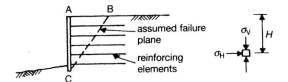

Fig. 4.34. Idealized reinforced soil structure

necessarily smaller than for drained conditions and depends upon, among other factors, the over-consolidation ratio, or the degree of compaction in the soil.

4.4.2 Vertically faced structures

4.4.2.1 Limit analyses

Vertically faced earth reinforced structures are designed in accordance with the principles of soil mechanics, Fig. 4.34. For convenience the analysis is usually considered in two parts.

(a) *Internal analysis.* This covers all areas relating to internal behaviour mechanisms, studies of stress within the structure, arrangement of the reinforcement, durability of the reinforcement and backfill properties. In design terms, internal analysis is associated essentially with adhesion and tension failure mechanisms (a), (b), in Fig. 4.35.

(b) *External analysis.* This covers the basic stability of the earth reinforced structure as a unit, including sliding, tilt/bearing failure, and slip within the surrounding subsoil or slips passing through the reinforced earth structure. Failure mechanisms of this nature are represented by (c), (d) and (e) in Fig. 4.35. In addition, stresses imposed upon the reinforced earth structure due to particular external conditions such as the creep of the subsoil have to be considered, Fig. 4.35(f).

Internal stability

The internal stability is concerned with the estimation of the number, size, strength, spacing and length of the reinforcing elements needed to ensure

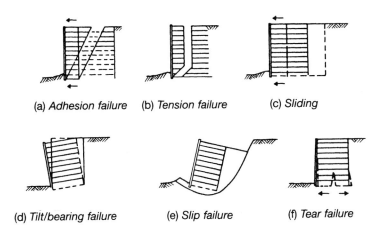

(a) *Adhesion failure* (b) *Tension failure* (c) *Sliding*

(d) *Tilt/bearing failure* (e) *Slip failure* (f) *Tear failure*

Fig. 4.35. Failure mechanisms

stability of the whole structure, together with the pressures exerted on the facing. Analyses to check for internal stability fall into two categories:

(a) those in which local stability is considered for the soil near a single strip or element of reinforcement, and

(b) those in which the overall stability of blocks or wedges of soil is considered.

Classical limit analysis methods have been proposed as design theories. Schlosser and Vidal (1969), Lee *et al.* (1973), Price (1975), Smith (1977) and the Department of Transport (1978) consider both types of analysis. Bolton and Choudhury (1976) consider the former, while Bacot (1974) considers the latter. Juran (1977) and Juran and Schlosser (1978) have developed a limit analysis method based upon a logarithmic spiral or circle.

The first method involves the consideration of the transfer of stress, from the soil to a single strip. Lee *et al.* and Bolton and Choudhury assume that each strip has to support a certain area of the skin (face) against which the soil exerts an active pressure, while Schlosser and Vidal assume that the reinforcement maintains the active earth pressure (K_a) in the soil. Both lead to the same force (T) in the reinforcement, i.e.

$$T = K_a \sigma_v S_h S_v \qquad (16)$$

where S_v is the vertical spacing and S_h the horizontal spacing of the reinforcement and σ_v the vertical stress caused by the soil overburden. Once the force in the strip is established, two modes of failure are considered. Firstly, the force may exceed the breaking stress of the reinforcement; secondly, it may not be possible for sufficient friction to occur between the strip and the soil to generate the force required, and failure occurs with a strip pulling out of the soil.

If the overall stability or equilibrium is considered, the same two failure mechanisms are normally assumed possible, i.e. reinforcement break or reinforcement slip. The method of calculating is to assume that the wedge is restrained by the reinforcement protruding through the wedge ABC into stable soil (Fig. 4.34). Either force or moment equilibrium is considered for the block ABC and the factor of safety can be calculated in two ways.

(a) Tension in the reinforcement (Lee *et al.*) is assumed to increase linearly with depth and the maximum necessary tension is compared with the maximum possible tension as in the local stability method.

(b) The maximum possible tension in each reinforcing element can be estimated. This is calculated from the pressure acting on the reinforcement (σ_v), the soil/reinforcement coefficient of friction, and the total surface area (top and bottom) of the reinforcement protruding beyond the failure plane. In the case of grid reinforcement the total surface area is the plan area of the grid. If the maximum possible tension in any reinforcement is found to exceed the breaking stress then the tension is set to the breaking stress. A factor of safety for the wedge can then be calculated equal either to the total restraining force divided by the total distributing force or to the total restraining moment divided by the total overturning moment.

From their work on models in a centrifuge, Bolton and Choudhury dispute the value of the equilibrium method and argue that for the reinforcement breaking mode it is reasonable to consider local equilibrium only, as once one layer of reinforcement breaks, the others will also break because of the dynamic effects caused by the sudden transfer of load.

4.4.2.2 Tension failures

Coulomb wedge theory. The reinforcement is assumed to be of sufficient length so as not to cause failure by lack of adherence. If the soil around the reinforcement is assumed to be in the limiting state equivalent to the active condition within the wedge, and assuming a linear distribution of tension in the layers of reinforcement, the maximum tension in the ith layer of reinforcement T_i is, Fig. 4.36:

$$T_i = \frac{n}{n+1} K_a \gamma H \Delta H \qquad (17)$$

where γ is the unit weight of the fill in the structure, H is the height of the structure, and ΔH is the zone of action of an individual layer of reinforcement.

Rankine theory. In 1967 the Laboratoire de Central des Ponts et Chaussées advocated the Rankine theory for maximum tension. The assumptions made with this approach include the following.

(*a*) The tension in the reinforcement is at maximum at the connection with the face unit.
(*b*) The direction of the principal stresses are either vertical or horizontal.
(*c*) The vertical overburden pressure near the vertical front face, $\sigma_v = \gamma H$.

Balancing the force developed within the bottom soil layer gives the maximum tensile force:

$$T_{\max} = K_a \gamma H \, \Delta H \qquad (18)$$

The limitations of the Coulomb and the Rankine methods lie with the assumption that the soil between the reinforcement is in a state of failure. A reinforced soil structure is usually constructed in layers and the distribution of stress and deformations will be different at each stage of construction with the

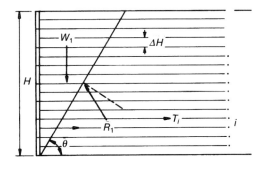
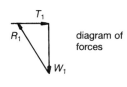

Fig. 4.36. Forces acting on a failure wedge

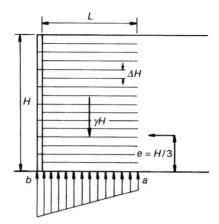

Fig. 4.37. Trapezoidal pressure distribution

soil in the upper region close to the at-rest (K_0) stress state, while lower down the active (K_a) stress state exists, Fig. 4.29.

Trapezoidal distribution. The effects of the backfill thrust (P) is neglected in the above arguments; the backfill thrust alters the state of stress within the block of reinforced soil and increases the vertical stress while increasing the tension in the reinforcement. Assuming a trapezoidal distribution of pressure under the base, Fig. 4.37.

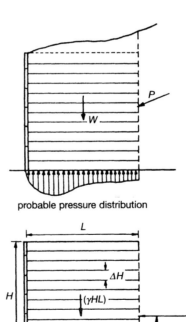

probable pressure distribution

Meyerhof distribution

Fig. 4.38. Probable and Meyerhof bearing pressure distribution

$$T_{max} = K_a \gamma H \ \Delta H \left[1 + K_a \left(\frac{H}{L} \right)^2 \right] \tag{19}$$

Meyerhof distribution. The probable pressure distribution is different from the trapezoidal, and a Meyerhof distribution has been suggested by Schlosser (1972), Fig. 4.38, which gives:

$$T_{max} = K_a \gamma H \ \frac{\Delta H}{\left[1 - 0 \cdot 3 K_a \left(\frac{H}{L} \right)^2 \right]} \tag{20}$$

Coulomb moment balance. By equating the moments due to the earth pressure and the reinforcement about the toe of a wall, the maximum tension in the bottom layer of reinforcement is defined as:

$$T_{max} = \frac{n^2}{(n^2 - 1)} K_a \gamma H \ \Delta H \tag{21}$$

Coulomb wedge. By assuming a Coulomb sliding wedge the sum of the tension components of the reinforcement (ΣT) may be expressed as:

$$\Sigma T = \left[\frac{FS - \tan \beta' \tan \phi'}{\cot \beta' \tan \phi' + 1} \right]^{1/2} \gamma H^2 \tag{22}$$

when FS = Factor of safety;

$$\tan \beta' = \sqrt{(\tan^2 \phi' + FS - \tan \phi')}$$

Elastic analysis. For a simple elastic analysis, using the finite element method, Banerjee (1975) found that the tension in a reinforcing element at a depth H is:

$$T_{max} = 0 \cdot 35 \gamma H \ \Delta H \tag{23}$$

Plotting the results from equations (17)–(23) shows the variation of the minimum cross-sectional areas of one row of strip reinforcing elements per metre width of a theoretical 5 m high wall for differing vertical spacing of the layers of reinforcement. It is apparent that the Coulomb wedge theory gives the minimum area of reinforcement and that the trapezoidal distribution gives the maximum, Fig. 4.39(a). Similarly, the following expression may be plotted to show the variation in factor of safety, FS, against failure due to lack of adherence for different vertical spacings of a strip reinforcing element, Fig. 4.39(b).

4.4.2.3 Adhesion failures

Where adhesion due to friction is relied upon to develop the combined behaviour mechanisms of soil and reinforcement, the bond length or length of adherence is critical.

Rankine theory (1). For a uniform normal stress $(\sigma_v = \gamma H)$ the factor of safety, FS, against an adhesion failure is:

$$FS = 2 \left(\frac{BL\mu}{K_a \Delta H} \right) \tag{24}$$

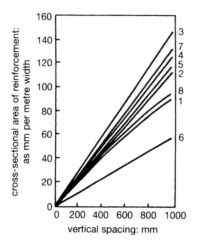
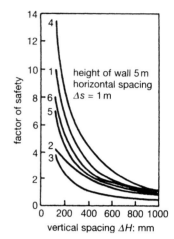

1,6,8 coulomb wedge
2 Rankine theory
3 trapezoidal distribution
4 Meyerhof distribution
5 coulomb moment balance
7 elastic analysis

1,6 coulomb wedge
2 Rankine theory
3 trapezoidal distribution
4 Meyerhof distribution
5 coulomb moment balance

Fig. 4.39. Comparison of different analytical methods

Meyerhof distribution. The vertical stress σ_v associated with the Meyerhof distribution differs from the uniform condition; consequently, the limiting length of adherence differs.

$$FS = \frac{L}{\dfrac{H^2 K_a}{3L} + \left[\dfrac{1}{1 - \frac{1}{3}K_a(H/L)^2}\right]\dfrac{K_a \Delta H}{2B\mu}} \tag{25}$$

Rankine theory (II). It can be argued that the part of the reinforcement which lies within the failure wedge may not be active in preventing failure by lack of adherence, in which case equation (24) can be modified to:

$$FS = \frac{2B\mu[L - H\tan(45 - \phi'/2)]}{K_a \Delta H} \tag{26}$$

Coulomb force balance. Considering the Coulomb theory, the overall factor of safety produced by taking balancing forces about the toe and equating is:

$$FS = \frac{4B\mu\Delta H}{K_a H^2}\sum_{i=N}^{n} i[L - (n-i)\Delta H\tan(45 - \phi'/2)] \tag{27}$$

where N is the number of the first layer of reinforcement to cross the theoretical failure line.

Coulomb moment balance. By taking moments, the Coulomb theory gives:

$$FS = \frac{12B\mu H^2}{K_a H^3}\sum_{i=N}^{n}(n-i)i$$

$$\times [L - (n-i)\Delta H\tan(45 - \phi'/2)] \tag{28}$$

Coulomb wedge. Resistance to the development of a Coulomb wedge failure using the adherence developed outside the wedge gives:

$$FS = \frac{2B\mu\gamma H\left[\dfrac{L}{2}(n+1) + H\tan\beta'\left(\dfrac{1-N^2}{6N}\right)\right]}{\frac{1}{2}\gamma H^2\left[\dfrac{FS - \tan\beta'\tan\phi'}{\cot\beta' + \tan\phi' + 1}\right]} \qquad (29)$$

As the height of a structure or wall increases, the adherence developed between the soil and the reinforcement will increase; as a result, for low walls at limiting factor of safety, the adhesion criteria rather than the tension criteria will normally be critical. Fig. 4.40 shows the relationship of spacing of strip reinforcing elements with heights of vertical structures and the point at which, in this particular case, tension or bond conditions become dominant. The values of the vertical spacing used in the figure are devised from the average results obtained from the expressions given above for tension and adhesion failure coupled with a factor of safety of three against failure by either mode.

The simplistic situation shown in Fig. 4.40 and the field performance of reinforced soil structures does not always agree. In particular, the above theories make no allowance for the effects of differing construction techniques, compaction or reinforcement possessing high adherence properties.

4.4.2.4 Logarithmic spiral

The classical limit analysis methods do not generally produce good agreement with observations on models and full-scale structures. In particular, the classical theories make no allowance for the presence of the reinforcing members and the restraint to lateral deformation that these engender. Juran (1977) concluded that the failure mechanism involves a rotation of a quasi-rigid block limited by a thin zone where the soil resistance to shearing is entirely mobilized. This failure zone separates the active zone and the resistant zone along the locus of the maximum tensile forces in the reinforcement, and is

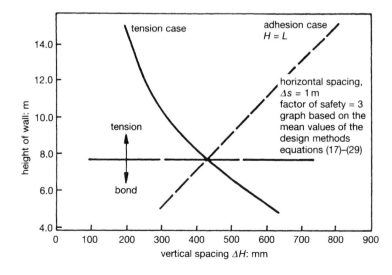

Fig. 4.40. Effect of wall height on failure criteria

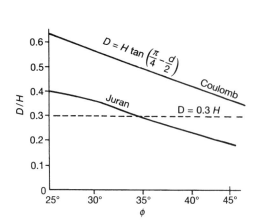 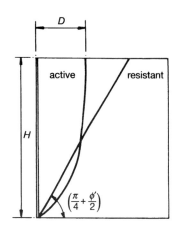

Fig. 4.41. Logarithmic spiral failure plane (after Juran, 1977)

orthogonal to the vertical free face to comply with the kinematic conditions of zero lateral displacement at the top of the wall or structure. A logarithmic spiral passing through the toe fits these conditions, Fig. 4.41.

The tensile forces are determined considering overall equilibrium of the active zone. The soil reaction along the failure surface is determined by integration of Köter's equation. By assuming that the horizontal shear stresses on each horizontal plane positioned between two layers of reinforcement are zero the tensile forces can be determined by the horizontal equilibrium of each soil layer having the reinforcement at its centre.

4.4.2.5 Elastic analysis

Analysis of the stress and deformation fields which develop in reinforced soil structures under normal conditions can be undertaken using elastic methods. During normal working conditions the state of stress within the structure is different from that prevailing at failure. The working condition state of stress may be equated with an elastic condition.

Finite element methods. Two finite element approaches are possible.

(*i*) The reinforced soil may be idealized as a unit cell or composite structure in which the reinforcement system is modelled as a locally homogeneous orthotropic material (Hermann and Al-Yassin, 1978). The composite material properties are given the equivalent properties of the soil matrix, the reinforcing elements and their composite interaction.

(*ii*) The reinforced soil is considered as a heterogeneous system in which the soil and the reinforcement are separately represented (Al Hussaini and Johnson, 1978).

An essential feature in the use of the finite element approach is that the analytical system should model accurately the following characteristics (Naylor, 1978):

(*a*) the longitudinal stiffness of the reinforcing elements
(*b*) the transfer of shear stress between the reinforcing elements and the soil

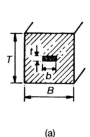
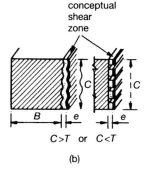

C>T or C<T

(a)   (b)

Fig. 4.42. Transformation from actual reinforced soil to equivalent material (after Naylor, 1978): (a) actual material; (b) equivalent material

(*c*) the transfer of shear through the soil in the vertical plane containing the reinforcement.

The longitudinal stiffness of the reinforcement may be incorporated as a ratio parameter, (*a*), defined in terms of the reinforcement cross-sectional area a_r, and the horizontal and vertical spacing, Fig. 4.42(a).

$$(a) = \frac{a_r}{S_v\,S_h} = \frac{Bt}{S_v\,S_h} \tag{30}$$

The transfer of shear stress between the reinforcement and the soil matrix may be incorporated by having the area of the equivalent reinforcement connected to the soil by the conceptual shear zone, C, the same as the surface area of the actual reinforcement, Fig. 4.42(b).

$$C = 2(B + t) \approx 2B \tag{31}$$

Therefore, $P = \dfrac{C}{S_v}$,

where P is a dimensionless bond area parameter.

Shear in the vertical plane may be accommodated by using a conceptual shear zone. (Note: Idealizing the reinforced soil system as a two-dimensional system, with the reinforcement as equivalent sheets, does not accommodate the transfer of shear through the soil in the vertical plane; any structure idealized in this mode would behave like a chest of drawers.)

4.4.2.6 Energy method

The energy method of analysis proposed by Osman (1977) is based on a consideration of the equilibrium of the external work due to earth pressure and the internal strain energy stored in the reinforcement. The following variables may be considered:

(*a*) the effect of reinforcement length on the magnitude of tension
(*b*) the variation in tension along a particular reinforcement and the distribution of tension with depth
(*c*) the deflected shape of the facing.

Figure 4.43 is a generalization of the earth pressure distribution and the

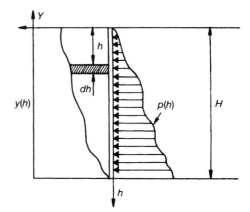

Fig. 4.43. Energy theory
parameters (after Osman et al.,
1979)

deflected shape of the structure. The total external work done by the earth pressure, U_{ext} per unit is given by the expression:

$$U_{ext} = \int_0^H p(h)\,y(h)\,\mathrm{d}h \qquad (32)$$

where $p(h) =$ earth pressure function; $y(h) =$ wall deflection function.

The energy method assumes that the external work done is stored in the reinforcement as elastic strain energy which may be calculated if the distribution of tension in the reinforcement is known. Assuming that:

(a) the distribution of tension along the reinforcement is linear, with the tension at the connection with the facing half the maximum tension
(b) the face deflection is parabolic and a function of the state of stress and the composite action of the soil and reinforcement
(c) the earth pressure distribution is hydrostatic,

the following relationships may be developed with regard to strip reinforcement:

Maximum reinforcement tension T' at depth h:

$$T' = \sqrt{\left(\frac{6K_a^{2.5}}{L}\right)} \gamma h S_v S_h \sqrt{(H - h)} \qquad (33)$$

Maximum reinforcement tension T_{max}:

$$T_{max} \sqrt{\left(\frac{8K_a^{2.5}}{9L}\right)} \gamma S_v S_h H^{1.5} \qquad (34)$$

Critical wall height, H_c:

$$H_c = \frac{P_{at}}{\gamma S_v S_h} \sqrt{\left(\frac{9L}{8K_a^{2.5}}\right)^{2/3}} \qquad (35)$$

Safety factor against reinforcement pullout, FS:

$$FS = \frac{2B\mu L^{1.5}}{S_v S_h \sqrt{[6K_a^{6.5}(H - h)]}} \qquad (36)$$

where b is the fill height above reinforcement, H is the total fill height above base of structure, S_v is the vertical reinforcement spacing, S_h is the horizontal reinforcement spacing, L is the reinforcement length, γ is the net weight of soil, K_a is the active pressure coefficient, P_{at} is the allowable tensile stress in the reinforcement, B is the strip reinforcement width, and μ is the reinforcement–soil friction coefficient.

4.4.2.6 Semi-empirical methods

Coherent gravity hypothesis
Studies on laboratory models and on full-scale structures show that the application of limit analysis methods does not produce results and performance characteristics consistent with experimental results. Semi-empirical methods have been introduced for practical design. The coherent gravity hypothesis relates to a reinforced soil structure constructed with a factor of safety and in a state of safe equilibrium. (Note: The design stresses relate to actual working stresses not to failure conditions. Thus the coherent gravity hypothesis relates to the serviceability limit state.) The coherent gravity hypothesis assumes that:

(*a*)   the reinforced mass has two zones, the active zone and the resisting zone, Fig. 4.44.
(*b*)   the state of stress in the fill, between the reinforcements is determined from measurements in actual structures constructed using well-graded cohesionless fill, Fig. 4.45.
(*c*)   an apparent coefficient of adherence, μ^*, between the soil and reinforcement is derived from an empirical expression developed from pullout tests on metallic strip reinforcements, Fig. 4.46.

Then, for a structure using strip reinforcement, the maximum tension per element at depth b, Fig. 4.47:

$$T_{max} = K\sigma_v \frac{\Delta H}{N} \tag{37}$$

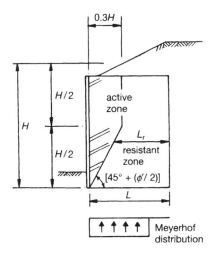

Meyerhof distribution

Fig. 4.44. Coherent gravity hypothesis

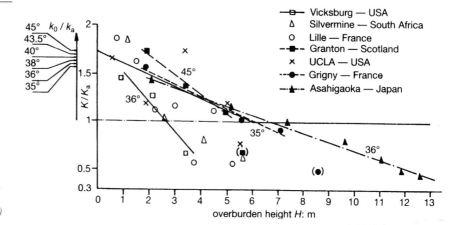

Fig. 4.45. Variation in earth pressure coefficient K (after Schlosser and Segrestin, 1979)

where $\sigma_v = \gamma h$, ΔH is the zone of action of the reinforcement, N is the number of reinforcements per area considered, and

$$K = K_0 \left(1 - \frac{h}{h_0} \right) + K_a \frac{h}{h_0}, \quad \text{for } h \leq (h_0 = 6 \text{ m}) \tag{38}$$

$$K = K_a, \quad \text{for } h > (h_0 = 6 \text{ m}) \tag{39}$$

Similarly, the maximum adhesion force per element of reinforcement, assuming a well-graded, cohesionless fill:

$$T_{ad} = 2B \int_{L-L_r}^{L} \mu^* \sigma_v \, dL \tag{40}$$

where

$$\mu^* = \mu_0 \left(1 - \frac{h}{h_o} \right) + \frac{h}{h_0} \tan \phi', \text{ for } h \leq (h_0 = 6 \text{ m}) \tag{41}$$

$$\mu^* = \tan \phi', \quad \text{for } h > (h_0 = 6 \text{ m}) \tag{42}$$

L_r is defined in Fig. 4.44. μ_0 for rough reinforcement is defined empirically as:

$$\mu_0 = 1 \cdot 2 + \log C_u \tag{43}$$

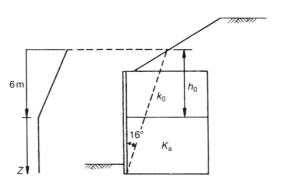

Fig. 4.46. Variation of apparent friction coefficient μ^* with depth (after Schlosser and Segrestin, 1979)

line of maximum tension

tension profile in reinforcement

ΔH

zone of action of reinforcement

Fig. 4.47. Zone of action of individual reinforcement

where

C_{u} is the coefficient of uniformity $\dfrac{D_{60}}{D_{10}}$

$\mu_0 = 0\cdot4$ for smooth reinforcement   (44)

Critically, the coherent gravity hypothesis does not consider a wedge stability check except in the case of unusual external loading or geometrics. There is evidence to suggest that this is a weakness in the method, which in some cases can result in failures (Lee *et al.*, 1994). (Additional consideration of coherent gravity hypothesis is given in Chapter 6, Design and analysis.)

4.4.2.7 Tie-back hypothesis

The tie-back hypothesis is based on the following design requirements for vertically faced structures.

(*a*) The design criteria is simple and safe.
(*b*) The design life of the structure is 120 years.
(*c*) The design procedure is consistent with the use of a wide range of potential fill materials, including frictional and cohesive–frictional soil. Any form of reinforcement may be used.

Internal stability considerations include:

(*a*) the stability of individual elements, Fig. 4.48
(*b*) resistance to sliding of upper portions of the structure
(*c*) the stability of wedges in the reinforced fill, Fig. 4.49
   (Note: Centrifuge studies indicate that at failure a wedge failure mechanism can develop. Therefore, the tie-back hypothesis relates to the ultimate limit state rather than the serviceability limit of the coherent gravity hypothesis.)

The following factors which influence stability are included in the design:

(*a*) the capacity to transfer shear between the reinforcing elements of the fill
(*b*) the tensile capacity of the reinforcing elements
(*c*) the capacity of the fill to support compression.

The state of the stress within the reinforced fill is assumed to be K_{a}. The at-rest K_0 condition measured in some structures, Fig. 4.30, is assumed to be a temporary condition produced by compaction during construction. The active state of stress is assumed to develop during the working life of the structure.

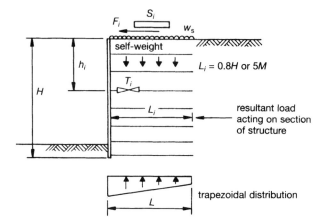

Fig. 4.48. Tie-back analysis—
local stability

Local stability. The maximum tensile force, $T_{\max}$ is obtained from the summation of the appropriate forces acting in each reinforcement:

$$T_{\max} = T_{hi} + T_{wi} + T_{si} + T_{fi} + T_{mi} \tag{45}$$

where T_{hi} = reinforcement tension due to fill above the reinforcement layer,

T_{wi} = reinforcement tension due to uniform surcharge,

T_{si} = reinforcement tension due to a concentrated load,

T_{fi} = reinforcement tension due to horizontal shear stress applied to the structure, and

T_{mi} = reinforcement tension due to bending moment caused by external loading acting on the structure.

Centrifuge studies have shown that at failure a wedge failure mechanism can occur. A Coulomb wedge failure is assumed to be possible within the reinforced soil structure. For design, wedges of reinforced fill are assumed to behave as rigid bodies of any size or shape and all potential failure planes are investigated (i.e. η and β' may both vary, Figs 4.49, 4.50). (Additional consideration of the tie-back hypothesis is given in Chapter 6, Design and analysis.)

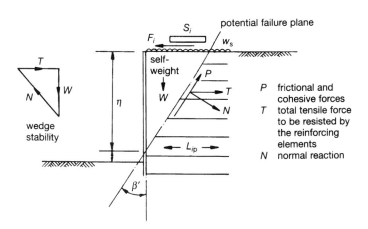

Fig. 4.49. Tie-back analysis—
wedge stability

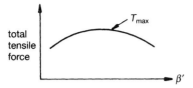

Fig. 4.50. Angle of failure plane

4.4.3 Overall stability—vertically faced structures

The overall stability of the reinforced soil structure is normally taken to be the same as for a mass or conventional retaining structure. Stability is checked for forward sliding at the base, bearing pressure failure beneath the toe and slip circle failure of the whole mechanism, Fig. 4.35(c),(d),(e). In good foundation conditions this approach is adequate but with poor subsoils the inherent ability of the reinforced soil mass to stabilize or alleviate the poor ground conditions is not utilized and in these circumstances a more realistic approach to the global stability problem is required. Thus, the consideration of the external stability is supplemented by a consideration of the overall settlement characteristics of the soil structure and any embankment it supports together with a consideration of the shear stress and the mobilized strength of the subsoil.

A soil structure founded on good subsoil usually complies with the tilt/bearing criteria, and the assumption of a trapezoidal or Meyerhof bearing pressure distribution is adequate. However, on weak subsoils the use of these empirical rules produces design difficulties, and it is possible to demonstrate in geometrical terms that conventional earth retaining structures are better suited to soft subsoils than soil structures, Fig. 4.51. This apparently paradoxical situation arises through a lack of appreciation of the overall design problem in which the internal design is only a part. The nature of the foundation type will have an influence on the overall behaviour of the structure.

4.4.3.1 Stiff foundation

In the case of a reinforced soil structure standing upon a rigid or very stiff foundation, it is reasonable to assume a trapezoidal bearing pressure distribution beneath the foundation similar to that depicted in Fig. 4.51. Numerous model tests have been undertaken on this condition and it is from the results of these that many design theories have evolved. Inherent in the acceptance of the trapezoidal or Meyerhof pressure distribution beneath the structure is the acquiescence that the reinforced earth mass and the front face of the structure

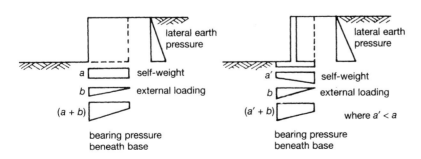

Fig. 4.51. Bearing pressures beneath a reinforced soil structure and a cantilever wall

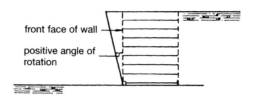

Fig. 4.52. Behaviour of a stiff–stiff material system

will tend to rotate about the toe with an active mechanism away from the fill. Because of its method of construction it is reasonable to describe an earth reinforced structure in these conditions as a stiff structure interacting with a stiff subsoil, i.e. a stiff–stiff material system, Fig. 4.52.

4.4.3.2 Soft foundation

The use of the concept of the trapezoidal pressure distribution, together with the assumption that the structure will rotate forward about its toe due to the loadings depicted in Fig. 4.51, does not hold in the case of a yielding foundation. With a soft foundation the self-weight of the stiff reinforced earth structure and the weight of the adjacent material, which it is supporting, may cause the structure to rotate in a negative sense. Thus, the behaviour of the stiff–soft system is fundamentally different to that of a stiff–stiff material system, Fig. 4.53.

The most significant conclusion that can be drawn from a study of the interaction of a stiff–soft material system is that the global failure criterion is not one of overturning about the toe, as the application of some design methods suggest, but that of a rotational slip through the retained embankment. The rotational behaviour of the reinforced soil mass in these circum-

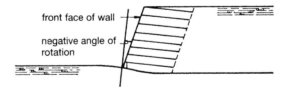

Fig. 4.53. Behaviour of a stiff–soft material system

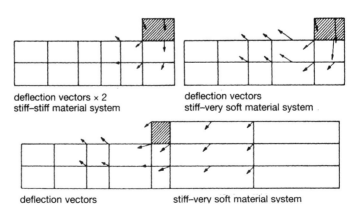

deflection vectors × 2
stiff–stiff material system

deflection vectors
stiff–very soft material system

deflection vectors

stiff–very soft material system

Fig. 4.54. Deflection vectors

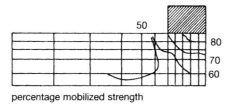

Fig. 4.55. *Deflection of original ground level; stiff–semi soft material system*

stances is similar to that experienced by the bridge abutments built on soft ground reported by Nicu *et al.* (1970) and Daniels (1973). The relative behaviour of stiff–soft material systems can be seen from the plots of the accumulated deflection vectors at the end of the individual incremental loading sequences following completion of construction, Fig. 4.54. The conclusion that can be gained is that it is the level of resistance to movement of the toe of the reinforced soil structure produced by the subsoil material which radically alters the validity of the assumption that the structure will rotate forward about the toe.

The difference between semi-soft cohesionless foundation and very soft cohesive foundation, each loaded with identical reinforced soil structures, can be seen by comparing the percentage of mobilized strengths (MS) depicted in the two subsoils, Figs 4.55 and 4.56. It is from a study of this parameter that the overall stability of the structure can be determined.

$$MS = \frac{(\sigma_1 - \sigma_3)}{(\sigma_1 - \sigma_3)_{ult}} = \frac{(\sigma_1 - \sigma_3)[1 + \sin \phi]}{2c' \cos \phi' + \sigma_3 \sin \phi'} \times 100 \qquad (46)$$

The difference between a narrow and a wide embankment can be seen by comparing Figs 4.56 and 4.57 which have identical foundation conditions and on which the reinforced soil structure is of an identical shape and stiffness.

The difference in the percentage mobilized strength in the case of the wide embankment is due principally to the influence of the weight of the retained embankment, behind the earth reinforced wall. It is the marked difference

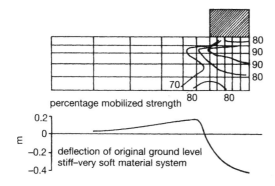

Fig. 4.56. *Deflection of original ground level; stiff–very soft material system*

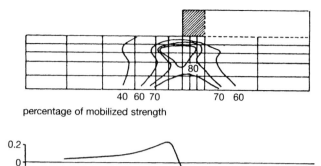

percentage of mobilized strength

Fig. 4.57. Deflection of original ground level; stiff–very soft material system

deflection of original ground level
stiff–very soft material system

between Figs 4.56 and 4.57 which demonstrates the necessity of a wholistic approach to the global analysis of a reinforced soil structure resting on a yielding foundation. As confirmation of the analysis produced by a study of the mobilized strength of the subsoil provided by the finite element method, equilibrium methods using the failure planes predicted by the mobilized strength plot can be used.

Inspection of the internal stresses in the bottom of the structure will indicate the probability of the potential failure mechanism shown in Fig. 4.35(f). One possible counter to this tear failure system is the use of high-strength reinforcement having a low elastic modulus, thereby providing an element of flexibility in the base of the structure to accommodate the consolidation strain which may occur.

4.4.4 Embankment and cuttings

4.4.4.1 Embankments

Reinforced embankments can take several forms depending on the nature of the problem to be solved. Reinforcement may be used for the following function, Fig. 4.58:

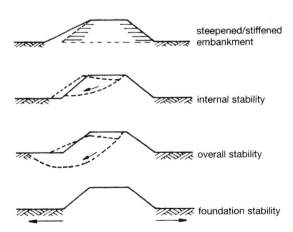

steepened/stiffened embankment

internal stability

overall stability

foundation stability

Fig. 4.58. Reinforcement functions in an embankment

(*a*) to permit increased compaction to stiffen the embankment
(*b*) to enable the embankment to be steepened
(*c*) to assist in internal stability problems
(*d*) to assist in overall stability problems
(*e*) to assist in foundation stability problems.

The mechanics of a reinforced embankment on a soft foundation can be likened to a bearing capacity problem with a surface load. An unreinforced

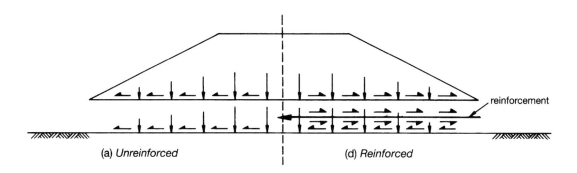

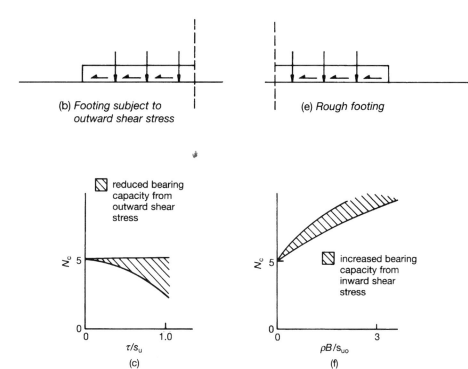

Reinforcement improves stability by:
1. carrying the outward (disturbing) shear stress
2. providing inward (resisting) shear stress

Fig. 4.59. The mechanics of a reinforced embankment

embankment exerts a worse loading on the embankment than a smooth footing, because the lateral thrust developed in the embankment fill creates outward shear stress which acts to reduce the bearing capacity of the foundation. The reinforcement at the base of the embankment has two functions:

(*i*) to support the outward thrust from the embankment fill; in this situation, the reinforcement is acting to reduce the forces causing failure

(*ii*) to restrain the surface of the foundation soil against lateral displacement; in this situation, the reinforcement is acting to increase the forces resisting failure, Fig. 4.59.

The general factors governing the behaviour of reinforced soil mentioned earlier may be applied to embankments. For design purposes, mathematical modelling techniques may be adopted (Sims and Jones, 1979; Jones, 1980), alternatively limit equilibrium methods of analysis may be employed. In the latter, the reinforcement is modelled as a thin, cohesive layer or a search is made for the failure mechanism that requires the maximum reinforcement force for stability, and a comparison made with the available reinforcement tensile strength (see also Chapter 6, Design and analysis).

4.4.4.2 Reinforced cuttings and soil nailing

In the case of cuttings, reinforcement may be used *in situ*, or placed in a layered construction, similar to embankment structures. In the latter condition an element of over excavation is required which is most likely to arise when using earth reinforcing techniques to reinstate or correct unstable slopes, Fig. 4.60. The use of reinforcement *in situ* as a form of soil stabilization in cuttings is possible only when it forms part of a well-developed technique used to form the cut. Techniques which have been developed are soil nailing and the lateral earth support system, Fig. 4.61.

Analytical techniques used for the reinforcement of cuttings range from the more rigorous methods of the slip-line method and limit analysis, to the limit equilibrium methods which produce approximate but reliable solutions of complex situations. The latter technique assumes a failure mechanism, a failure surface and the stress distribution along the failure surface, such that an expression of equilibrium in terms of stress resultants may be produced for any condition; solution is by statics. However, for a given set of soil parameters and soil stresses, the shape of the failure surface affects the solution of the

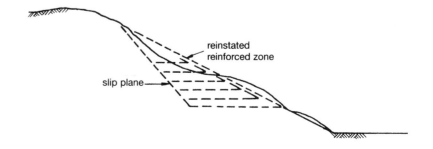

Fig. 4.60. Reinforcement of unstable slope

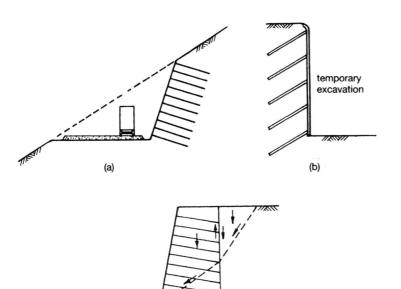

Fig. 4.61. Reinforcement placed
in situ: (a) soil nailing;
(b) lateral earth support system

Fig. 4.62. Failure mechanism in
soil nailed structure

problem and it is important that the failure surface used in the analysis is a close
approximate of the true failure surface of the soil structure. Thus, in the case of
a failed cutting slope the slip plane is known. In the case of soil nailing,
experimental evidence (Gässler and Gudehus, 1981), indicates that with a
cohesionless soil block, failure accompanied by rotation or translation is
probable, Fig. 4.62, although other mechanisms are possible, Fig. 4.63.

The potential failure mechanism for lateral earth support systems has been
studied using centrifuge techniques which suggest that these soil structures are
relatively rigid and coherent and that they are capable of sustaining large
deformations. Failure of the lateral support system is assumed at the initiation
of surface cracks rather than total collapse, a failure surface represented by a
parabola passing through the toe is the expected mechanism, Fig. 4.64. (See
also Chapter 6, Design and analysis.)

4.4.5 Composite reinforcing systems

In many reinforced soil systems a single consistent reinforcing material is used

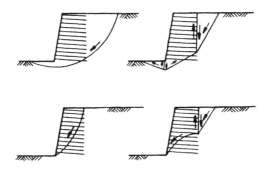

Fig. 4.63. Failure mechanisms

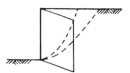

Fig. 4.64. Failure passing through toe

in conjunction with a uniform soil. It is possible, and it may be advantageous, to use two or more different reinforcing materials or systems in the same structure, Fig. 4.65.

When two separate reinforcing materials are used, each is likely to have a different response in terms of reinforcing the soil in keeping with the variables considered in Table 4.1. For practical purposes, assuming that the location, orientation and spacing of the different reinforcements are compatible, the effects of different longitudinal stiffness may be accommodated by equating the contribution of the lower modulus material in terms of an equivalent material

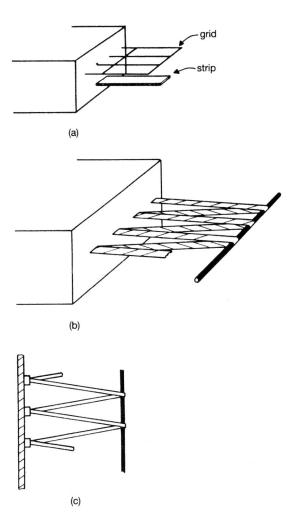

Fig. 4.65. Composite reinforcing systems: (a) grid and strip; (b) and (c) strip and anchor

to that of the stiffer reinforcement. Thus in Fig. 4.65(a) if grid (ar_g, Er_g) and strip (ar_s, Er_s) are combined:

where ar_g is the cross-sectional area of the longitudinal elements of the grid,
   ar_s is the cross-sectional area of the strip,
   Er_g is the elastic modulus of the grid reinforcement,
   Er_s is the elastic modulus of the strip reinforcement, and
   $Er_g = 10Er_s$.

Equivalent reinforcing material

$$= ar_g Er_g + ar_s \frac{Er_g}{10}$$

$$= Er_g \left(ar_g + \frac{ar_s}{10} \right) \tag{47}$$

Similarly in Fig. 4.65(b), the adhesion resistance or resistance to pullout, T_{ad} per metre width of composite reinforcement of length L at a depth h_i, made up of strip reinforcement and transverse anchors, may be derived from equations (5) and (8).

$$T_{ad} = 2LP_i \gamma h_i \mu + N_w d N_q \gamma h_i \tag{48}$$

where P_i is the total width of the strip reinforcement per metre width,
   N_w is the number of transverse anchors,
   N_q is a bearing capacity coefficient,
   d is the diameter of the anchor.

4.5 Seismic effects

During an earthquake, acceleration of the material mass of the backfill of a soil structure may occur; in time these may cause additional forces to develop in the reinforcing elements. The total force in each reinforcement can be assumed to be the sum of the static forces before the seismic event, plus the dynamic forces generated during the earthquake activity. At the end of the earthquake it is assumed that reinforcement forces will return to the initial static condition. It has been established that two internal failure modes can exist with reinforced soil; failure of the reinforcement in tension or failure of adhesion between the soil and the reinforcement. During an earthquake, failure by breakage of reinforcement could be catastrophic, accordingly a failure mechanism based upon loss of adhesion is preferred. In this situation distortion of the structure may occur during seismic conditions but stability is restored once this ceases. The observed deformation mechanism is one of forward rotation about the toe, Fig. 4.66.

4.5.1 Seismic designs

Seismic design methods are based on two approaches:

(a) prediction of dynamic reinforcement tensions resulting from the ground motion
(b) selection of reinforcement adhesion failure as the design criteria, coupled

with the determination of the probable deformation and a check that the latter is within the serviceability limits.

4.5.1.1 Prediction of dynamic tension method

Richardson and Lee (1975) have suggested an empirical method for estimating the dynamic forces in each reinforcing element, based upon a technique of spectral analysis of models. The total lateral dynamic force is assumed to be proportional to a design spectral acceleration, A_{des}:

$$A_{des} = \Gamma_1 S_{a1} + \Gamma_2 S_{a2} \tag{49}$$

where Γ_1 and Γ_2 are the first and second modal participation factors and S_{a1} and S_{a2} are the first and second spectral accelerations. S_{a1} and S_{a2} are functions of the natural frequencies and damping of the reinforced soil structure. It has been established that the density of reinforcement influences the distribution of dynamic reinforcement tension. Tests on full-sized structures have shown that the low strain first and second fundamental periods (T_1, T_2) of reinforced soil walls, height H m, to be:

$$T_1 = \frac{H}{38}, \quad T_2 = \frac{H}{100} \tag{50}$$

Richardson (1976) has developed an empirical stiffness factor, I based upon a conceptual stiffness coefficient for the structure. The stiffness coefficient is the second moment of the ultimate tensile forces resisting deformation about the base, Fig. 4.67. In the upper part of the structure the ultimate tensile force will be governed by adhesion criteria lower in the structure and in tall structures tensile forces will be dependent upon the tensile strength of the reinforcement.

The stiffness factor, I, is defined as relating to a structure with a factor of safety of one against reinforcement failure under static forces. With $I < 2.0$, peak dynamic strain is inversely proportional to structural stiffness. The distribution of dynamic forces, D_F within the reinforcement system as a function of wall stiffness is shown in Fig. 4.68.

The effective mass, M_{eff} of a soil structure affected during an earthquake may be defined by the empirical expression:

$$M_{eff} = 0.75 K_0 \gamma H^2 / g \tag{51}$$

and the total dynamic force, D_F is given by:

$$D_F = (S_{a1} + 0.2 S_{a2}) M_{eff} \tag{52}$$

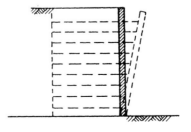

Fig. 4.66. Deformation mechanism under seismic conditions

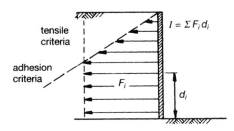

tensile criteria

adhesion criteria

$I = \Sigma F_i d_i$

F_i

d_i

Fig. 4.67. Ultimate tensile forces resisting deformation

If the design soil structure stiffness, I' is in the range

$$0 \cdot 9I < I' < 1.1I \qquad (53)$$

where I is the initially assumed normalized stiffness $= 1 \cdot 0$, then design is complete. If $I' \gg I$ the design is conservative, but if $I' \ll I$ the lateral dynamic earth pressures are underestimated.

Note: The above method of analysis is based on an empirical correlation between the displacements measured in blast tests on a full-scale vertical structure and that predicted by a computer program, developed to provide an initial estimate of the acceleration amplification and displacement due to horizontal shearing of structures (Idriss and Seed, 1968). The strain correlation between the program and the test structure is dependent upon the density of reinforcement, the orientation of the reinforcement and the gravity of the wall and backfill. Variations of the geometry of the backfill can produce considerable variations in dynamic lateral stress, Fig. 4.69. Therefore, the above method would appear to be appropriate to conventional vertical structures, but not necessarily to embankment structures having sloping sides.

4.5.1.2 Displacement method

The displacement method assumes that the construction will fail by reinforcement pullout with movement occurring on a known failure surface. By

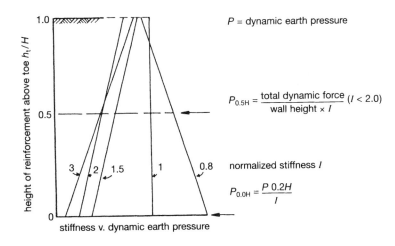

height of reinforcement above toe h_t/H

1.0

0.5

0

3 2 1.5 1 0.8

stiffness v. dynamic earth pressure

P = dynamic earth pressure

$$P_{0.5H} = \frac{\text{total dynamic force}}{\text{wall height} \times I} \ (I < 2.0)$$

normalized stiffness I

$$P_{0.0H} = \frac{P\ 0.2H}{I}$$

Fig. 4.68. Influence of wall stiffness

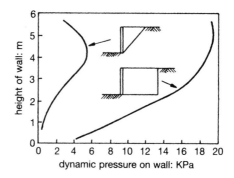

Fig. 4.69. Dynamic pressures on conventional retaining walls for different configurations of backfill (after Bracegirdle, 1979)

determining the distribution of tension in the reinforcement a critical horizontal acceleration, K_c applied as a pseudo static force can be calculated to produce failure on the predetermined slip plane. The design is undertaken using a further assumption of an average seismic design coefficient, K_m. If $K_m > K_c$ slippage will occur and displacements of the reinforced structure may be determined. Failure within the reinforced structure may occur within the reinforced mass (a contained failure) or may extend into the backfill (an uncontained failure).

(a) Contained failure
Summing vertically and horizontally, Fig. 4.70:

$$K_c - \Sigma T'_{max} + R\cos(\theta + \phi) = 0 \tag{54}$$

$$W - R\sin(\theta + \phi) = 0 \tag{55}$$

$$\text{Therefore, } K_c = \frac{2\Sigma T'_{max}}{\gamma(H - h_t)^2 \tan\theta S_h} - \cot(\theta + \phi) \tag{56}$$

The lowest value of K_c is obtained when the failure plane passes through the toe of the structure ($h_t = 0$). Assuming a uniform distribution of strip reinforcement ($S_h = S_v$):

$$T'_{max} = 2BL\gamma[H - iS_v]\mu$$

$$= 2B\gamma\mu(H - iS_v)(L - nS_v\tan\theta) \tag{57}$$

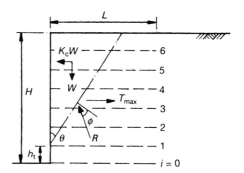

Fig. 4.70. Seismic design–displacement method (contained)

Therefore $\displaystyle\sum_{i=1}^{N} T'_{\text{max}} = \gamma B\mu(NHL - NH^2)$

$$+ N\frac{(2N+1)}{3(N+1)}H^2 \tan\theta \tag{58}$$

where

$$20 = \beta' + \cos^{-1}\left[\frac{(\alpha''-1)}{(\alpha''^2 - 2\alpha''\cos 2\phi + 1)^{1/2}}\right]$$

$$\alpha = 2\beta\frac{NL\mu}{HS_{\text{v}}}$$

and

$$\tan\beta'' = \frac{\alpha'' \sin 2\phi}{(-\alpha'' \cos 2\phi)}$$

By determining (T'_{max}) from equation (58), substitution into equation (56) yields K_{c}. The method is valid for any contained slip surface at any height, h_{t} within the wall.

(b) Uncontained failure
From area (1), Fig. 4.71

$$Q = W_1 \frac{K_{\text{c}} \sin(\theta_1 + \phi_1) + \cos(\theta_1 + \phi_1)}{\sin(\delta + \theta_1 + \phi_1)} \tag{59}$$

where Q is the resultant inter-slice force.

From area (2),

$$Q = \frac{(\Sigma T'_{\text{max}} - K_{\text{c}}W_2)\sin(\theta_2 + \phi_2)}{\sin(\delta + \theta_2 + \phi_2)} \tag{60}$$

from which

$$K_{\text{c}} = \frac{\Sigma T'_{\text{max}} \sin(\theta_2 + \phi_2) - \cos(\theta_2 + \phi_2) - \Omega\frac{W_1}{W_2}\cos(\theta_1 + \phi_1)}{\Omega\frac{W_1}{W_2}\sin(\theta_1 + \phi_1) + \sin(\theta_2 + \phi_2)} \tag{61}$$

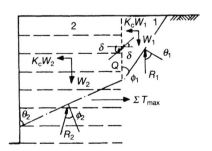

Fig. 4.71. Seismic design–displacement method (uncontained)

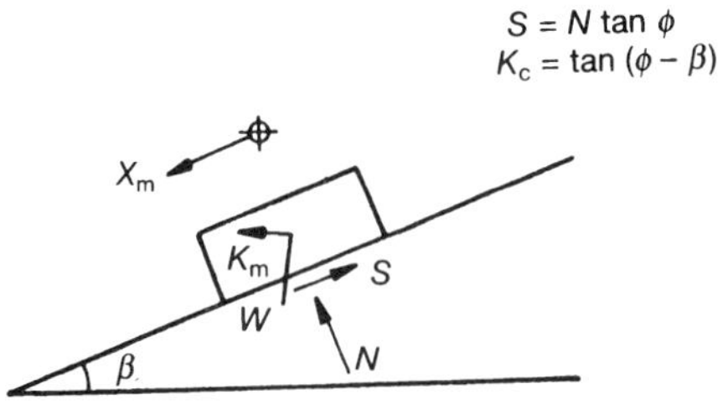

$$S = N \tan \phi$$
$$K_c = \tan (\phi - \beta)$$

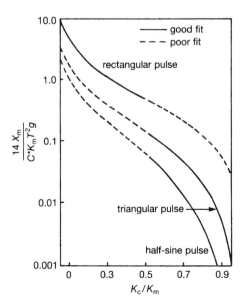

Fig. 4.72. Sliding block method of analysis

where

$$\Omega = \frac{\sin(\theta_2 + \phi_2 + \delta)}{\sin(\theta_1 + \phi_1 + \delta)} \text{ and } 0 \leq \delta \leq \phi$$

The minimum value of K_c in equation (61) may be obtained by differentiation.

(c) Average seismic coefficient, K_m
The average seismic coefficient, K_m is the maximum inertial force contained within the volume of soil defined by the free surface and the slip surface. If the reinforced structure acts as a rigid block then K_m is the maximum base acceleration, an exception being where a large surcharge exists when model participation factors may be derived (Bracegirdle, 1979).

(d) Displacement
When K_m exceeds K_c, displacement X_m may be determined using the sliding block method developed to calculate the displacement of earth dams (Newmark, 1975), Fig. 4.72.

Fig. 4.73. Sliding block mechanism

From Fig. 4.72

$$D - R = \frac{W \cos(\phi - \beta)}{\cos \phi}(K_m - K_c) \qquad (62)$$

where D is the driving force, R is the resisting force, and $K > K_c$.

Displacements X_m, with respect to time, may be obtained from equation (62) using Newton's law of motion. These have been obtained by Sarma (1975), for rectangular, triangular and half-sine pulses of period T, Fig. 4.73. In practice, T is the predominant period of the ground motion and the term C^*, Fig. 4.73:

$$C^* = \frac{\cos(\phi - \beta)}{\cos \phi} \qquad (63)$$

from which X_m may be obtained for values of K_c and K_m.

4.6 Mining subsidence

The mining of coal or other minerals results in earth movements in the vicinity of the excavated area. These movements, known collectively as subsidence, are three-dimensional in nature, any affected point having components of displacement along all three axes of a general Cartesian co-ordinate system (Jones and Bellamy, 1973). The displacements are imposed on any structure in the affected zone and may result in damage or even collapse unless adequate safeguards have been made in the design of the structure. The displacements of the ground surface subjected to mining subsidence are illustrated in Fig. 4.74.

In many conventional building structures both the tensile and compressive ground strain may cause damage. With conventional earth-retaining structures the tensile strain phase does not usually produce a problem, but the compressive phase may result in a significant increase in the lateral earth pressures. Reinforced soil structures are unique in that their susceptibilities to damage from compressional strain is very low, however, the tensile strain in the ground beneath a reinforced soil structure can increase the tensile stress in the bottom layers of the reinforcement and a tear failure mechanism must be guarded against, Fig. 4.35(f).

A review of the effects of mining subsidence on reinforced soil undertaken by the UK Transport and Road Research Laboratory confirmed that reinforced soil walls and bridge abutments in France and the UK have survived mining subsidence events without any distress. The relevance and desirability of using the K_0 state of stress when structures in areas of mining subsidence are being designed has not been resolved (Jones, 1989).

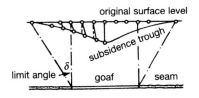
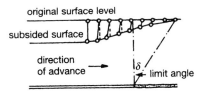

Fig. 4.74. Surface displacement due to mining subsidence

4.7 Repeated loading

Many reinforced soil structures will be subjected to repeated or cyclic surcharge loading; in addition, some may also be subjected to repeated direct loading of the reinforcing elements. Heavy traffic loading will produce cyclic surcharge loading, and wave action may cause repeated direct loading of the reinforcement.

Little attention has been given to the long-term effects of repeated loading on reinforced earth structures. Murray *et al.* (1979) conducted pullout tests on model reinforcing strips under static and vibrating loads which showed a significant loss of pullout resistance under the effect of vibration. Richardson and Lee (1975), Richardson and Lee (1976), and Bracegirdle (1979) have studied the effect of dynamic loading associated with seismic work. Al-Ashou (1981) in a large-scale study, established the following criteria related to repeated loading on strip and anchor reinforcements.

(*a*) The behaviour of reinforcement under repeated loading has an initial stable state representing the major part of the loading period. If failure occurs, this will happen following a period of accelerated slip.

(*b*) Loading amplitude is the critical element in the loading-life of strip reinforcement for all types of repeated loading.

(*c*) No slippage will occur under static surcharge and cyclic loading when the loading amplitude is $< 0.25 P_u$ (P_u = ultimate pullout resistance of the reinforcement).

(*d*) The effect of repeated loading is to cause a redistribution of load along the reinforcement and breakdown of the frictional resistance at the soil/reinforcement interface.

(*e*) Depth of surcharge level has no influence on the behaviour of reinforcing strip other than to affect the length of the initial stable stage.

(*f*) In the case of strip reinforcement a residual load of 13–16 per cent of the peak amplitude is locked into soil adjacent to the reinforcement.

(*g*) The ultimate pullout capacity of strip reinforcement may be reduced by 20–30 per cent after repeated loading.

(*h*) Application of cyclic surcharge improves pullout capacity and extends the life of the reinforcement under repeated loading.

(*i*) Under a static tensile load and cyclic surcharge pressure the reinforcement may experience an initial displacement of 1–3 mm, followed by a very stable state.

(*j*) No slippage will occur under cyclic surcharge and cyclic loading when the loading amplitude is $< 0.5 P_u$.

(*k*) Anchors significantly increase the pullout resistance of strip reinforcement; a 20 mm anchor increases pullout 23 per cent, and a 40 mm anchor increases pullout 45 per cent.

(*l*) Reinforcement formed with anchors has a gradual failure mode.

4.8 References

AL-ASHOU M. O. (1981). *The behaviour of Reinforced Earth under repeated loading*. Ph.D. Thesis, Sheffield University.

AL-HUSSAINI M. M. and JOHNSON L. D. (1978). Numerical analysis of Reinforced Earth Wall. *Proc. ASCE Symp. Earth Reinforcement*, Pittsburgh.

ANDRAWES K. Z. and McGOWN A. (1978). Alteration of soil behaviour by the inclusion of materials with different properties. *Ground Engineering*, Sept., 35–42.

BACOT J. (1974). *Etude théorique et experimentale de soutenements réalisés en terre armée.* D.Eng. Thesis, Claude Bernard University, Lyon, France.

BANERJEE P. K. (1975). Principles of analysis and design of Reinforced Earth Retaining Walls. *J. Inst. Highway Engs*, January.

BASSETT R. H. and HORNER J. N. (1977). *Centrifugal model testing of the approach embankment to the M180, Trent Crossing.* Report to North Eastern Road Construction Unit, London University.

BASSET R. H. and LAST N. C. (1978). Reinforcing Earth below Footings and Embankments. *Proc. ASCE Spring Conv.*, Pittsburgh.

BINQUET J. and LEE K. L. (1975). *Bearing capacity of strip footings on Reinforced Earth Slabs.* Report to Nat. Sc. Foundation, Grant No. GR38983, May.

BIRGISSON G. I. (1979). *Horizontally and inclined Reinforced Earth Structures.* M.Sc. Thesis, Heriot-Watt University, Edinburgh.

BISHOP J. A. and ANDERSON L. (1979). *Performance of a welded wire retaining wall.* Master's thesis, Utah State University.

BOLTON M. D. and CHOUDHURY S. P. (1976). *Reinforced Earth.* UMIST/TRRL Research Contract, First Report.

BOLTON M. D. (1986). The strength and dilatancy of sands. *Géotechnique*, **36**, no. 1, 65–78.

BRACEGIRDLE A. (1979). *Reinforced Earth Walls static and dynamic considerations.* M.Sc. Dissertation, Imperial College, London University.

CHANG J. C., HANNON J. B. and FORSYTH R. A. (1977). *Pull resistance and interaction of earthwork reinforcement and soil.* Transportation Research Board 56th Annual Meeting, Washington DC.

COYNE M. A. (1927). Murs de soutènement et murs de quai à échelle. *Le Génie Civil*, Tome XCI, no. 16, 29 October, Paris.

COYNE M. A. (1945). Murs de soutènement et murs de quai à échelle, extrait du *Génie Civil*, 1–15 May, Paris.

DANIELS P. (1973). *Pile group analysis.* M.Sc. Dissertation, Leeds University.

DEPARTMENT OF TRANSPORT (1978). *Reinforced Earth Retaining Walls and Bridges Abutments for Embankments.* Tech. Memo (Bridges) BE3/78.

DYER M. R. (1985). Observations of the stress distribution in crushed glass with applications to soil reinforcement. D.Phil.Thesis, University of Oxford.

FORSYTH R. A. (1978). Alternative earth reinforcements. *ASCE Symp. Earth Reinforcement*, Pittsburgh, 358–370.

GASSLER G. and GUDEHUS G. (1981). Soil nailing—some aspects of a new technique. *Proc. 10th Int. Conf. SMFE*, Stockholm.

HARRISON W. J. and GERRARD C. M. (1972). Elastic theory applied to Reinforced Earth. *Journal of the Soil Mechanics and Foundations Division of the American Society of Civil Engineers*, **98**, SM12, December.

HERMANN L. R. and AL-YASSIN Z. (1978). Numerical analysis of Reinforced Earth Systems. *ASCE Symp. Earth Reinforcement*, Pittsburgh.

HESHMATI S. (1993). *The action of geotextiles in providing combined drainage and reinforcement to cohesive soil.* Ph.D. Thesis, University of Newcastle upon Tyne.

IDRISS I. M. and SEED H. B. (1968). Seismic response of horizontal soil layers. *J. SMF Div.*, ASCE, SM4, July.

JABER M. B. (1989). *Behaviour of reinforced soil walls in centrifuge model tests.* Ph.D. Thesis, University of California, Berkeley.

Theory

JEWELL R. A. (1980a). *Some effects of reinforcement on the mechanical behaviour of soils.* Ph.D. Thesis, University of Cambridge.

JEWELL R. A. (1980b). *Some factors which influence the shear strength of reinforced sand.* Cambridge University Engineering Department, Technical Report No. CUED/D–SOIL-S/TR.85.

JEWELL R. A. (1990). Strengths and Deformations in Reinforced Soil Design. *4th Int. Conf. on Geotextiles and Related Products*, The Hague, 77.

JONES C. J. F. P. (1979). Lateral earth pressures acting on the facing units of Reinforced Earth structures. *C.R. Coll. Int. Reinforcement des Sols*, Paris, 445–451.

JONES C. J. F. P. (1980). Computer applications relating to Reinforced Earth. *Symp. Computer Applications to Geotechnical Problems in Highway Engineering*, Cambridge.

JONES C. J. F. P. (1989). *Review of effects of mining subsidence on reinforced earth.* Transport and Road Research Laboratory, Crowthorne, Contract Report 123, 43.

JONES C. J. F. P. and BELLAMY J. B. (1973). Computer prediction of ground movements due to mining subsidence. *Géotechnique*, **23**, no. 4, 515–30.

JONES C. J. F. P. and SIMS F. A. (1975). Earth pressures against the abutments and wingwalls of standard motorway bridges. *Géotechnique*, **25**, no. 4.

JONES C. J. F. P. and ZAKARIA N. A. (1994). The use of geosynthetic reinforcement as a construction aid over soft super soils and voids. *Proc. 2nd Geotechnical Eng. Conf.*, Cairo, **II**.

JURAN I. (1977). *Dimensionement interne des ouvrages en terre armée.* D. Ing. Thesis, Laboratoire Central des Ponts et Chaussees.

JURAN I. and SCHLOSSER F. (1978). Theoretical analysis of failure in Reinforced Earth Structures. *Proc. ASCE Symp. Earth Reinforcement*, Pittsburgh, 528–55.

JURAN I. and SCHLOSSER F. (1979). Étude théorique des efforts de traction développés dans les annatures des annrages en terre armée. *C.R. Coll. Int. Reinforcement des Sols*, Paris.

LEE K. L., ADAMS B. D. and VAGNERON J. J. (1973). Reinforced Earth retaining walls. *J. SMFE, Div. Proc. ASCE*, **99**, no. SM10, 745–64. Discussion, vol. 100, no. GT8, 958–66.

LEE K., JONES C. J. F. P., SULLIVAN W. R. and TROLINGER W. (1994). Failure and deformation of four reinforced soil walls in Eastern Tennessee. *Géotechnique*, **44**, no. 3, 397–426.

MAIR R. and HIGHT D. (1983). Private Communication, Geotechnical Consulting Group, London.

MANUJEE F. (1974). *Reinforced Earth.* M.Sc. Thesis, Leeds University.

MCGOWN A., ANDRAWES K. Z. and AL-HASANI M. M. (1978). Effect of inclusion properties on the behaviour of sand. *Géotechnique*, **28**, no. 3, 327–46.

MCGOWN A., MURRAY R. T. and ANDRAWES K. Z. (1987). *Influence of wall yielding on lateral stresses in unreinforced and reinforced fills.* Transport and Road Research Laboratory, Crowthorne, Research Report 113, 14.

MCKITTRICK D. P. (1978). Reinforced Earth—application of theory and research to practice. *Proc. Symp. Soil Reinforcing and Stabilizing Techniques*, NSWIT/MSW. Univ. (Separate volume).

MCKITTRICK D. P. and WOJCIECHOWSKI L. J. (1979). Examples of design and construction of Seismically Resistant Reinforced Earth Structures. *Proc. Int. Conf. Soil Reinforcement*, Paris.

MILLIGAN G. W. E. (1974). *The behaviour of rigid and flexible retaining walls in sand.* Ph.D. Thesis, Cambridge.

MURRAY R. T., CARDERS D. R. and KTAWCZYK J. V. (1979). Pull-out tests on reinforcements embedded in uniformly graded sand subjected to vibration. *Seventh Eur. Conf. Design Parameters Geot. Engng*, London, **3**, 115–20.

NAYLOR D. J. (1978). A study of Reinforced Earth Walls allowing slip strip. *Proc. ASCE Spring Conv.*, Pittsburgh.

NEWMARK N. M. (1975). *Seismic design criteria for above ground facilities other than pipelines and pumpstation buildings, Trans-Alaskan pipeline systems*. Report to Alaska Pipeline Service Com., March.

NICU N. D., AUTES D. R. and KESSLER R. S. (1970). *Field measurements on instrumented piles under an overpass abutments*. Report to the Committee for Foundations in Bridges and Other Structures, New Jersey Department of Transportation.

OSMAN M. A. (1977). *An analytical and experimental study of Reinforced Earth Retaining Walls*. Ph.D. Thesis, Glasgow University.

OSMAN M. A., FINDLEY T. W. and SUTHERLAND H. B. (1979). The internal stability of Reinforced Earth Walls. *C.R. Coll. Int. Reinforcement des Sols*, Paris.

PRICE D. I. (1975). Aspects of Reinforced Earth in the UK. *Ground Engineering*, **8**, no. 2, 19–24.

RICHARDSON G. N. (1976). *The seismic design of Reinforced Earth walls,* Rep. no. UCLA-ENG-7586, University of California, Los Angeles.

RICHARDSON G. N., FEGER D., FONG A. and LEE K. L. (1977). Seismic testing of Reinforced Earth Walls. *J. Geot. Eng. Div., ASCE*, **103**, GT1, January.

RICHARDSON G. N. and LEE K. L. (1975), Seismic design of Reinforced Earth Walls. *Jour. SMFD, ASCE*, **101**, GT2, February.

RICHARDSON G. N. and LEE K. L. (1976). The seismic design of Reinforced Earth Walls. *Soil Mech. Lab Report 7699*, University of California, Los Angeles.

ROSCOE K. H. (1970). The influence of strains in soil mechanics. Tenth Rankine Lecture, *Géotechnique*, **20**, no. 2, 129–170.

ROWE P. W. (1962). The stress dilatancy relation for the static equilibrium of an assembly of particles in contact. *Proc. Royal Society*, London, **269**, 500–27.

SARMA S. K. (1975). Seismic stability of earth dams and embankments. *Géotechnique*, **25**, no. 4, 743–761.

SCHLOSSER F. (1972). La terre armée—recherches et réalisations. *Bull. de Liaison des Laboratoires Central des Ponts et Chaussées*, no. 62, 79–92.

SCHLOSSER F. and ELIAS V. (1978). Friction in Reinforced Earth. *Proc. ASCE Symp. Earth Reinforcement*, Pittsburgh.

SCHLOSSER F. and SEGRESTIN P. (1979). Local stability analysis method of design of reinforced earth structures. *C.R. Coll. Int. Reinforcement des Sols*, Paris, 157–162.

SCHLOSSER F. and VIDAL H. (1969). Reinforced Earth. *Bull. de Liaison des Laboratoires Routiers Ponts et Chaussées*, no. 41, France.

SHEN C. K., BANG S., HERRMANN L. R. and ROMSTAD K. L. (1978). A Reinforced Lateral Earths Support System. *ASCE Symp. Earth Reinforcement*, Pittsburgh.

SIMS F. A. and JONES C. J. F. P. (1974). Comparison between theoretical and measured earth pressures acting on a large motorway retaining wall. *J. Inst. Highway Eng.*, Dec. 26–9.

SIMS F. A. and JONES C. J. F. P. (1979). The use of soil reinforcement in highway schemes. *C.R. Coll. Inst. Reinforcement des Sols*, Paris.

SMITH A. K. C. S. (1977). *Experimental and computational investigations of model Reinforced Earth retaining walls*. Ph.D. Thesis, Cambridge University.

SMITH G. N. (1977). Principles of Reinforced Earth Design. *Symp. Reinforced Earth and other Composite Soil Techniques*. TRRL and Heriott-Watt University, Edinburgh.

SMITH G. N. and BIRGISSON G. L. (1979). Inclined strips in Reinforced Earth Walls. *Civil Engrng*, June, 52–63.

VIDAL H. (1963). Diffusion restpeinte de la terre armée. Patent No. 1 069 361. Patent Office, London.

VIDAL H. (1966). La terre armée. *Annales de L'Institut Technique du Bâtiment et des Travaux Publics*, **19**, nos. 223–4, July–August, France.

VIDAL H. (1969a). La terre armée (réalisations récentes). *Annales de L'Institute Technique du Bâtiment et des Travaux Publics*, no. 259–60, July–August.

VIDAL H. (1969b). The principle of Reinforced Earth. *Highway Research Record*, No. 282, USA.

VIDAL H. (1972). Reinforced Earth 1972. *Annales de L'Institut Technique du Bâtiment et des Travaux Publics,* Supplement No. 299, November.

WESTERGAARD H. M. (1938). *A problem of elasticity suggested by a problem in soil mechanics: soft material reinforced by numerous strong horizontal sheets.* Contributions to the mechanics of solids dedicated to Stephen Timoshenko by his friends on the occasion of the sixtieth birthday anniversary, the Macmillan Company, New York.

5

Materials

5.1 Introduction

There are three basic materials or material composites required in the construction of any reinforced soil structure. They are:

(a) soil or fill matrix
(b) reinforcement or anchor system
(c) a facing if necessary.

In addition, other materials are required to cover associated elements such as the foundations, drainage, connecting elements and capping units and to act as barriers and fencing. There is usually an interrelationship between the various materials used, the choice being based upon design considerations relating to theoretical need, material availability, material properties, relative costs and delivery restrictions.

5.2 Soil/fill

The shear properties of soil can be improved as theoretically any soil could be used to form an earth-reinforced structure. For practical purposes, only a limited range of soils are likely to be used, particularly in vertically faced reinforced soil structures, although marginal material may be used in embankments. The choice of which soil or fill material is used will depend upon the technical requirements of the structure in question and also upon the basic economics associated with the scheme. In general, indigenous or waste material would be the most economic choice although these soils are likely to have inferior properties. The use of indigenous material may prove difficult for a variety of reasons especially with regard to long-term durability of the reinforcing elements and the ease with which the soil can be handled.

It is possible to differentiate between conventional vertically faced reinforced soil structures and sloping soil structures with regard to fill or soil employed. With faced reinforced soil structures a better quality fill is likely to be specified in contrast to embankment structures where the whole object of the reinforcing concept may be to improve existing marginal fill.

The soil used in long-term conventional structures is usually a well-graded cohesionless fill (granular backfill) or, alternatively, a good cohesive frictional fill, although purely cohesive soils have been used with success. The advantages of cohesionless fills are that they are stable, free draining, not susceptible to frost and relatively non-corrosive to reinforcing elements. The main disadvantage is that it would usually be imported material and therefore, might , be costly. With the cohesive soils the main advantage is availability, but there may be long-term durability problems together with distortion of the structure. The limitations to the fill material used are essentially those imposed by the design codes or specifications and relate principally to considerations of long-term stability and durability.

Cohesive frictional fill can be a convenient compromise between the technical benefits of cohesionless soil and the economic advantages of cohesive fill.

5.2.1 Cohesionless fill

Cohesionless fill (frictional fill, granular backfill) is defined as good quality, well-graded non-corrosive material usually possessing a good angle of internal friction. Examples include crushed rock, river sand or gravel. In addition, 'frictional fill' is in the UK defined (Department of Transport, 1978, BE 3/78) as a material in which no more than 10% passes a 63 μm BS sieve, and 'granular backfill', the term used in France when constructing earth reinforced structures, refers to fill in which no more than 15% (by weight) is smaller than 15 μm. This criterion is chosen as representing the point where intergranular contact of the material skeleton breaks down causing loss of internal friction.

5.2.1.1 Material properties

Knowledge of the following material properties is required for the selection of cohesionless fill:

(a) density
(b) grading
(c) uniformity coefficient, C_u
(d) pH value, pH
(e) *chloride ion content, Cl^-
(f) total SO_3 content, SO_3
(g) resistivity, P_a
(h) *redox potential, E_r
(i) angle of internal friction under effective stress conditions, ϕ'
(j) coefficient of friction between the fill and the reinforcing element, μ.

* May not be required for non-metallic reinforcement.

Density. Reinforced soil structures are gravity structures, the density of the parent material has a direct effect upon internal and external stability.

Grading. See Table 5.1.

Examples of gradings suitable and unsuitable for use as frictional fill are shown in Fig. 5.1.

Angle of internal friction. In the United Kingdom the effective angle of internal friction of cohesionless soil, $\phi' \geq 25°$. In France the angle of internal friction, ϕ

Table 5.1. Grading limitations for frictional fill

Sieve size	% passing
125	100
90	85–101
10	25–100
600 μm	10–65
63 μm	0–10
2 μm	0–10

of saturated consolidated frictional fill must be $> 25°$. However, measurement of the internal friction is not required if no more than 15% (by weight) of fill is smaller than 80 μm. Frictional fill should not normally be used in alternate layers with cohesive friction fill.

Uniformity coefficient. The uniformity coefficient, C_u is the ratio of the maximum size of 60 per cent of the sample to the effective size. The effective size is the maximum particle size of the smallest 10 per cent of the sample:

$$C_u = \frac{D \text{ passing } 60\%}{D \text{ passing } 10\%}$$

where D is particle diameter.

The coefficient of uniformity for cohesionless soil, $C_u \geq 5$. (In UK using cohesive frictional fill.)

pH value, chloride ion, SO_3 content, resistivity and redox potential. These soil/fill properties are associated with the durability of the reinforcing materials used. Some reinforcing materials are more durable than others and acceptable limitations are shown in Table 5.2.

Friction between the fill and reinforcing elements. See section 5.3.

Design criteria. Because of contractual arrangements the designer may not know what material will be used during construction. The following design properties can be assumed for a normal frictional fill:

In UK $\phi' = 30°$, $\gamma = 19 \text{ kN/m}^3$
In France $\phi' = 32°$, $\gamma = 19{\cdot}6 \text{ kN/m}^3$

If an elastic finite element analysis is being considered, the following material parameters relating to the soil backfill may be used, Table 5.3.

(*Note*: Bolton (1986) has demonstrated that $\phi'_{cv} = 32°$ for a granular fill. The implication is that a design value of $\phi' = 30\text{–}32°$ is conservative.)

At-rest pressure K_0 may be defined as: $K_0 = 1 - \sin \phi'$ or from elasticity

$$K_0 = \frac{\nu}{1 - \nu}$$

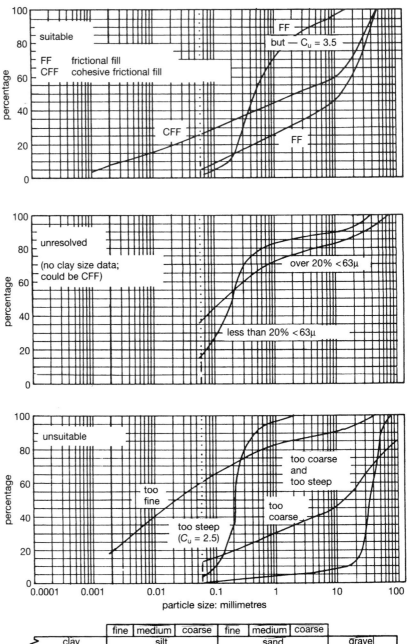

Fig. 5.1. Examples of suitable and unsuitable gradings for frictional or cohesive frictional fill

Table 5.2. Acceptable limitations for reinforcing materials

Reinforcing material	pH value		Maximum chloride ion content: %	Maximum total SO_3	Minimum resistivity (ohm–cm)	Minimum redox potential (volts)		Maximum microbial activity index	Maximum organic content: %
	min.	max.				Frictional fill	Cohesive frictional fill		
Aluminium alloy	6	8	0·05	0·5	3000	0·4	0·43	5	0·2
Copper	5	9	0·05	0·5	2000	0·25	0·25	5	0·28
Hot-dip galvanized steel	5	10	0·02*	0·1*	1000*	0·4	0·43	5	0·2
Stainless steel	5	10	0·025*	0·1*	3000	0·35	0·35	5	0·2
Fibretain (FRP)	4	9	2·0	1·0	1000	not applicable	not applicable	not applicable	not applicable
Paraweb (polyethylene)						not applicable	not applicable	not applicable	not applicable
Tensar (geogrid)						not applicable	not applicable	not applicable	not applicable

*When the structure is submerged in fresh water these values are reduced—see BS 8006: 1995.

Table 5.3. Material parameters for soil backfill

		Elastic modulus E_r: MN/m^2	Poisson ratio ν
Clay	soft	3	0·4
	medium	7	0·3
	hard	14	0·25
	sandy	36	0·25
Sand	loose	15	0·2
	dense	80	0·3
Sand/Gravel	loose	100	0·2
	dense	150	0·3

The active pressure,

$$K_a = \frac{1 - \sin \phi'}{1 + \sin \phi'}$$

5.2.2 Cohesive frictional fill

Cohesive frictional fill (CFF) can be defined as material with more than 10% passing 63 μm BS sieve (BE 3/78). The main advantage of cohesive frictional fill is better availability when compared with frictional fill. This may represent an economy. Cohesive frictional fill is specified in the UK design memoranda, including the UK Code of Practice for Reinforced Soil, BS 8006: 1995, but is not permitted in some other specifications, Table 5.4.

5.2.2.1 Material properties

Knowledge of the following material properties is required for the selection of cohesive frictional fill:

(*a*) density
(*b*) grading
(*c*) uniformity coefficient, C_u
(*d*) pH value, pH
(*e*) *chloride ion content, Cl^-
(*f*) *total SO_3 content, SO_3
(*g*) resistivity, P_a
(*h*) *redox potential, E_r
(*i*) angle of internal friction under effective stress conditions, ϕ'
(*j*) coefficient of friction between the fill and the reinforcing elements, μ
(*k*) cohesion under effective stress conditions, c'
(*l*) adhesion between the fill and the reinforcing elements under effective stress conditions, c'_r
(*m*) liquid limit, LL
(*n*) plasticity index, PI
(*o*) consolidation parameters.

* May not be required for non-metallic reinforcement.

Table 5.4. Grading limitations for cohesive frictional fill

Sieve size	% passing
125	100
90 mm	85–100
10 mm	25–100
600 μm	11–100
63 μm	11–100
2 μm	0–10

Table 5.5. Materials used to construct a reinforced earth structure

	LL	PI	Sand (%)	Silt (%)	Clay (%)
Material (A)	42	21	7	65	28
Material (B)	30	17	51	39	10

Grading. See Table 5.4.

Coefficient of uniformity: $C_u* \geq 5$ (unless agreed)

Examples of gradings suitable and unsuitable for use as cohesive frictional fill are shown in Fig. 5.1.

Angle of internal friction. The effective angle of internal friction, $\phi' \geq 20°$.

Liquid limit and plasticity index. For material $< 52 \mu$m. Liquid limit, LL $\leq 45\%$. Plasticity index, PI $\leq 20\%$.

The maximum limit of 10 per cent clay is related to the need for the fill to be sufficiently free draining to provide stability during construction. The clay content of soil can be difficult to measure and large variations are common. The liquid limit and the plasticity index may give a more direct indication of mechanical properties, in which case a measure of the clay content may be omitted. As an example, Table 5.5 illustrates materials which have been used to construct a reinforced earth structure (Boden *et al.*, 1979).

pH value, chloride ion, SO_3 content, resistivity and redox potential. The durability criteria associated with cohesive frictional fill is the same as for frictional fill, Table 5.2.

Moisture content. Minimum moisture condition values in the range 6–10 will normally produce satisfactory conditions relating to stability and handleability. Alternatively, a value of 1·2–1·3 times the plastic limit of the soil may be used.

5.2.3 Cohesive fill

Cohesive soils can be reinforced and may be economical to use. The use of cohesive soil falls into two separate categories:

(a) when the cohesive soil is used as the fill to a vertically faced reinforced soil construction

(b) when reinforcement is provided to improve the mechanical properties of the soil, as in the case of a reinforced embankment constructed of marginal material on top of a weak subsoil.

5.2.3.1 Vertically faced structures

The main reasons why fine graded and cohesive soils are generally held to be unsuitable for vertically faced reinforced soil construction are short-term stability and durability.

(a) Short-term stability: the bond between cohesive soil and strip reinforcement is poor and subject to reduction if positive pore water pressures develop.

(b) Some fine-grained cohesive soils are significantly more aggressive than cohesionless soils. It is known that clay materials such as illite accelerate metal corrosion.

It is thought that long-term deformation may occur when plastic soils are reinforced. However, many widespread benefits and applications arise if suitable reinforcements and construction techniques can be adapted to use cohesive fill, particularly in areas where cohesionless fill is in short supply.

Cohesive soils may be susceptible to frost, and in faced structures can lead to additional earth pressures being generated behind the facing units; these have to be accommodated by the reinforcement/facing connections.

Cohesive soil will normally require comprehensive drainage and may be difficult to place, especially in wet conditions.

Permanent geosynthetic-reinforced soil retaining walls and bridge abutments, built using cohesive fill, are used in Japan for railway works. These structures have been shown to be very stable and capable of surviving large seismic forces such as the 1995 Kobe earthquake (Tateyama and Musata, 1994; Tatsuoka *et al.*, 1992, and Tatsuoka *et al.*, 1995).

5.2.3.2 Embankment structures

The use of cohesive soil in reinforced embankment structures is accepted procedure; however, the forms of the reinforcement used would normally preclude strip reinforcement in favour of fabrics and geogrids.

5.2.4 Chalk

Chalk with a saturation moisture content (SMC) $\geq 29\%$ may be used as fill for permanent reinforced soil structures. Chalk with a SMC $> 29\%$ may be used for temporary structures. Particles should be < 125 mm in size.

5.2.5 Waste materials

The use of waste materials as fill for reinforced soil structures is attractive from an environmental as well as economic viewpoint. The following waste products are produced in quantity.

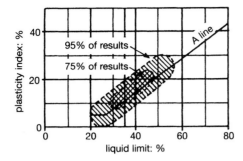

Fig. 5.2. General range of plasticity characteristics for coarse discards and lagoon deposits as found in existing mine waste tips (after McKechnie Thompson et al., 1973)

5.2.5.1 Mine waste

The quantity of mine waste produced is increasing and disposal is a cause of concern. Waste has been used for embankment construction. By improving the strength of mine waste by reinforcing, the range of civil engineering uses can be improved.

The range of particle size distribution in mine waste materials is very variable and depends upon many factors, including the method of handling and placement. Typically the materials are predominantly fine grained but include sand and gravel-sized particles. As the particle size characteristics are variable, so are also the plasticity characteristics. The general ranges of index properties for coarse discard and lagoon deposits are shown on the plasticity chart Fig. 5.2.

There are strong similarities between the mechanical properties of mine waste and inorganic clays of medium plasticity. Mine waste has been successfully used to construct earth reinforced structures using strip and grid reinforcements. Because of the concern for the durability of the construction elements used with mine waste, reinforcing materials formed from materials which have high corrosion and degradation resistance are preferred (Jewell and Jones, 1981).

Mine waste can be very susceptible to moisture and, therefore, proper drainage and shaping of any proposed structure is essential. The material can be satisfactorily compacted at a moisture content close to the optimum but may prove to be slightly more difficult to place than the conventional filling materials.

5.2.5.2 Pulverized fuel ash

The use of pulverized fuel ash as a lightweight fill in embankment construction, is an established practice. The material can also be used as a lightweight fill for earth reinforced structures. Typical properties of pulverized fuel ash are given in Table 5.6.

Pulverized fuel ash is relatively easy to place with compaction by vibrating rollers or footpath compactors, giving an optimum moisture content of approximately 19 per cent, with 10 per cent air voids. Because of its fine structure, grid reinforcement may prove the most satisfactory form of reinforcement.

Reinforcement resistant to corrosion is essential and care must be taken with regard to drainage, as pulverized fuel ash is particularly sensitive to the effects

Table 5.6. Typical properties of pulverized fuel ash

Sieve size	% passing		
10 mm	100	Bulk density, γ	12 kN/m^3
600 μm	96	SO$_3$	1·0%
300 μm	95	ϕ'	25–30°
63 μm	84	c$'$	5–15
			kN/m^2

of uncontrolled water. Many pulverized fuel ash materials display pozzolanic properties and can develop cohesive shear strengths in excess of 40 kN/m^2 (Table 5.7). It is essential to test the strength parameters of the pulverized fuel ash material proposed prior to use as the self-hardening properties are time dependent; this can be undertaken using the procedures detailed in BS 1377: 1990. Both conditioned hopper ash and lagoon ash can have self-hardening properties. Some codes of practice, including BS 8006: 1995, limit the value of effective cohesion c' developed by pulverized fuel ash materials to ≤ 5 kN/m^2. The rate of development of cohesion is compatible with the construction rate of unformed soil structures. Even assuming conservative values for c', the implications for cohesion on reinforced soil design is significant (see Chapter 9, Costs and economics), Fig. 5.3.

5.2.6 Improved fills

The implication of cohesion on reinforced soil structures has been considered by Güler (1990). The addition of 5–10% lime to cohesive fill with an effective angle of friction $\phi' = 19°$ results in increased workability (easier compaction),

Table 5.7. Self-hardening characteristics of pulverized fuel ash

		Tilbury Power Station			
		Direct shear test (c–d)			
Elapsed time		0	7	14	28
Stockpiled					
Unsoaked	C_d (kN/m^2)	14·9	19·7	24	26·6
	ϕ_d (deg.)	44	43·6	41	43
Soaked	C_d	1·7	2·7	5	7·8
	ϕ_d	39·5	40	42	40
Conditioned					
Unsoaked	C_d	28·2	29·7	29·7	38·2
	ϕ_d	43	41	43	44
Soaked	C_d	2·4	8	8·7	28
	ϕ_d	41	41	42	43

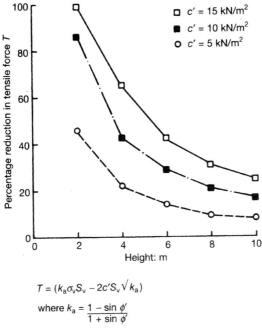

$$T = (k_a\sigma_v S_v - 2c'S_v \sqrt{k_a})$$

where $k_a = \dfrac{1 - \sin \phi'}{1 + \sin \phi'}$

S_v = reinforcement spacing
σ_v = vertical stress

Fig. 5.3. Influence of cohesion on required tensile force in reinforcements

increased permeability and the development of cohesive strengths, $c' = 150–180$ kN/m². An alternative method of improving the soil properties is to add cement (Morrison and Crockford, 1990). The reported strength of the resultant fill is high, $c' \approx 180$ kN/m² but the material is brittle.

5.3 Reinforcement

A variety of materials can be used as reinforcing materials. Those that have been used successfully include steel, concrete, glass fibre, wood, rubber, aluminium and thermoplastics. Reinforcement may take the form of strips, grids, anchors and sheet material, chains, planks, rope, vegetation, and combinations of these or other material forms.

5.3.1 Types of reinforcing material
5.3.1.1 Strips

These are flexible linear elements normally having their breadth b greater than their thickness t. Dimensions vary with application and structure, but are usually within the range $t = 3–20$ mm, $b = 30–100$ mm. The most common strips are metals. The form of stainless, galvanized or coated steel strips are either plain or have several protrusions such as ribs or grooves to increase the friction between the reinforcement and the fill. Strips can also be formed from aluminium, copper, polymers and glass fibre reinforced plastic (GRP). Reed and bamboo reinforcements are normally categorized as strips, as are chains. Mild steel rebar is also used as strip reinforcement.

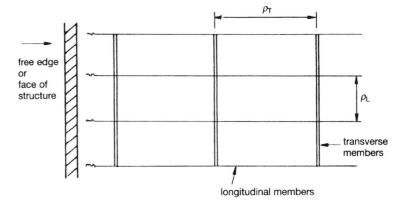

Fig. 5.4. *Diagrammatic representation of grid reinforcement*

5.3.1.2 Planks

Planks are similar to strips except that their form of construction makes them stiff. Planks can be formed from timber, reinforced concrete or prestressed concrete. The dimensions of concrete planks vary; however, reinforcements with a thickness, $t = 100$ mm and breadth, $b = 200$–300 mm have been used. They have to be handled with care as they can be susceptible to cracking.

5.3.1.3 Grids and geogrids

Reinforcing elements formed from transverse and longitudinal members, in which the transverse members run parallel to the face or free edge of the structure and behave as abutments or anchors, Fig. 5.4. The main purpose of the longitudinal members is to retain the transverse members in position. Since the transverse members act as an abutment or anchor they are often stiff relative to their length; however, this is not essential. The longitudinal members may be flexible having a high modulus of elasticity not susceptible to creep. The pitch of the longitudinal members, ρ_L is determined by their load-carrying capacity and the stiffness of the transverse element. The pitch of the transverse elements, ρ_T depends on the internal stability of the structure under consideration. A surplus of longitudinal and transverse elements is of no consequence provided that the soil or fill can interlock with the grid.

Fig. 5.5. Two different geogrid structures

Grids can be formed from steel in the form of plain or galvanized weldmesh, or from expanded metal. Grids formed from polymers are known as 'geogrids'

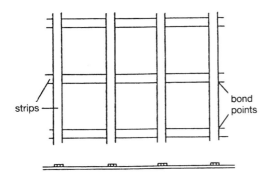

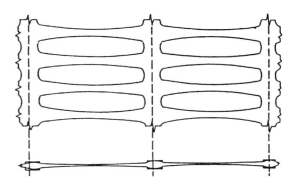

  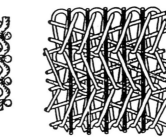

(a) *Nonwoven structure*   (b) *Woven structure*   (c) *Stitch-bonded structure*

Fig. 5.6. Different conventional geotextile constructions

and are normally in the form of an expanded proprietary plastic product, or manufactured as a composite material, Fig. 5.5.

5.3.1.4 Sheet reinforcement (geotextiles)

The most usual sheet materials used as reinforcement are geotextiles. Geotextiles can be divided into two categories, namely, conventional geotextiles and specials. *Conventional geotextiles* are products of the textile industry and include nonwoven, woven, knitted and stitch-bonded textiles, Fig. 5.6.

Nonwoven geotextiles consist of a random arrangement of fibres bonded together by heat (melt bonded) or physical entanglement (needle punched). The fibres used can be in the form of either stable (short lengths) or continuous filaments. The structure of nonwoven geotextiles is illustrated in Fig. 5.6(a). Woven geotextiles consist of fibres arranged essentially at right angles to one another in varying configurations, the general structure is shown in Fig. 5.6(b). Alternative configurations are identified on the basis of the cross-sectional shape of the constituent fibres. Monofilament wovens are manufactured from fibres with circular or elliptical cross-sections. Multifilament and fibrillated tape wovens result from a gathering of fibres in parallel arrays along the length and across the width of the geotextile. Tape wovens are made from fibres with a flat cross-section. Knitted geotextiles consist of fibres which are inter-looped. This process produces two different structures, i.e. weft knitted and warp knitted geotextiles. Stitch-bonded geotextiles are formed by the stitching together of fibres or yarns, Fig. 5.6(c). Fabric reinforced retaining walls have proved to be economic but are somewhat utilitarian in appearance, and the larger use of geotextile fabrics has proved to be in the areas of separation, filtration and drainage. Knitted geotextiles in the form of open grids are being used extensively for reinforcement.

5.3.1.5 Composite reinforcement

Reinforcement can be in the form of combinations of materials and material forms such as sheets and strips, grids and strips, or strips and anchors, depending on the requirements. In a soil reinforcement context, *geocomposites* generally consist of high strength fibres set within a polymer matrix or encased within a polymer skin. The fibres provide the tensile properties for the material while the matrix or skin provides the geometrical shape and protects the fibres from damage. There are two common types of geocomposite structure, strips

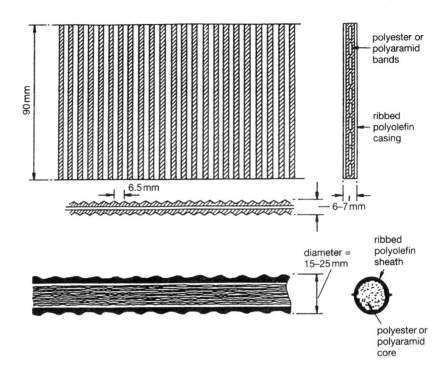

Fig. 5.7. Composite polymeric
reinforcements

and bars, Fig. 5.7. In addition to these special geotextiles, knitted grid structures are also encased within a polymer skin used to provide protection for the tensile members.

The form of any geotextile required to reinforce a structure depends on the nature and life of the structure itself. The properties required for permanent reinforcements are likely to be different from those required for temporary structures. In order to achieve maximum reinforcement efficiency, the load-carrying elements of the geotextile are laid flat and in a highly directional alignment within the geotextile structure; in this manner, the tensile characteristics of the load-carrying elements determine the tensile characteristics of the material as a whole. Typical properties of polymer reinforcements are listed in Table 1.

5.3.1.6 Anchors

Flexible linear elements having one or more pronounced protrusions or distortions which act as abutments or anchors in the fill or soil. They may be formed from steel, rope, plastic (textile) or combinations of materials such as webbing and tyres, steel and tyres, or steel and concrete, Fig. 5.8.

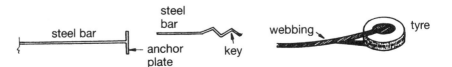

Fig. 5.8. Composite strip and
bar structures

Table 5.8. Properties of sheet and strip materials—frictional fill

Material	Maximum thickness to which stresses apply: mm	Basic permissible stresses			Coefficient α
		Axial tensile strength: N/mm^2	Shear strength: N/mm^2	Bearing strength: N/mm^2	
Aluminium alloy BS 4300/8: NS51, H4	6	120	72	180	0·46
Copper BS 2870: C101, $\frac{1}{2}$H: C102, $\frac{1}{2}$H	10	108	65	163	0·46
Carbon steel BS EN 10025: 1993 S235 JR	16	340	205	340	0·50
Carbon steel BS EN 10025: 1993 S255 JR	16	490	295	490	0·50
Stainless steel BS 1449: Pt 2: 316S31/33	10	510	305	510	0·46
Stainless steel BS 1449: Pt 2: 316S31, CR	6	650	390	650	0·46
Fibretain	(see Tables 5.14 and 5.15)				

5.3.2 Properties

The principal requirements of reinforcing materials are strength and stability (low tendency to creep), durability, ease of handling, a high coefficient of friction and/or adherence with the soil, together with low cost and ready availability. The properties of metallic reinforcements used in some specifications are shown in Table 5.8. Steel reinforcement which is to be galvanized should have a silicon content compatible with the requirements of BS 729: 1986. In the UK, the average zinc coating $> 1000 \ gm/m^2$. Other specifications require coatings $> 700 \ gm/m^2$. The particular properties of some reinforcing materials in use are illustrated in Table 5.9.

5.3.2.1 Coefficient of friction

The coefficient of friction or adherence between the reinforcement and the soil can be obtained from shear box tests. For frictional fill, an assumed value of the coefficient of friction μ for strips, grids or sheets can be obtained from:

$$\mu = \alpha' \tan \phi', \text{ where } \alpha' \text{ is a coefficient.}$$

The values of α' for some strip reinforcements complying with UK design memoranda, are shown in Table 5.8.

Table 5.9. Comparison of different materials for linear strip reinforcing elements

Material	Tensile strength: N/mm^2	Yield or 0·2% proof stress: N/mm^2	Specific gravity	Young modulus: kN/mm^2	Permissible tensile stress: N/mm^2	% extension at working load	Relative weight per unit force
Aluminium alloy NS51—H4 to BS 4300/8	270	200	2·68	70	120	0·17	27
Copper C101—$\frac{1}{2}$H to BS 2870	245	180	8·93	125	110	0·09	97
Galvanized mild steel KHR 34/20P to BS 1449: Part 1	340	200	7·85	200	120	0·06	100
Cold rolled stainless steel 316 S16 to BS 1449: Part 2	540	400	7·96	200	220	0·11	44
Glass-fibre reinforced plastic (GRP)	354	—	—	40	80	0·20	37
Polymer fibre strips	10–100	—	—	—	Factor of safety = 6	1·80	19
Aluminium alloy NS51—H8	400	285	2·68	70	170	0·24	19
Galvanized high-yield steel KHR 54/35P to BS 1449: Part 1	540	350	7·85	200	190	0·10	69
Hard-rolled stainless steel 316 S16 to BS 1449: Part 2	1150	800	7·96	200	440	0·22	24

For cohesive frictional fill and cohesive fill, a shear box text may be used to ascertain the friction and adhesion between the reinforcement and the soil. Examples of values obtained for various strip reinforcing materials are shown in Table 5.10. The soil–geotextile frictional behaviour of geosynthetic reinforcements can be determined using the method detailed in BS 6906: 1991.

5.3.2.2 Durability

The reinforcement must be durable and maintain its integrity over the life of the structure. Design life of earth-reinforced structures varies from 20 to 120 years. With the longest life structures, the use of sacrificial thicknesses on metallic components is usually necessary. Examples of the sacrificial thickness needed are illustrated in Table 5.11 (see also BS 8006: 1995). Glass fibre and polyethylene are durable although the latter is sensitive to ultraviolet light and must be stored under cover before use (see also Chapter 10, Durability).

Note: Where metallic reinforcements overlap, as at the re-entrant corners of abutments, or retaining walls, the figures in Table 5.11 should be increased by 25% for frictional fill and 20% for cohesive frictional fill. The zone of the increase should be taken as the plan area covering the overlapping elements. In addition, to assist in durability, it is good practice to ensure that the materials used are electrolytically compatible.

5.3.2.3 Properties of polymeric reinforcing materials

Polymeric materials used as soil reinforcement have four main requirements. They must be strong, relatively stiff, durable and bond with the soil. Of critical importance is that the strength of the reinforcement is sufficient to support the force required to achieve stability of the structure. The magnitude of the required force will vary depending on the application.

In a steep slope strengthened by a geotextile reinforcing layer, each reinforcement might have to support a force of 10–40 kN/m; alternatively, a single geotextile reinforcing layer in an embankment of soft soil may be required to support a tensile force of 100–400 kN/m to ensure stability.

The requirement of the geotextile to be stiff is so that the required force can be mobilized at a tensile strain which is compatible with the deformation of the soil. The concept of strain compatibility between the reinforced soil and the soil is implicit in any reinforced soil structure (Jewell, 1992). The allowable tensile strain depends on the application and, in the case of a reinforced slope on soft soil, the allowable extension can vary from 5–10 per cent. In the case of a reinforced soil wall, the design allowable tensile extension of the reinforcement is unlikely to exceed 2–4 per cent, with a limitation of < 1 per cent strain occurring after construction.

Durability of the polymeric reinforcement is influenced by time and has to be considered together with the environment conditions. With permanent structures, durability is the dominant consideration of the designers.

The mechanical requirement for bond between reinforcement and the soil is important, but often a function of the form of the polymer reinforcement. Geogrids and conventional geotextiles in the form of sheets provide good bond with the soil attributable either to the large surface offered by the

Table 5.10. *Frictional properties for various strip materials when embedded in two types of clay fill (derived from shearbox tests) (reference Table 5.5)*

Type of reinforcement	Angle of friction without reinforcement (ϕ)	Type of fill					
		Coefficient of friction between fill and reinforcement (μ)					
		Galvanized mild steel	Stainless steel	Glass-fibre reinforced plastic (GRP)	Aluminium coated mild steel (Aludip)	Plastic coated mild steel	Polyester filaments in polyethylene (Paraweb)
Effective stress range 0–40 kPa LL42, PI21	37	0·38	0·40	0·53 to 0·64	0·51 to 0·58	0·36	0·42
Effective stress range 0–100 kPa LL30, PI17	37	0·36	0·39	0·53 to 0·64	0·51 to 0·58	0·37	0·40

*GRP (Glass fibre reinforced plastic).

Table 5.11. Corrosion allowance for metallic components exposed to various environments (after Department of Transport BE 3/78)

| | Sacrificial thickness to be allowed for on each surface exposed to corrosion (mm) | | | |
| | Atmospheric environment | | Buried in fill | |
	Urban industrial, industrial coastal	Other	Frictional fill	Cohesive frictional fill
Aluminium alloy	—	—	0·15	0·5
Copper	—	—	0·15	0·3
Galvanized steel	0·85	0·3	0·75	1·25
Stainless steel	0	0	0·1	0·2

geotextile or to soil/reinforcement interlock in the case of geogrids. In the case of strip or bar reinforcement, bond can become a critical consideration particularly in the top of a reinforced soil structure (Hassan, 1992).

Strength-stiffness of the polymer
The tensile strength and extension characteristics of geotextiles are a function of the tensile properties of the constituent materials and the geometrical arrangement of the elements within the geotextile, Table 5.12. The tensile characteristics of a range of geotextiles are shown in Fig. 5.9(a) which shows that the strength of polyaramide fibres can be greater than that of prestressed steel tendons. Polyaramide fibres are seldom used for geotextiles because of cost, and alternatives are formed from polyester fibres, polypropylene tapes and high density polyethylene (HDPE) grids which all exhibit good tensile characteristics at relatively low costs. All of these materials have been shown to be well suited for reinforced soil applications. High density polyethylene grids can be manufactured in a form which is immediately suitable for use as reinforcement, while polyester fibres can be produced as specific constructions to enable their easy installation into soils. The influence of geometrical structure on the resultant geotextile stress–strain characteristics is shown in Fig. 5.9(b). For maximum efficiency, it is desirable that the geotextile reinforcement should be able to reproduce as closely as possible the character-istics of the constituent load-carrying elements. Reference to Fig. 5.9(b) indicates why woven, stitch-bonded, geogrid and geocomposite structures are preferred for reinforced soil applications.

Effect of long-term loads on strength-stiffness
For many polymeric based materials, ambient operating temperatures coincide with their visco-elastic phase, thus creep becomes a significant consideration in assessing their long-term load-carrying capacity. Creep is the increase in extension of a material under a constantly applied load. The stress–strain time characteristics (at constant temperature) of geotextile reinforcements can be visualized in terms of a three-dimensional body with stress, strain and time comprising the three axes, Fig. 5.10 (Lawson, 1991). By projecting the three-

Table 5.12. Representative properties of geosynthetics

Geotextile construction	Tensile strength: kN/m	Extension at max. load: %	Apparent opening size: mm	Water flow: l/m²per s*	Unit weight: g/m²
Conventional geotextiles					
Nonwovens					
Melt-bonded	3–25	20–60	0·02–0·35	25–150	70–350
Needle-punched	7–90	50–80	0·03–0·20	30–200	150–2000
Wovens					
Monofilament	20–80	9–35	0·07–2·5	25–2000	150–300
Multifilament**	40–800	9–30	0·20–0·9	20–80	250–1350
Flat tape	8–70	10–25	0·07–0·15	5–20	90–250
Knitteds					
Weft	2–5	300–600	0·2–1·2	60–2000	
Warp	20–120	12–15	0·4–5	100–2000	
Stitch-bonded	30–1000	8–30	0·07–0·5	30–80	250–1200
Special geotextiles					
Geogrids					
Cross-laid strips	25–200	3–20	50–300	NA	300–1200
Punched sheets	10–200	11–30	40–150	NA	200–1100
Geocomposites					
Strips	20–150†	3–20	NA	NA	NA
Bars	20–500†	3–20	NA	NA	NA
Link structures	100–4000	3–20	NA	NA	600–4500

* Normal to the plane of the geotextile with 10 cm constant head.
** Fibrillated tapes are included in this category.
† Measured in kN (not kN/m).
NA not applicable.

dimensional body into each of three phase planes, three sets of curves are obtained which can be used to describe creep behaviour:

(*a*) isochronous creep curves (projecting on to the stress–strain plane)
(*b*) isostrain creep curves (projecting on to the stress–time plane)
(*c*) isostress creep curves (projecting on to the strain–time plane).

Isochronous creep curves depict the change in the stress–strain curve of the material at different points in time.

Isostrain creep curves depict the changes in the load-carrying capacity of the material at different points of time and at different strain levels. The stress rupture (or creep rupture) curve is used to predict the expected lifetime of the load-carrying element. The stress rupture curve plots the time to rupture of the material when loaded at different stress levels. It also depicts stress relaxation, which is the complementary relationship to creep. Greenwood (1990) has found reasonable agreement between comparable isochronous stress–strain

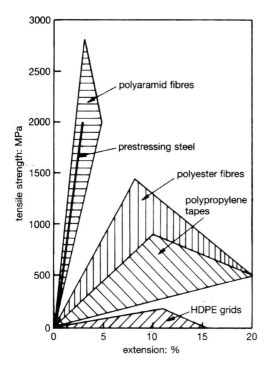

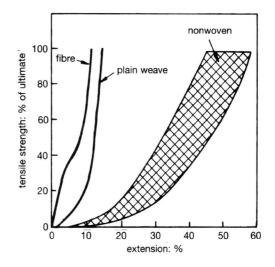

curves obtained from creep measurements with isochronous stress–strain curves from stress relaxation tests.

Isostress creep curves depict the change in strain of the material at different points in time and at different stress levels. Two additional sets of curves are derived from isostress creep curves: Sherby–Dorn plots and creep coefficients. Sherby–Dorn depict the rate of change in strain of the material versus its total strain for different stress levels. The long-term creep characteristics of geotextile reinforcements cannot be obtained from Sherby–Dorn plots as there is no reference to time in these curves. Creep coefficients can be determined by linearizing the isostress creep curves on to strain versus log time axes. When plotted in this manner, most polymeric materials approximate to a linear relationship. The equation for the total strain of the materials is:

$$\epsilon_t = \epsilon_0 + b \log t \qquad (1)$$

where ϵ_t is the total extension in the material after the time period, t
ϵ_0 is the initial extension at time, $t = 0$
b is the creep coefficient.

The creep coefficient b is dependent on the level of stress applied and is a measure of the rate of creep.

The creep curves of most practical use for geotextile reinforcements are the stress–rupture curves, the isochronous creep curves and the creep coefficient curves. The stress–rupture curves are used to predict the lifetime over which the geotextile reinforcement can carry a specific load. The isochronous creep curves and the creep coefficient curves are used to estimate both the total extension and the creep extension of the geotextile reinforcement over

Fig. 5.9. Load/extension characteristics of geotextiles and influence of construction: (a) tensile strength/extension characteristics of various geotextile load-carrying elements and that of prestressing steel; (b) effect of geotextile construction on resulting extension characteristics using polyester fibres

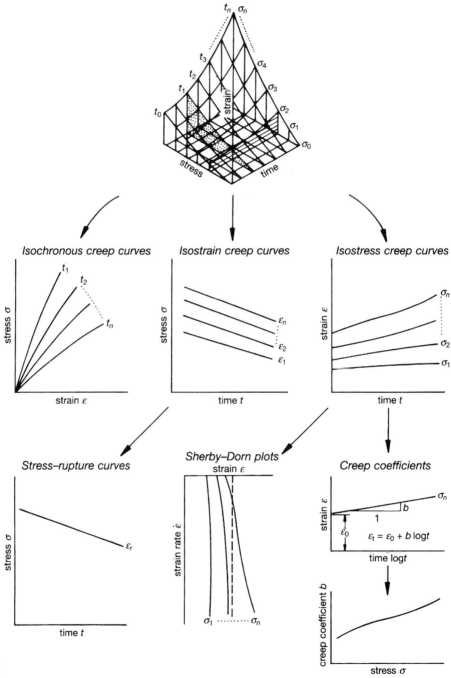

Fig. 5.10. *Various ways of representing creep data (at constant temperature)*

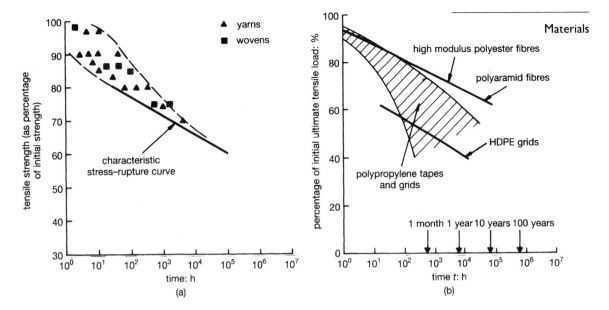

different design lives and stress levels. In order to describe the creep behaviour of geotextiles (in terms of stress–strain time), it is necessary to present both the stress–rupture curve for the material and either the isochronous creep curve or the creep coefficient curve. A characteristic stress–rupture curve for a high modulus polyester yarn is shown in Fig. 5.11(a). Also shown in Fig. 5.11(b) is the stress–rupture properties for various structural materials used in polymeric reinforcements. It has been established that high modulus polyester fibres, polyamide fibres and proprietary high density polyethylene grids conform to well-defined stress–rupture patterns, while polypropylene fibres and grids exhibit varying stress–rupture properties depending on the processing carried out during the manufacture (Hollaway, 1990).

As well as quantifying the long-term strength capability of polymer reinforcement, the long-term extension and stiffness also need to be determined. Fig. 5.12(a) shows the isochronous creep curves for a commercially available high modulus polyester fibre-based geostrip. The shape of the curve indicates that there is little change in the load extension curve with time for load levels below 40 per cent of the initial tensile strength (less than 1 per cent creep extension). The difference in behaviour of a polypropylene geotextile is shown in Fig. 5.12(b). At working stress level of 20 per cent for the initial ultimate tensile strength in the material, long-term creep extensions could be expected to be 5–6 per cent.

Creep coefficients provide a convenient means of comparing the rate of creep of different polymeric materials. Fig. 5.12(c) shows the distribution of creep coefficients for various structural materials used in geotextile reinforcements. It can be seen that the creep coefficient increases for increasing applied load for polymers, although processing techniques can alter significantly the rate of creep of a particular polymer. An example is the wide range of creep coefficients seen for polypropylene.

Fig. 5.11. Characteristic stress–rupture curve and general stress–rupture properties of various geotextile reinforcements: (a) characteristic stress–rupture curve of 10⁵h duration for high modulus polyester geotextiles at 23°C; (b) stress–rupture behaviour of various geotextile load-carrying elements at 23°C (after Hollaway, 1990)

119

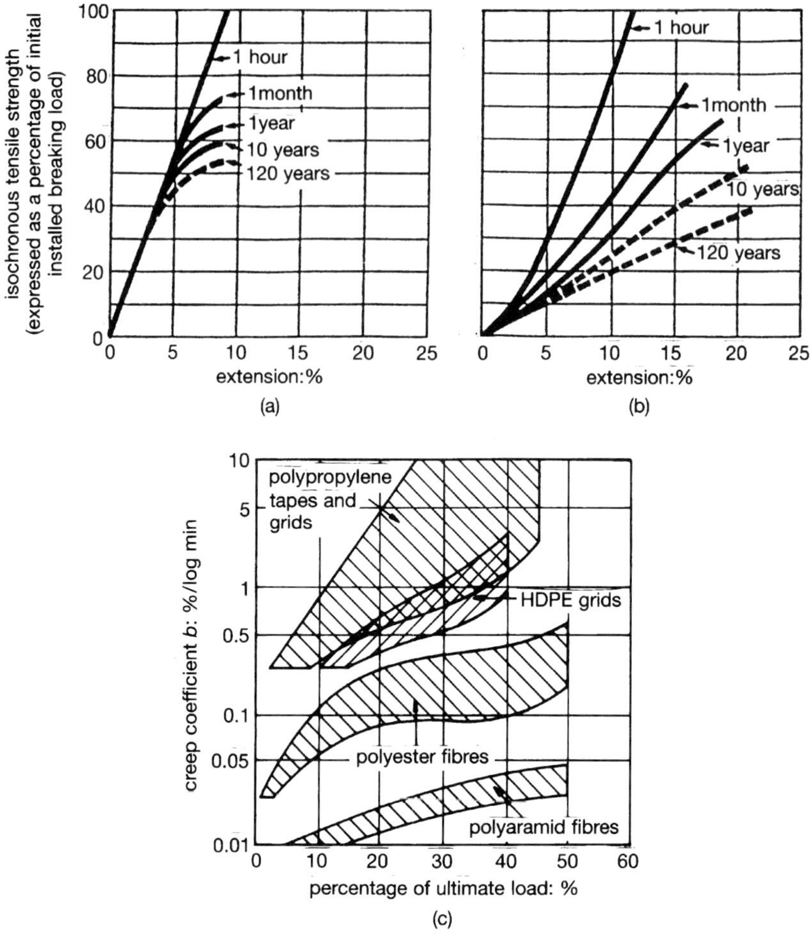

Fig. 5.12. Isochronous creep curves for two commercially available geotextiles and comparison of the rate of creep of various geotextile reinforcements: (a) isochronous creep curves for high modulus polyester fibre geostrips at 23°C (after ECGL, 1989); (b) isochronous creep curves for high modulus polypropylene tape woven geotextiles at 23°C (after ECGL, 1989); (c) creep coefficient versus percentage applied load for various geotextile reinforcing elements at 23°C (after Hollaway, 1990)

Effect of temperature on strength and stiffness

The temperature at depth within a soil mass remains constant; however, near the surface the ambient temperature may vary depending on the external temperature and the environment, Fig. 5.13. At depth the constant soil temperature may range from 10°C in temperate climates to 20°C in tropical climates. The temperature profile of a reinforced soil structure in a desert environment is shown in Fig. 5.14 which shows that the temperature of the soil immediately behind the concrete soil reinforcing units could reach 35°C during the summer months.

For those polymeric materials whose operating temperatures coincide with their visco-elastic phase, a change in operating temperature can affect their strengths/stiffness characteristics particularly in relation to creep, Table 5.3. At operating temperatures below its glass transition temperature, a material behaves in an elastic manner with a relative small plastic (creep) component. At operating temperatures between the glass transition temperature and its melting point, a material behaves in an essentially visco-elastic manner with a significant creep component when load is applied. If changes in operating

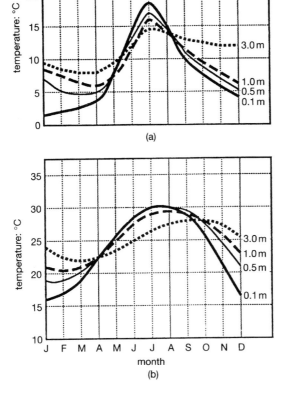

Fig. 5.13. In-soil temperatures in the UK and Hong Kong: (a) mean monthly in-soil temperatures at different depths in Britain (Murray and Farrar, 1988); (b) mean monthly in-soil temperatures at different depths in Hong Kong (based on data from Royal Hong Kong Observatory, after Howells and Pang, 1989)

temperature are confined to a region below the materials glass transition temperature, there is an insignificant change in the materials behaviour under load. Thus for polyester reinforcements, changes in ambient operating temperatures in the range of 10–40°C would not be expected to alter the creep characteristics. However, changes in the same temperature range would be expected to alter the creep characteristics of polypropylene tapes and high density polyethylene.

A family of creep curves for high modulus polyester yarns and high modulus polypropylene tapes at 23°C and 40°C is shown in Fig. 5.15. In the case of the polyester yarns there is no apparent difference in creep between the two temperature environments. For the polypropylene fibres there is an increase in the rate of creep with an increase in temperature, although the increase is not particularly significant. This increase in the rate of creep with temperature has been used by researchers to provide accelerated creep data at ambient temperatures for specific products (Bush, 1990).

As temperature affects the rate of creep and the stress–rupture characteristics of many polymer reinforcements, this should be taken into account if the creep data to be used in design are obtained at different operating temperatures from those occurring in service. Where the creep data have been derived at

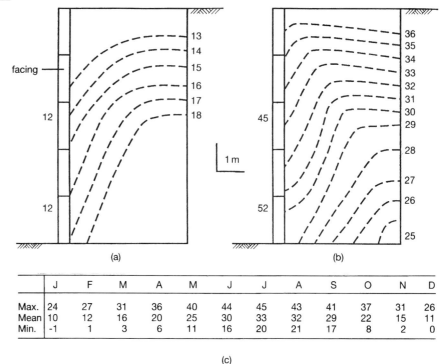

Fig. 5.14. Temperature conditions in desert environment A and B show in-soil temperatures (°C) for a reinforced soil wall near Tucson, Arizona: (a) winter (desert); (b) summer (desert); (c) air temperature in Phoenix, Arizona, USA

	J	F	M	A	M	J	J	A	S	O	N	D
Max.	24	27	31	36	40	44	45	43	41	37	31	26
Mean	10	12	16	20	25	30	33	32	29	22	15	11
Min.	-1	1	3	6	11	16	20	21	17	8	2	0

(c)

higher ambient temperatures than those expected in service, conservative predictions will result if the data are used in the calculation of long-term design strength and extensions. Alternatively, if the creep data have been derived at lower ambient temperatures than those expected in service, unsafe predictions may result. In the majority of tests for creep, a test temperature of 20–23°C has become the industrial standard; reference to Fig. 5.13 shows that the majority of this information is directly applicable to many reinforced soil structures.

5.3.3 Proprietary reinforcing materials

With the increasing importance and use of earth reinforcing systems, a range of proprietary materials developed specifically for reinforced soil structures has

Table 5.13. Glass transition temperatures and melting points for various geotextile elements

Geotextile element	Glass transition temperature, T_g (°C)	Melting point, T_m (°C)
Aramid fibres		370
Polyester fibres	90 to 110	260
Polypropylene tapes	−20	170 to 180
High density polyethylene	−120 to −90	130

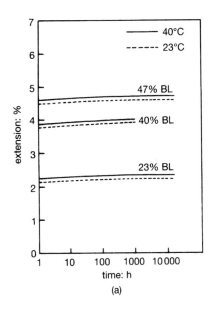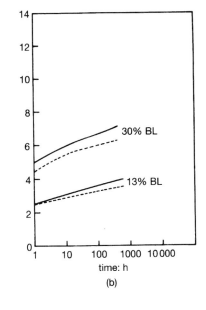

Fig. 5.15. Effect of temperature on creep: (a) high modulus polyester: (b) high modulus polypropylene

Table 5.14. Aligned glass-fibre reinforcing strip (Fibretain) in the form of a hairpin

Type	Length	Nominal width: mm	Nominal thickness: mm	Load capacity: kN	Nominal strength, glass: MN/m^2	Anchor load capacity: kN
48	3 m to 10 m	40	2	16	354	8
96/1	Other	80	2	32	354	16
192/1	as	160	2	64	354	32
192/2	special	80	4	64	354	32
240/1		161	2·5	80	354	40

Table 5.15. Frictional characteristics of glass-fibre reinforced plastic strip reinforcements

Frictional value at horizontal displacement	Bunter sand versus Bunter sand	GRP versus Bunter sand	Glacial sand versus Glacial sand	GRP versus Glacial sand	PFA versus PFA * (Staythorpe)	GRP versus PFA
tan ϕ'	0·73	0·61	0·73	0·54	0·71	0·53
ϕ'	30°	31°	36°	28°	35°	28°

123

Table 5.16. Tensar product data

Geogrid	Quality control strength: kN/m*	Load from QC test: k/Nm*		Polymer	Typical aperture pitch: mm	Roll dimensions: m
		at 2% strain	at 5% strain			
SS20	20/20	7/7	14/14	Polypropylene	39/39	50 × 4
SS30	30/30	10·5/10·5	21/21	Polypropylene	39/39	50 × 4
SS40	40/40	14/14	28/28	Polypropylene	33/35	30 × 4

* Determined in accordance with BS 6906, Part 1 and as a lower 95% confidence limit in accordance with ISO 2602 1988 (BS 2846, Part 2, 1981). Figures are transverse direction/longitudinal direction.

Geogrid	QC strength: kN/m	Approx peak strength: %	Creep limited strength: kN/m*		Polymer	Typical aperture pitch: mm	Roll dimensions: m
			10°C**	20°C**			
SR55	55†	11·2	22	20·5	High density polyethylene	160/22·4	30 × 1
SR80	80†	11·2	32·5	30·5	High density polyethylene	160/22·4	30 × 1
SR110	110†	11·2	45	42	High density polyethylene	150/22·5	30 × 1
55RE	55†	11·5	23	21	High density polyethylene	235/22	50 × 1·3
80RE	80‡	11·5	33	31	High density polyethylene	235/22	50 × 1·3
120RE	120‡	11·5	50	46	High density polyethylene	235/22	50 × 1·3
160RE	160‡	11·5	66	60	High density polyethylene	230/22	30 × 1·3

* Determined by the application of standard extrapolation techniques to creep data obtained in accordance with BS 6906, Part 5, for a strain not exceeding 10% in 120 years. The strengths obtained allow for extrapolation of data and variations in product manufacture.
** In-soil temperature.
† Determined by Netlon Limited QC test method and as a lower 95% confidence limit in accordance with ISO 2602 1980 (BS 2846, Part 2, 1981).
‡ Determined in accordance with BS 6906, Part 1 and as a lower 95% confidence limit in accordance with ISO 2602 1980 (BS 2846, Part 2, 1981).

Table 5.17. Paragrid product data

| Construction | | | High modulus polyester fibre core, polyethylene sheath | | | |
| Polymer composition | | | High modulus polyester fibre core, polyethylene sheath | | | |
Product grade			50S/25S	50S/50S	100S/25S	100S/100S
MECHANICAL PROPERTIES—Characteristic values						
Tensile strength						
Nominated breaking load (NBL)—length	BS 6906, ASTM D4595	kN/m	50	50	100	100
Extension at NBL—length	BS 6906, ASTM D4595	%	12	12	12	12
Working modulus—length	BS 6906, ASTM D4595	kN/m	500	500	1000	1000
Nominated breaking load—across	BS 6906, ASTM D4595	kN/m	25	50	25	100
Extension at NBL—across	BS 6906, ASTM D4595	%	12	12	12	12
HYDRAULIC PROPERTIES—Mean values						
Permeability—10 cm constant head		$l/m^2 s$	NA*	NA	NA	NA
Apparent opening size—O_{90}		μm	NA	NA	NA	NA
PRODUCT DIMENSIONS—Nominal						
Weight per m^2	ISO 3801, ASTM D1910	g	400	550	550	850
Thickness—under load 2 kN/m^2		mm	1·5	1·5	1·5	1·5
Width		m	4·5	4·5	4·5	4·5
Standard length		m	50	50	50	50
Roll diameter		m	0·35	0·35	0·35	0·35
Roll weight (gross)		kg	105	140	140	210

* Not applicable.

125

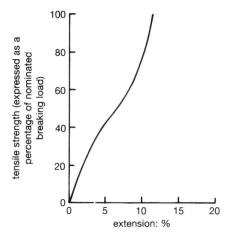

Fig. 5.16. Load/extension characteristics of Paragrid

been developed. The following have become established and are used as examples of the various materials and material forms available.

5.3.3.1 Glass-fibre reinforced plastic (GRP)

Glass-fibre reinforced plastic strips have been developed by Pilkington Brothers. They are formed from a continuous filament of E-glass roving embedded in a thermo-setting polymer. The materials are combined to form an aligned fibre reinforcing strip in the form of a hairpin, the end connection being formed at the loop. The diameter of the anchor hole varies with requirement. Typical properties are illustrated in Table 5.14.

The ultimate strength is the level below which no individual anchor load should fall in 100 years. The long-term load capacity values are the loads to which the appropriate factor of safety should be applied to arrive at working loads. GRP does not exhibit plastic deformation. The frictional characteristics of GRP are illustrated in Table 5.15.

5.3.3.2 Geogrid—Tensar

Tensar is a polymer grid material developed as a soil reinforcing material. Various forms are available dependent upon application. Typical properties of the material are illustrated in Table 5.16. The soil frictional characteristics of geogrids are superior to other forms of reinforcement due to the interlocking of the soil with the grid members. A coefficient of friction between the soil and the geogrid of unity may be assumed, i.e. $\alpha' = 1 \cdot 0$. Tests may indicate that a higher value is possible (see Chapter 4, Theory). (The properties of geogrids vary widely and are being developed very rapidly; the reader is referred to specialist literature.)

5.3.3.3 Paragrid

Paragrid is a geogrid formed from high modulus polyester fibre encased in a polyethylene skin. This form of structure is illustrated in Fig. 5.8. The fibres provide the load-carrying capacity of the material while the polyethylene skin

Table 5.18. *ParaLink product data*

| Construction | | High modulus polyester fibre core, polyethylene sheath | | | | |
| Polymer composition | | High modulus polyester fibre core, polyethylene sheath | | | | |
Product grade		200S	300S	700S	1000S	1250S
MECHANICAL PROPERTIES—Characteristic values						
Tensile strength						
Nominated breaking load (NBL)—length	kN/m BS 6906, ASTM D4595	200	300	700	1000	1250
Extension at NBL—length	% BS 6906, ASTM D4595	9	9	9	9	9
Working modulus—length	kN/m BS 6906, ASTM D4595	2200	3300	7500	11 000	13 500
Nominated breaking load—across	kN/m BS 6906, ASTM D4595	—	—	—	—	—
Extension at NBL—across	% BS 6906, ASTM D4595	—	—	—	—	—
HYDRAULIC PROPERTIES—Mean values						
Permeability—10 cm constant head	l/m^{2}s	NA*	NA	NA	NA	NA
Apparent opening size—O$_{90}$	μm	NA	NA	NA	NA	NA
PRODUCT DIMENSIONS—Nominal						
Weight per m^2	g ISO 3801, ASTM D1910	1100	1600	2600	3600	4500
Thickness—under load 2 kN/m^2	mm	4	5	6	8	10
Width	m	4·5	4·5	4·5	4·5	4·5
Length: (roll length can be made to order if required)	m	100	100	50	50	50
Roll diameter	m	0·6	0·72	0·56	0·61	0·62
Roll weight (gross)	kg	570	800	570	790	990

* Not applicable.

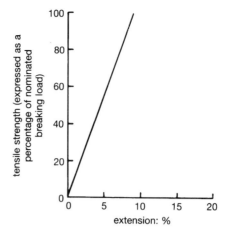

Fig. 5.17. Load/extension characteristics of ParaLink

maintains the shape of the geogrid and protects the polyester fibres from installation damage and against the effects of hydrolysis. The material is formed by bonding strips of the material to form a grid, Fig. 5.5. The ultimate tensile strength of Paragrid ranges from 50 kN/m to 100 kN/m in the transverse direction. The properties of the material are illustrated in Table 5.17. The load/extension characteristics of the material are shown in Fig. 5.16.

5.3.3.4 ParaLink

ParaLink is a geocomposite soil reinforcing material formed from high modulus polyester material encased in a polyethylene skin. ParaLink is a uni-directional material, having tensile properties in one direction only. To create a bi-directional reinforcement, two layers of the material are cross laid.

ParaLink is manufactured to any ultimate tensile strength ranging from 100 kN/m to 1250 kN/m. Typical properties are illustrated in Table 5.18. The load-extension characteristics of the material are shown in Fig. 5.17.

5.3.3.5 Engtex

Engtex soil reinforcing materials are woven geogrids formed from high modulus polyester fibres covered with a protective coating. The ultimate strength of the material ranges from 25 kN/m to 100 kN/m. The properties of the materials are illustrated in Table 5.19.

5.4 Facings

For vertical structures a facing is required. The function of the facing is to stop erosion of the fill and to provide a suitable architectural treatment to the structure. In achieving these objectives it must be compatible with the basic requirements of the particular construction system that is being employed.

Various materials can be used to form the facing, many will have particular advantages and disadvantages related to the particular application, scale of structure, shape adopted and material used, Table 5.20.

Table 5.19. *Engtex product data*

Construction		High strength woven geogrid						
Polymer composition		High tenacity polymer core with PVC coating						
PRODUCT		S35·30	S50·25	S50·50	S75·25	S75·75	S100·25	S100·100
Tensile strength (BS 6906) warp	kN/m	37	56	50	75	84	105	100
weft		34	27	50	25	77	25	100
Extension at break warp	%	12	12	12	12	12	12	12
weft		12	12	12	12	12	12	12
Aperture size warp	mm	28	28	28	28	30	28	30
weft		29	29	29	57	31	57	30
Weight per m²	g	220–260	210	280	188	250	220	340
Width	mm	2100	2100	2100	2100	2100	2100	2100
Roll length	m	50 100	50 100	50 100	50 100	50 100	50 100	50 100

Creep tensile strength based upon BS 6906, 40, 60 or 80% of tensile strength.

Table 5.20 (pages 130–133). Advantages and disadvantages of various facing materials

Materials	Potential advantages	Potential disadvantages	Comment
Aluminium	Durable Easily transported and handled, very versatile for erection purposes Can be cast	Requires experience in extrusion techniques for manufacturer Aesthetics (surface finish may deteriorate or distort) May encourage electrolytic action	Little used to date
Brick or masonry	Good material common to industry Good aesthetics Very durable	Produces a stiff facing unsuitable for soft foundations or where differential movements are likely Not particularly suited to tall structures	Very suitable for small structures
Concrete precast slabs	Technique available to industry, no need for further development Finishes good Relatively inexpensive, although s.steel lugs can be expensive Particularly suited to large construction where cost of formwork is low	Care required over durability, air entrainment may be necessary Shape of unit critical to rate of production and cost Rate of production and cost dependent on the lead time available for manufacture	Most common form of facing
Concrete pressed slabs	Durability and finish very good Once system developed, individual units have very low cost Weight of proposed units can be within the lifting capacity of one man	Difficult to reinforce Potential staining problem from reinforcement Fixing of holes in unit difficult Shape of unit is critical to production technique	Little used to date

Table 5.20—continued

Materials	Potential advantages	Potential disadvantages	Comment
Cast concrete (Full height)	Very stable and durable Reduces the potential failure planes to those passing under the toe	Placed after the reinforced soil structure has been constructed	Used extensively in Japan Shown to be stable during earthquake
King post and panel composite (Full height)	Constructed with readily available materials Very stable and durable Reduces potential failure planes to those passing beneath the toe Architectural finishes can be added after construction	King posts need to be propped during part of the construction process	Probably the most stable and durable facing available
Concrete prestressed	Readily available, if adopted from existing construction elements, e.g. Double Tee Very durable Easily transported and erected Good aesthetics	Best suited to small and medium-height structures	Most durable material

Table 5.20—continued

Materials	Potential advantages	Potential disadvantages	Comment
Fabric/ textiles/ geogrids	Very lightweight Very flexible Good for temporary structures Good for military structures	Aesthetics Durability (UV, rodents and vandals)	Covering face with soil or vegetation eliminates problem Concrete cladding system developed in France for bridge abutments. Full height cast concrete facing added in Japan
Glass-fibre reinforced cement, GRC	Very resistant to impact Finish good and economic in material Fixing of lugs presents no problems Factory production if spray-up manufacture used	New material which may not be completely understood Premix manufacture in early stages of development, manufacturers cost can be high	Little used because of cost
Glass-fibre reinforced plastic, GRP	Durable and strong, very resistant to impact Finish is good Very light and easily shaped Rate of unit output high	Colour affected by ultra-violet light Susceptible to damage from intense heat	Used in maintenance
Plastics (PVC, ABS)	Very lightweight, strong Finish good Easily shaped Rate of unit output high and easily transported	Melts at relatively low temperatures Fairly new materials, life expectancy unknown Some materials susceptible to ultra-violet light Creep characteristics not known Durability	Used in maintenance

Table 5.20—continued

Materials	Potential advantages	Potential disadvantages	Comment
Steel, galvanized	Relatively inexpensive Easily transported, rate of production high Easily shaped Good for industrial structures	Durability Aesthetics	Normally used with industrial environment
Steel, stainless (316)	Durable Easily transported, rate of production high Easily shaped	Special surface treatments increase cost significantly Can be expensive Thin sections lead to great flexibility	Little used
Steel, weather resistant	Relatively inexpensive Interesting architectural finish	Overlapping thin plates corrode	Little used
Timber (planks)	Readily available Particularly suitable for short life or temporary structures Suitable for Third World countries	Aesthetics Susceptible to termites, etc.	Roman structures recently discovered
Timber (plywood)	Readily available Suitable for short life/temporary structures Suitable for small-scale structures	Not suitable for long life Durability	Used in forestry applications

5.5 References

BARBER E. C., JONES C. J., KNIGHT P.G.K. and MILES M. H. (1972). PFA *Utilization*, CEGB.

BODEN J. B., IRWIN M. J. and POCOCK R. G. (1979). *Construction of experimental reinforced earth walls at the Transport and Road Research Laboratory*. TRRL Supplementary Report 457, 162–194.

BOLTON M. D. (1986). The strength and dilatancy of sands. *Géotechnique*, **36**, no. 1, 65–78.

BRITISH STANDARDS INSTITUTION (1983). *BS 1449: Part 1: Specification for carbon and carbon-manganese plate, sheet and strip; Part 2: Specification for stainless and heat resisting steel plate, sheet and strip*. BSI, London.

BRITISH STANDARDS INSTITUTION (1986). *BS 729: Specification for hot dip galvanized coatings on iron and steel articles*. BSI, London.

BRITISH STANDARDS INSTITUTION (1990). BS 1377: *Methods of test for soils for civil engineering purposes: Part 7: shear strength tests (total stress); Part 8: shear strength tests (effective stress)*. BSI, London.

BRITISH STANDARDS INSTITUTION. *BS 6906: Part 1: Determination of tensile properties using a wide width strip (1987); Part 5: Determination of creep (1991); Part 8: Determination of sand–geotextile frictional behaviour by direct shear* (1991). BSI, London.

BRITISH STANDARDS INSTITUTION (1995). *BS 8006: Code of practice for strengthened/reinforced soils and other fills*. BSI, London.

BUSH D. I. (1990). Variation of long-term design strength of geosynthetics in temperatures up to 40°C. *Proc. 4th Int. Conf. on Geotextiles*, The Hague. Balkema, **2**, 673–676.

CEN STANDARDS (1993). *BS EN 10025: Hot rolled products of non-alloy structural steels*. BSI, London.

CLARKE B. G., TRI UTOMO S. H. and JONES C. J. F. P. (1990). *Geotechnical properties of Tilbury Power Station Ash*. University of Newcastle, Research Report.

DEPARTMENT OF TRANSPORT (1978). *Reinforced Earth Retaining Walls and Bridge Abutments for Embankments*. Tech. memorandum (bridges), BE 3/78.

EXXON CHEMICAL GEOPOLYMERS LTD (ECGL) (1989). *Designing for soil reinforcement*, Exxon Chemical UK.

GREENWOOD J. H. (1990). The creep of geotextiles. *Proc. 4th Int. Conf. on Geotextiles*, The Hague. Balkema, **2**, 645–650.

GÜLER E. (1990). D. Holt (ed.). Lime stabilized cohesive soil as a fill for geotextile reinforced retaining structures. *Geotextiles, geomembranes and related products*. Balkema, **1**, 39–44.

HASSAN C. A. (1992). *The use of flexible transverse anchors in reinforced soil*. PhD thesis, University of Newcastle upon Tyne.

HOLLAWAY L. (1990). *Polymers and polymer composites in construction*. Thomas Telford, London, 275.

HOWELLS D. J. and PANG P. L. R. (1989). Temperature considerations in the design of geosynthetic reinforced fill structures in hot climates. *Proc. Symp. on Application of Geosynthetics and Geofibres in South East Asia*, Petaling Jaya, Malaysia, 1.1–1.7.

JEWELL R. A. and JONES C. J. F. P. (1981). Reinforcement of clay soils and waste materials using grids. *X. Int. Conf. for Soil Mechanics and Foundation Engineering (SMFE)*, Stockholm, **3**, 701.

JEWELL R. A. (1992). Ochiai, Hayashi and Otani (eds). Keynote Lecture: Links between the testing, modelling and design of reinforced soil. *Earth reinforcement practice*, Balkema, **2**, 755–772.

KOENER R. M. and WELSH J. P. (1980). *Construction and Geotechnical Engineering Using Fabrics*, John Wiley.

LAWSON C. R. (1991). *Use of geotextiles in reinforced soil retaining structures*. Sydney Technological University, Australia, 44.

McGOWN A., ANDRAWES K. Z. and AL-HASANI M. M. (1978). Effect of inclusion properties on the behaviour of sand. *Géotechnique*, **28**, 3, 327–46.

McKECHNIE THOMPSON G., ECCLES D., RODIN S. and WEBB S. P. (1973). Spoil heaps and lagoons. National Coal Board Technical Handbook, London.

MORRISON D. V. and CROCKFORD W. W. (1990). Lambe and Hanson (eds.). Cement stabilized soil retaining walls. *Design and construction of earth retaining structures*. ASCE, New York, Special Geotechnical Publication 25, 307–321.

MURRAY R.T. and FARRAR D. M. (1988). Temperature distributions in reinforced soil retaining walls. *Int. J. Geotextiles and Geomembranes*, **6**.

RANKILOR P. R. (1981). *Membranes in Ground Engineering*, John Wiley.

RISSEEUW P. and SCHMIDT H. M. (1990). Hydrolysis of HT polyester yarns in water at moderate temperatures. *Proc. 4th Int. Conf. on Geotextiles*, The Hague. Balkema, **2**, 691–696.

TATEYAMA M. and MURATA O. (1994). Permanent geosynthetic-reinforced soil retaining walls used for bridge abutments. *13th Int. Conf. SMFE*, New Delhi.

TATSUOKA F., MURATA O. and TATEYAMA M. (1992). W. Wu (ed.). Permanent geosynthetic-reinforced soil retaining walls used for railway embankments in Japan. *Proc. Geosynthetic-Reinforced Soil Retaining Walls*. Balkema, 101–130.

TATSUOKA F., TATEYAMA M. and KOSEKI J. (1995). Performance of geogrid-reinforced soil retaining walls during the Great Hanshin-Awaji Earthquake. *1st Int. Conf. Earthquake Geotechnical Engineering*, IS-Tokyo '95.

WATTS G. R. A. and BRADY K. C. (1990). Site damage trials on geotextiles. *Proc. 4th Int. Conf. on Geotextiles*, The Hague. Balkema, **2**, 603–608.

WISSE J. D. M., BROOS C. J. M. and BOELS W. H. (1990). Evaluation of the life expectancy of polypropylene geotextiles used in bottom protection structures around the Ooster Schelde Storm Surge Barrier. *Proc. 4th Int. Conf. on Geotextiles*, The Hague. Balkema, **2**, 697–702.

6

Design and analysis

6.1 Idealization

The development of modern soil reinforcing techniques has been rapid. Even so the benefits to be gained from their use have been demonstrated not only in the financial savings achieved but also in their ability to produce novel solutions to construction problems.

Owing to the extensive lead times involved in civil engineering schemes, it is probable that the first consideration of soil reinforcing systems by many designers will be as an alternative to a conventional solution. The disadvantages of substitution can be considerable: contractors inexperienced in the technique may tender high; short lead times for material delivery can cause logistical problems; and the lack of knowledge relating to specific subsoil conditions may create design problems. The fact that reinforced soil can frequently provide financial benefits when used as a late alternative to a conventional design suggests that greater benefits could be obtained if the use of soil strengthening systems were considered at the conceptual design stage of any scheme (Jones, 1994).

6.2 Conceptual design

The full benefit of soil reinforcing methods can be obtained only if the engineer is aware of the advantages and limitations of the technique, and has access to the necessary analytical testing and estimating procedures required for design. An essential requirement is a comprehensive soil survey which must be planned with the understanding that soil reinforcing techniques could form part of the design solution. In particular, if finite element techniques are to be used in the analysis, the conventional soil survey may need to be supplemented to provide information relating to the initial stresses in the subsoil.

Recent experience suggests that some of the more beneficial applications lie in the following areas; others are illustrated in Chapter 3, Application areas:

(a) retaining walls, bridge abutments or retained embankments
(b) a solution for environmental or special problems
(c) reinforced embankments and cuttings, either as an aid to construction or as a means of reducing land requirements (soil nailing)

(d) industrial bulk storage structures
(e) military structures
(f) foundations for structures, embankments and roads
(g) control of coastal erosion or wave walls
(h) as retaining structures in areas of seismic activity
(i) tension membranes supporting structures and embankments over voids
(j) as a construction aid over areas of very soft (super soft) soil.

6.2.1 Walls, abutments and retained embankments

The design and construction of conventional reinforced soil walls are now established, although the action of the complex mechanisms involved are not properly understood. When viewed at the conceptual design stage of a highway scheme, reinforced soil walls present few problems, although their cost effectiveness may suggest vertical and horizontal alignments which could not be contemplated with conventional structures.

The general motorway or trunk road bridge, if constructed using abutments, will have a significant proportion of the cost of the structure invested in the substructure. Split costs of decks and abutments on small span bridges have indicated substructure costs rising to 50–70 per cent of overall cost. Since reinforced soil has been shown to produce economies in abutment costs, significant reductions in total bridge costs are possible (see Chapter 9, Costs and economics). The use of reinforced soil abutments cannot be accomplished without some change to the deck design, the span of which will almost certainly be increased at a cost dependent on span and skew, Fig. 6.1.

The possibility of differential settlement across the width of reinforced soil abutments raises concern over potential difficulties in articulation of the deck. However, experience with bridges constructed in areas of active mining show that this problem can be resolved with the use of a low torsion beam and slab deck, provided that the twist does not exceed 1 in 80 (representing a differential settlement of 300 mm across the abutment of a typical two-lane overbridge). An acceptable design concept in these circumstances for reinforced soil abutments is to adopt the procedure used in mining areas, where piled foundations cannot be used owing to the problem of differential subsoil

Fig. 6.1. Deck costs relative to span and skew

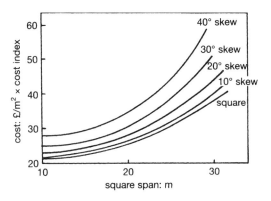

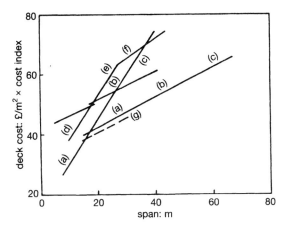

Fig. 6.2. Deck cost comparisons: (a) reinforced concrete slab; (b) post-tensioned voided slab; (c) post-tensioned cellular; (d) pre-tensioned inverted T; (e) post-tensioned inverted T; (f) post-tensioned T; (g) pre-tensioned M + low torsion slab

strain caused by a moving subsidence wave or where past mining activities have resulted in the presence of migrating lens cavities (i.e. cavities which over a period of many years move to the surface), Fig. 6.2.

One solution is to provide a substantial bearing pad, up to 7 m thick, of compacted granular material under the abutment, and to accept any residual differential settlement. The use of a thinner reinforced soil foundation formed as an integral part of the reinforced approach embankment is a practical alternative which has the advantage of minimizing the incident of differential settlement which occurs frequently behind conventional abutments (Walkinshaw, 1975). The same concept can be used under central piers of a two-span structure, although a degree of sophistication may be required in the analysis to permit the settlements of the abutments and the pier to be of the same order.

Sutherland (1973) has shown that the use of retained embankments in place of viaducts may offer considerable financial benefits, although in an urban environment the local severance caused by such structures may cause problems of access, Fig. 6.3. In rural conditions the use of viaducts is synonymous with bad ground conditions on which a conventional embankment cannot be constructed. In these conditions the use of earth reinforcing systems to improve the bearing capacity coupled with a reinforced soil embankment warrants special consideration.

On weak foundations, bridge abutments are frequently supported on piles; analytically, the abutments, piles and the adjacent embankments are considered in isolation. If the abutment is subjected to vertical loads from the deck and the piles to lateral loads from the embankment, a frequent conclusion of this separate analysis is that the abutment will move away from the embankment and the piles should be raked forward. A global assessment of the behaviour of the abutment, piles and embankment, in which the ground movements caused by the placing of the embankment are considered, may show an alternative behaviour. Abutment rotation may be towards the embankment, introducing an increase in moment in the abutment stress due to increased lateral pressures (i.e. the interaction of a stiff/soft material, Chapter 4, Theory). An alternative idealization of the role of the pile reinforcement is possible. Instead of providing compressive reinforcement in the form of conventional piles, tensile

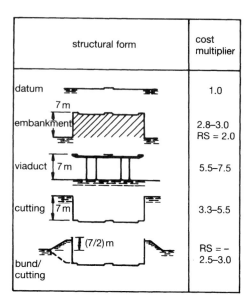

	structural form	cost multiplier
datum		1.0
embankment	7 m	2.8–3.0 RS = 2.0
viaduct	7 m	5.5–7.5
cutting	7 m	3.3–5.5
bund/ cutting	(7/2) m	RS = – 2.5–3.0

Fig. 6.3. Cost alternative of structural forms

reinforcement located in the tensile strain field may be used and the abutment becomes a reinforced soil structure forming one end of the embankment, Figs 6.4(a) and (b).

Practice indicates that the pile lengths associated with the idealization of Fig. 6.4(a) may be substantial, > 30 m on occasions; whereas the reinforcement length needed with the condition of Fig. 6.4(b) may be substantially less. An additional benefit of the reinforced soil solution is that this idealization effectively eliminates the problem of differential settlement between the abutment and adjacent embankments which is a common feature with piled abutments.

Fig. 6.4. (a) conventional idealization abutment construction on weak soil; (b) alternative idealization

6.2.2 Environmental or special problems

Civil engineering design standards cater not only for the user but also for the community and the environment. In urban environments the problem of noise

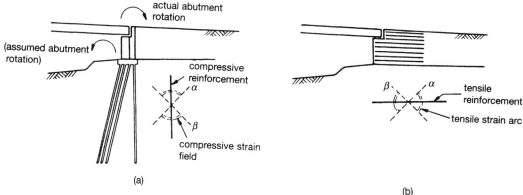

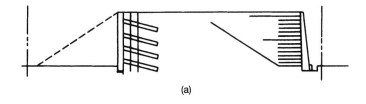

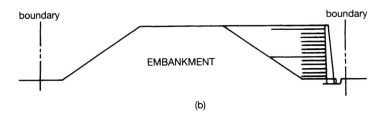

Fig. 6.5. Reinforced soil used to: (a) reduce width of embankment; (b) widen embankment, without the need for additional land

pollution must be taken seriously, particularly with highways. Dropping the highway into a cutting is one way of reducing noise pollution. The relative cost of a highway in a cutting against that at datum can be high and a compromise solution is the use of reinforced soil environmental bunding in conjunction with a partially sunken road. With this solution, not only is the noise pollution eased but also the severance problem is less acute as bridge crossings of the sunken highway require relatively short climbs for elderly pedestrians, Fig. 6.3.

The potential for using earth reinforcing techniques as a method of solving particular technical problems is extensive. The following example illustrates a unique quality with regard to bridge abutments in areas of mining subsidence. Bridge and other structures in an active mining area are subjected to both compressive and tensile ground strains (Jones and Bellamy, 1973). To design bridges to resist these strains is usually uneconomic, the solution is to permit the abutments to move together during the compressive phase of the subsidence wave (Sims and Bridle, 1966). Designing the abutments to accommodate the lateral pressures needed to cause this movement can add 25 per cent to the cost. Reinforced soil structures can tolerate compressive strains and stresses without any major change in design, although care must be taken to ensure that the tension phase of the mining wave does not cause tension failure of the reinforcement (Jones, 1989).

Reinforced soil structures have been shown to provide economic solutions to the problems associated with the widening of highways or railway embankments. A major design problem in these cases is the lack of available land. This is particularly the case in many urban areas and in densely populated areas such as parts of Japan. It is possible to widen a cutting or embankment without the need for additional land take by using reinforced soil techniques, Figs 6.5(a) and (b). Solutions using reinforced soil structures have an added bonus, in that they are very resistant to earthquakes. A number of structures of this type, constructed using geosynthetic reinforcement and rigid facings, survived the Kobe earthquake in Japan which measured 7·2 on the Richter scale (Tateyama *et al.*, 1995).

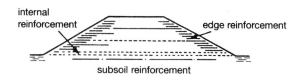

Fig. 6.6. Reinforced soil embankment

6.2.3 Reinforced embankments

Reinforcement in an embankment can take several forms depending on the nature of the problem to be solved. In Japan, the reinforcement of the edge of railway embankments, using geogrids to combat the acute climatic conditions, has been reported by Iwasaki and Watanabe (1978). The use of reinforcement close to the face permits the operation of heavy compaction plant near the shoulder of the slope and encourages uniform compaction throughout the embankment. Adoption of this construction technique on highway schemes could ensure the stability of the edge of the hard shoulder which is known to cause some problems, Fig. 6.6.

Similarly, the use of geogrids throughout the embankment permits higher compaction to be obtained. The latter has been demonstrated on the Uetsu Railway in Japan, where the recorded stiffness of unreinforced earth embankments, using standard penetration values (N-values), averaged N4 with a peak of N7·5. The equivalent readings for a reinforced embankment of N30 with a peak of N60 permits a reduction of the embankment width by steepening the slopes. The benefit of reducing the width can be substantial, in terms of both land take and materials, Fig. 6.7. Reinforcement of low-grade fill which would normally be unusable produces similar benefits, but is only practical when the concept of adding reinforcement to strengthen the weak soil is accepted.

A further use for earth reinforcement is beneath embankments situated on weak subsoils. The objective in this application is either to permit construction to take place by creation of an artificial subsoil crust, or to enable the erection to proceed at a faster pace than the dissipation of the pore water pressures would normally allow; to this end the use of subsoil reinforcement may be used as an alternative or adjunct to expensive subsoil drainage.

6.3 Analysis
6.3.1 Cornerstones of analysis

The cornerstones for the analysis of reinforced soil structures are that the soil strains, soil stresses, soil reinforcement interaction and gravity, reinforcement and boundary forces are all interconnected. This has been illustrated by Bolton (1991), Fig. 6.8. Equilibrium in a reinforced soil structure is reached when the strain in the reinforcement and that in the adjacent soil are compatible, Fig. 6.9. From the concept of strain compatibility illustrated in Fig. 6.9, it is possible to deduce that:

Fig. 6.7. Reinforcement used to reduce the width of an embankment

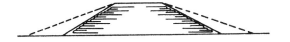

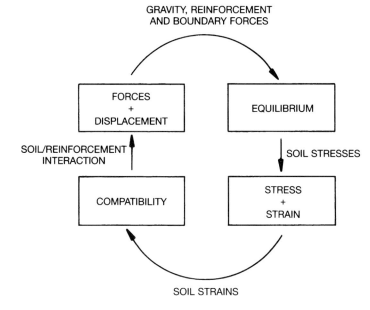

Fig. 6.8. Four cornerstones of analysis (after Bolton, 1991)

(a) stress in different materials (soil/reinforcement) is based on strain level
(b) stiff reinforcing materials will attract stress
(c) reinforcing materials prone to creep will lose stress.

The stiffness of the reinforcement has a fundamental influence on the behaviour and performance of reinforced soil structures. Axially stiff reinforcement will take up little strain before taking up load. Stress in the reinforcement can accumulate rapidly and may occur at lower strains than those required to mobilize peak soil strength. By contrast, extensible reinforcements require greater deformation before they take up the stresses imposed by the soil. This may lead to higher strains and the peak shear strength of the soil may be approached or exceeded.

The pattern of plastic soil strains associated with movements of a cantilever structure rotating about the toe shows lines of zero strain are referred to as velocity characteristics. These lines are oriented at angles of $45° + \psi/2$ either

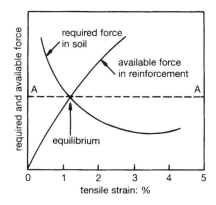

Fig. 6.9. Compatibility curve for steep reinforced slopes and walls (Jewell, 1985)

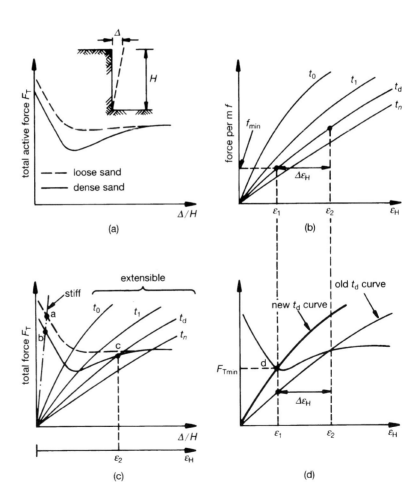

Fig. 6.10. *Velocity characteristic for reinforced soil structure with cantilever deformation of facing*

Fig. 6.11. *System compatibility of strains and forces for reinforced soil structures: (a) active earth force; (b) isochronous curves for single reinforcement; (c) system compatibility diagram; (d) new* t_d *curve*

side of the direction of maximum tensile strain, where ψ is the angle of dilatancy of soil, Fig. 6.10 (see Chapter 4, Theory). Similar velocity characteristics can be developed for steep slopes.

A simple model for reinforced soil deformation is to assume that the wall rotates about its base generating a displacement pattern similar to that in Fig. 6.10. Horizontal tensile strains will accumulate in proportion to wall rotation, and the horizontal strain, ϵ_H, in the reinforcement can be expressed as $\epsilon_H = (\Delta/H)\tan(45° + \epsilon/2)$, where Δ/H is the displacement of the wall.

Figure 6.11(a) shows the active earth force developed by a retaining wall, supporting loose and dense sand, as a function of wall rotation $\Delta/_H$. Fig. 6.11(b) shows isochronous curves for a single extensible reinforcement. Fig. 6.11(c) shows the system compatibility diagram in which the active earth force and the isochronous curves are superimposed. Equilibrium occurs when the curves representing reinforcement tension intersect the curve representing the active earth force which reflects the mobilized frictional resistance of the soil. For stiff reinforcements, this intersection may occur at low levels of strain, points a and b (equivalent to an at-rest condition of earth pressure, K_0), Fig. 6.11(c). For extensible reinforcements, the active earth force will approach a limit state, point c (equivalent to the active K_a, condition and ϕ'_{cv}).

6.3.2 Vertical walls and abutments

Walls and abutment structures are normally constructed using horizontal reinforcement, and they take the form illustrated in Fig. 6.12. The vertical spacing of the reinforcement may remain constant throughout the depth, but the density is likely to be greater near the base.

Structural layout. The simplest layout is a uniform distribution of identical reinforcing elements throughout the length and height of the structure. A

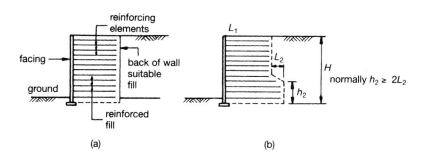

(a)   (b)

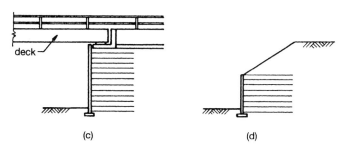

(c)   (d)

Fig. 6.12. (a) Wall elements; (b) stepped wall; (c) abutment; (d) part height wall

more economical layout may be achieved using reinforcement of different properties or by dividing the structure into different zones. The minimum length of reinforcement is often taken to be $0.7H$ or 5 m; this requirement relates to strip reinforcement and relates to the required bond length; grid reinforcements may not require these lengths. Bridge abutments frequently have a minimum base length of 7 m.

Current design methods. Many methods used to design reinforced soil structures use proprietary reinforcements and do not cover the methods used in the general design of reinforced soil in parts of Europe, the USA and the UK. These can be classified as those based on the coherent gravity hypothesis or the tie-back hypothesis, Fig. 6.13. The coherent gravity method is an empirical technique which has been described by Mitchell and Villett (1987) and the Ministeres des Transports (1979). It was developed to cater for structures reinforced with steel strip (inextensible) reinforcements. The tie-back method was developed by the UK Department of Transport (1978) and is based on limit equilibrium methods. It is independent of the reinforcement material, and is used with both inextensible reinforcements and with anchors. The UK Department of Transport (1978) tie-back method was revised in 1986.

Limit state design. The most recent innovation in the design of reinforced soil structures in the UK has been the introduction of the British Standard BS 8006 (1995). The British Standard (BS 8006) is written in a limit state format which

Fig. 6.13. (a) Coherent gravity hypothesis; (b) tieback hypothesis

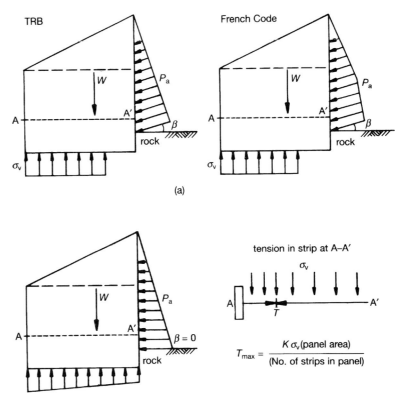

covers all forms of reinforced soil identified as internally stabilized systems with the exception of soil dowels, reticulated micropiles and special materials. The Code covers the design of walls, bridge abutments, steepened slopes, basal reinforcement and reinforcement over voids. Hybrid systems including tailed gabions, tailed masonry and also the gravity-faced soil-retaining structures used in Japan described by Tatsuoka (1992) are covered by the new Code of Practice. In the case of vertical structures and bridge abutments, both the coherent gravity method and the tie-back wedge are accepted, the selection of the appropriate analytical model being dependent on whether extensible or inextensible reinforcement is being used.

A particular development of BS 8006 is the use of the limit state concept. The two limit states considered are the *ultimate limit state* and the *serviceability limit state* which are defined as:

(a) *Ultimate limit state* at which collapse mechanisms form in the ground or the retaining structure, or when movement of a retaining structure leads to severe structural damage in other parts of the structure or in nearby structures or services.
(b) *Serviceability limit state* at which movements of the retaining structure affect the appearance or efficient use of the structure or nearby structures or services which rely upon it.

The limit states may be defined in terms of *limit modes*. Design evaluation involves ensuring the stability of the structure against failure by the limit modes appropriate for the circumstances. In the case of vertical walls and abutments, six limit modes are considered, Fig. 6.14.

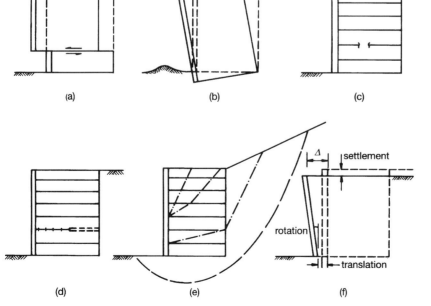

Fig. 6.14. Limit modes of failure of reinforced soil structures: (a) limit mode 1: sliding; (b) limit mode 2: bearing; (c) limit mode 3: element rupture; (d) limit mode 4: element pullout; (e) limit mode 5: wedge/slip circle stability; (f) limit mode 6: deformation

Limit mode 1: sliding failure of the structure:
  (*i*) on the interface between the reinforced fill and the subsoil.
  (*ii*) within the reinforced fill on a soil/soil interface.
Limit mode 2: bearing failure of the structure
Limit mode 3: tensile failure of the reinforcement
Limit mode 4: pullout failure of reinforcing elements
Limit mode 5: wedge/slip circle stability failure
Limit mode 6: deformation.

Other limit modes may be appropriate in certain circumstances and have to be checked accordingly; e.g.

(*i*)   where 3D effects could influence the structure
(*ii*) where the structure could be subject to seismic loadings
(*iii*) where the structure could be subject to cyclic loadings
(*iv*) where the structure could be subject to accidental flooding.

Table 6.1. *Partial factors used in the design of vertical walls and abutments*

	Partial factor	Ultimate limit state	Serviceability limit state
Loads	Soil unit weight (forming or on top of the structure)	$\gamma_{es} = 1.5*$ $= 1.0\dagger$ $= 1.0\ddagger$	$\gamma_{es} = 1.0*$ $= 1.0\dagger$ $= 1.0\ddagger$
	External dead loads (e.g. bridge deck loading)	$\gamma_{ff} = 1.2*$ $= 1.0\dagger$ $= 1.0\ddagger$	$\gamma_{ff} = 1.2*$ $= 1.0\dagger$ $= 1.0\ddagger$
	External live loads: traffic loading	$\gamma_{q} = 1.5*$ $= 1.5\dagger$	$\gamma_{q} = 1.5*$ $= 1.5\dagger$
	bridge loading	$\gamma_{eq} = 1.5\ (HA)*$ $= 1.3\ (HA+HB)*$ $\gamma_{eq} = 1.5\ (HA)*$ $= 1.3\ (HA+HB)\dagger$	$\gamma_{eq} = 1.5\ (HA)*$ $= 1.3\ (HA+HB)*$ $\gamma_{eq} = 1.5\ (HA)*$ $= 1.1\ (HA+HB)\dagger$
	temperature effects	$\gamma_{eq} = 1.3$	$\gamma_{eq} = 1.3$
	Earth pressure behind structure	$\gamma_{es} = 1.5*$ $= 1.5\dagger$ $= 1.0\ddagger$	$\gamma_{es} = 1.0*$ $= 1.0\dagger$ $= 1.0\ddagger$
	Horizontal loads due to shrinkage and creep	$\gamma_{ef} = 1.2*$ $= 1.2\dagger$ $= 1.0\ddagger$	$\gamma_{ef} = 1.2*$ $= 1.2\dagger$ $= 1.0\ddagger$

* When checking maximum reinforcement tension and foundation pressure.
† When checking maximum overturning loads, pullout resistance and sliding on base.
‡ When considering dead loads only, foundation settlement and the reinforcement tension at the serviceability limit state.

6.3.2.1 Analytical procedure

In common with other design codes, modern reinforced soil design methodology adopts a limit state partial factor approach in which a structure is shown to be safe against failure at the ultimate or serviceability levels by the application of partial factors. Partial factors are applied to the loads (γ_{fl}), reinforcing materials (γ_m), soil materials (γ_{ms}), soil/reinforcement interaction factors (sliding and pullout) (γ_s, γ_p), foundation bearing capacity (γ_{bc}) and horizontal sliding on a soil/soil interface (γ_{ss}). An additional partial factor (γ_n) related to the economic ramifications of failure is adopted by the UK Code BS 8006: 1994. This additional factor could be applied to either the material factors (γ_m) or the load factors (γ_{fl}). The application of increased (factored) external loads to an earth-retaining structure is not always unfavourable, as increased stress in a frictional soil can result in an enhanced shear strength. In addition, larger factors are generally applied to live loads than to dead loads as, in the case of reinforced soil structures, the superimposed live loads are often small in comparison to the self-weight dead loads. As a result, the application of γ_n to the reinforcement design strength is more likely to result in a satisfactory margin of safety due to increased ramifications of failure than if it were applied to the loads.

Values of partial factors which have been used in the design of vertical walls and bridge abutments are shown in Table 6.1. The values stated in Table 6.1 reflect conditions in the UK and may not be appropriate in other countries.

Table 6.1—continued

	Partial factor	Ultimate limit state	Serviceability limit state
Materials	Soil materials:		
	applied to $\tan'\phi_{des}$	$\gamma_{ms} = 1\cdot0$	$\gamma_{ms} = 1\cdot0$
	applied to c'	$\gamma_{ms} = 1\cdot6$	$\gamma_{ms} = 1\cdot0$
	applied to C_u	$\gamma_{ms} = 1\cdot0$	$\gamma_{ms} = 1\cdot0$
	Reinforcement	Varies with the form of reinforcement (See example 4, Ch. 11)	Varies with the form of reinforcement (See example 4, Ch. 11)
Soil/reinforcement interaction	Sliding on surface or reinforcement pullout	$\gamma_s = 1\cdot3$ $\gamma_p = 1\cdot35$	$\gamma_s = 1\cdot0$ $\gamma_p = 1\cdot0$
Bearing capacity	Foundation	$\gamma_{bc} = 1\cdot35$	—
Sliding	Base of structure or soil/soil interface	$\gamma_{ss} = 1\cdot2$	—
Economic ramifications of failure	Structure of strategic importance (e.g. motorway bridge or dam)	$\gamma_n = 1\cdot1$	$\gamma_n = 1\cdot1$
	Other structures (e.g. industrial structures)	$\gamma_n = 1\cdot0$	$\gamma_n = 1\cdot0$

At the ultimate limit state, calculations are needed to evaluate stability. At the serviceability limit state, calculations may not be necessary and requirements may be satisfied by reference to similar structures and details.

Limit mode 6 covering deformation has to be considered. This is used to check the serviceability of any structure and is also used to determine the stress state applicable in the analysis. The analytical model used is sensitive to the form of reinforcement used in that an extensible reinforcement will lead to greater structural deflection during and post construction than if inextensible (stiff) reinforcement is specified. The use of stiff reinforcement can result in additional stress being attracted to the reinforcement, a point implicitly acknowledged in the coherent gravity hypothesis where the K_0 stress state is used in the analysis.

In the limit mode method of analysis, the correct stress state of the soil for use in the analysis is obtained by the following analytical sequence:

(*i*)   The geometry of the structure is chosen.
(*ii*)   The stress state of the soil K_{des}, is assumed equal to K_a (the active condition).
(*iii*)   Limit modes 3 and 4 are checked, and the quantity of reinforcement needed to satisfy these conditions is identified.
(*iv*)   Limit mode 6 is checked.

If $\Delta \geq 001H$ or $\theta_{max} \geq 0{\cdot}002H$, where Δ is a rigid body translation of the reinforced soil structure, θ_{max} is a rigid body rotation (forward) of the wall or structure, the selection of $K_{des} = K_a$ is considered to be justified and the design proceeds with consideration of the remaining limit modes.

If $\Delta < 001H$ or $\theta_{max} < 0{\cdot}002H$, $K_{des} \neq K_a$ but at some value between the active and the at-rest condition. In this case, the reinforcement must be redesigned using the at-rest pressure within the reinforced zone ($K_{des} = K_0$). It is recognized that in tall structures the stress state of the soil in the lower part of the structure will equate to the active condition (i.e. in the top of the structure $K_{des} = K_0$, in the lower part of tall wall $K_{des} \rightarrow K_a$). In this condition, the adoption of the coherent gravity hypothesis as the analytical model is justified.

Design angle of friction of the soil fill (ϕ'_{des})
The mobilized shear strength of a soil subject to compressive loads increases with increasing axial and lateral strain, until a peak shear strength is mobilized, ϕ'_p, Fig. 6.15. In elastoplastic soils, shear strength continues to be mobilized at the peak values when strains are required to produce the peak shear strength. In the case of strain softening soils, the mobilized shear strength decreases as strains exceed those required to mobilize peak strength. At large strains, the mobilized shear strength achieves a constant minimum value, ϕ'_{cv}. In frictional fills subject to plane strain conditions, the strain needed to mobilize ϕ'_p is small, typically 1·5% for dense sands. The selection of ϕ'_{des} depends on the quality of the fill used in design. The British Standard BS 8006: 1994 adopts $\phi'_{des} = \phi'_p$ for walls and abutments constructed with good frictional fill, in the case of slopes and embankments on weak foundations $\phi'_{des} = \phi'_{cv}$.

Note: At the time of design, the source of the fill that will be used in a

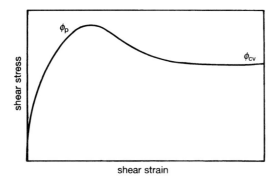

Fig. 6.15. Typical stress–strain curve for cohesionless soil

structure may not be known and a conservative value of ϕ'_{des} is often adopted, a typical value being 32°. In a good quality fill a 32° value for the angle of shearing resistance under effective conditions would be equivalent to the ϕ'_{cv} value.

6.3.2.2 General design case—tie-back analysis

Limit mode 1—sliding

The stability of the structure against forward sliding is checked at the base of the structure or at any level within the structure. Sliding can occur on a soil–soil interface or on a soil–reinforcement interface; both need to be considered. Resistance to movement should be based on the properties of the subsoil on the reinforced fill whichever is the weaker.

(*i*)   Long-term stability on a soil–soil interface

$$\gamma_{ss} P \leq W \frac{\tan \phi'_{des}}{\gamma_{ms}} + \frac{c'}{\gamma_{ms}} L \tag{1}$$

(*ii*)   Long-term stability on a soil–reinforcement interface

$$\gamma_s P \leq \alpha' \frac{\tan \phi'_{des}}{\gamma_{ms}} + \alpha'_{sr} \frac{\tan \phi'_{des}}{\gamma_{ms}} L \tag{2}$$

(*iii*)   Short-term stability on a soil–soil interface

$$\gamma_{ss} P \leq \frac{c_u}{\gamma_{ms}} L \tag{3}$$

(*iv*)   Short-term stability on a soil–reinforcement interface

$$\gamma_s P \leq \frac{\alpha'_{sr} c_u}{\gamma_{ms}} L \tag{4}$$

where P is the horizontal factored disturbing force
$= \frac{1}{2} K_{des} \gamma_{es} \gamma H^2$ for the simple case of backfill pressure only

W is the vertical factored resistance force
$= \gamma_{es} \gamma H L$ for the simple case of no surcharge loading and $\beta = 0$, Fig. 6.13

ϕ'_{des} is the design angle of shearing resistance under effective stress conditions

c' is the cohesion of the soil under stress conditions
c_u is the shear strength of the soil
α' is the interactive coefficient relating soil/reinforcement bond angle with $\tan \phi'_{des}$
α'_{sr} is the cohesive coefficient relating soil cohesion to soil/reinforcement bond.

Limit mode 2—bearing and tilt
This should be checked for the following conditions:

(*a*) Horizontal forces are at a *maximum* and vertical forces are at a *minimum*.
(*b*) Both horizontal and vertical forces are at a *maximum*.

Bearing. For structures situated on good subsoils a trapezoidal pressure distribution beneath the structure may be assumed, Fig. 6.16 (on poor/weak subsoils the pressure distribution may equate to a uniform condition). The imposed bearing pressure under the toe *a* should be compared with the ultimate bearing capacity of the foundation soil

$$a \leq \frac{q_{ult}}{\gamma_{bc}} + \gamma D_m \tag{5}$$

where *a* is the factored bearing pressure acting under the toe

$$= \frac{1}{L}\left(W + \frac{6Pe}{L}\right)$$

$$= \gamma_{es}\,\gamma H \left[1 + K_a \left(\frac{H}{L}\right)^2\right]$$

e is the eccentricity of line of action of horizontal thrust $H/3$ in the simple case
D_m is the embedment depth
γ_{bc} is the partial factor for bearing capacity of the foundation
q_{ult} is the ultimate bearing capacity of the foundation soil
γ is the foundation soil density.

Tilting. The overturning movement with respect to the toe must be less than the restoring movement:

$$\frac{WL}{2} > Pe \tag{6}$$

Fig. 6.16. (a) tilting; (b) slip

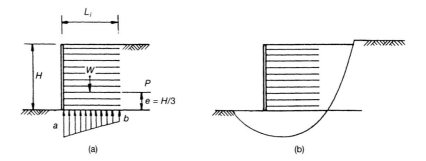

(a)   (b)

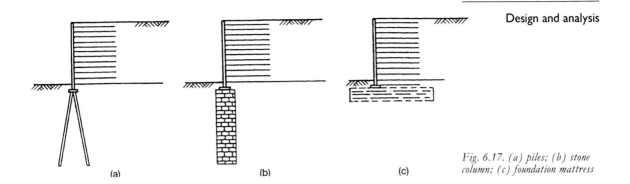

Fig. 6.17. (a) piles; (b) stone column; (c) foundation mattress

where P is the horizontal factored disturbing force (see equation (1))

W is the vertical factored resistance force (see equation (1)).

If the limit mode 2 condition is not satisfied, the stability of the structure can be increased by widening the base width of the structure, Fig. 6.12(b). With weak foundation soils, widening the base width may not be sufficient to satisfy the ultimate bearing capacity criteria. In this case, support for the base may be provided by external means, Fig. 6.17. Alternatively, the overall stability of the structure and the surrounding soil may be considered on a global basis using finite element or other continuum methods.

Limit mode 3—element rupture
Local stability of a layer of reinforcing elements. Consider a row i, of reinforcing members in Fig. 6.18. The maximum ultimate limit state tensile force T_i, at a

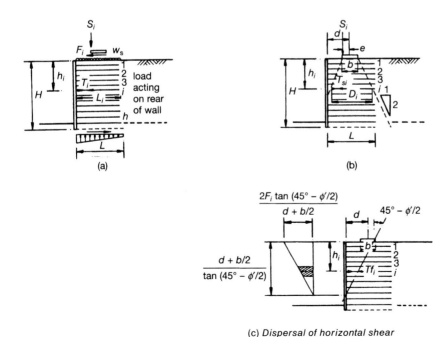

(c) *Dispersal of horizontal shear*

Fig. 6.18. Tie-back analysis: local stability

153

depth of h_i below the top of the structure is given by:

$$T_i = T_{hi} + T_{wi} + T_{fi} + T_{mi} + T_{si} \tag{7}$$

where   (*i*) T_{hi} is the design tensile force due to the self-weight of the reinforced fill above the *i*th layer of reinforcements

$$= (K_{des}\, \gamma_{ms}\, \gamma h_i - 2\, \gamma_{ms}\, c'\sqrt{K_a})\, V \tag{8}$$

(*ii*) T_{wi} is the uniform surcharge on top of the structure

$$= K_{des}\, \gamma_{ff}\, w_s\, V \tag{9}$$

where γ_{ff} is the partial factor for external dead loads

       w_s is uniformly distributed surcharge on top of structure.

[The combined effect of height and fill and uniform surcharge for cohesive-frictional fill only:

$$T_{hi} + T_{wi} \geq 5V\left(h_i + \frac{\gamma_{es} w_s}{\gamma}\right)]$$

(*iii*) T_{si} is the vertical loading applied to a strip contact area of width *b* on top of the structure

$$= K_{des} V\, \gamma_q\, \frac{S_i}{D_i}\left(1 + \frac{6e}{b}\right) \tag{10}$$

where γ_q is the partial load factor for external live loads

       S_i is vertical loading on top of structure

       d is load contact area

where: $D_i = h_i + b$ if $h_i \leq 2d - b$

         $D_i = d + \dfrac{h_i + b}{2}$ if $h_i > 2d - b$

         If $h_i > 2b$

$$T_{si} = K_{des}\, V\, \gamma_q\, \frac{S_i}{D_i} \tag{11}$$

(*iv*) T_{fi} is the horizontal shear applied to a strip contact area of width *b* on top of the structure

$$= 2V\, \gamma_q F_i\, Q\, (1 - h_i\, Q) \tag{12}$$

where F_i is the horizontal shear on top of structure

where: $Q = \dfrac{\tan(45° - \phi'_{des}/2)}{d + \dfrac{b}{2}}$

(*v*) T_{mi} is the bending movements caused by external loading

$$= 6\frac{K_{des}\, V\, M_i}{L_i^2} \tag{13}$$

where M_i is the bending moment arising from the external loading

       L_i is the length of reinforcement at the *i*th layer.

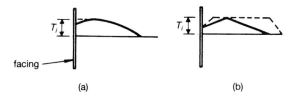

facing

(a)   (b)

Fig. 6.19. Assumed distribution of tensile force in reinforcement: (a) strip reinforcement; (b) grid reinforcement

Rupture. The tensile strength of the ith layer of reinforcing elements required to satisfy load and stability is:

$$\frac{T_D}{\gamma_n} \geq T_i \qquad (14)$$

where T_D is the design strength of the reinforcement

γ_n is the partial factor for ramifications of failure.

Limit mode 4—element pullout
Adherence. The perimeter P_i of the ith layer of reinforcing elements required to satisfy local stability requirements is:

$$P_i \geq \frac{T_i}{\mu L_{ei} \dfrac{(\gamma_{es}\gamma_{hi} + \gamma_{ff}w_s)}{\gamma_n\gamma_p} + \dfrac{a_{sr}c'L_{ei}}{\gamma_{ms}\gamma_n\gamma_p}} \qquad (15)$$

where T_i is the maximum value of the design tensile force calculated for limit mode 3—element rupture

P_i is the total horizontal width of the top and bottom faces of the reinforcing element at the ith layer per metric 'run'

μ is the coefficient of friction between the fill and the reinforcing elements

$$= \frac{\alpha' \tan \phi'_{des}}{\gamma_{ms}}$$

L_{ei} is the length of reinforcement in the resistant zone outside the assumed failure wedge at the ith layer of reinforcements.

Distribution of tensile force in strip reinforcement is assumed to be similar to Fig. 6.19(a). (Force at the connection between the reinforcement and the facing is taken as T_i. The distribution of tensile force in grid reinforcement is taken as Fig. 6.19(b). The individual values of T_i for strip and grid reinforcement are

Fig. 6.20. Potential wedge failure planes

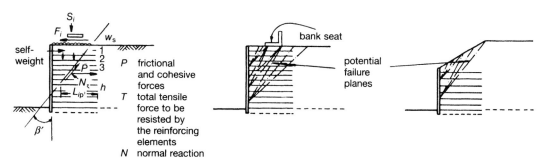

self-weight

S_i
F_i w_s
1
2
$P = 3$ P frictional and cohesive forces
N h
$L_{ip'}$ T total tensile force to be resisted by the reinforcing elements
β'
N normal reaction

bank seat

potential failure planes

unlikely to coincide. British Standard BS 8006: 1995 permits a reduction in the value of the connecting force at the top of structures with some structural details.)

Limit mode 5—wedge/slip circle stability
Wedges are assumed to behave as rigid bodies and may be of any size and shape. Stability of any wedge is maintained when frictional forces acting on the potential failure plane in connection with the tensile resistance/bond of the group of reinforcing elements embedded in the fill beyond the plane are able to resist the applied loads tending to cause movement, Fig. 6.20. The following loads and forces are considered in the analysis:

(*i*)   self-weight of the fill in the wedge
(*ii*) any uniformly distributed surcharge, w_s
(*iii*) external vertical loading, S_i
(*iv*) horizontal shear from any external loading, F_i
(*v*)   frictional and cohesive forces acting along the potential failure plane
(*vi*) the normal reaction on the potential failure plane.

In British Standard BS 8006: 1995 it is assumed that no potential failure plane passes through a bridge abutment bank seat. The wedge stability check ensures that the calculated frictional resistance does not exceed the tensile capacity, and vice versa, of any layer of elements. (When the facing consists of a structural element formed in one piece, potential failure planes passing through the facing may be neglected or the shear resistance offered by the rupture of the facing may be considered. This can have a major influence on stability.)

Fig. 6.21. Tie-back analysis: wedge stability

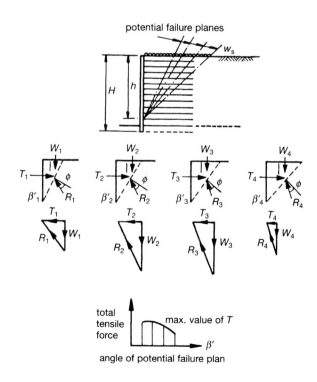

Table 6.2. Serviceability limits for vertical walls and bridge abutments

	Limit on post construction internal strains %
Bridge abutments	0·5
Walls	1·0

Various potential failure planes are considered, and a graphical search made for the maximum value of T, Fig. 6.21. For a regular level structure, with or without a uniform overcharge, the potential failure plane may be taken as:

$$\beta' = (45° - \phi'/2)$$

Wedge stability check. The resistance provided by any layer of reinforcing elements is taken as the smaller value of:

(*i*) the adherence or frictional resistance of the elements embedded in this fill outside the failure plane

(*ii*) the tensile resistance of the reinforcement elements.

The total resistance of the layers of elements anchoring the wedge is obtained from:

$$\sum_{i=1}^{n} \left[\frac{T_{Di}}{\gamma_n} \ \text{or} \ \frac{P_i L_{ip}}{\gamma_p \gamma_n} \left(\mu \ \gamma_{es} \gamma_{hi} + \mu \gamma_{ff} w_s + \frac{\alpha_{sr} c'}{\gamma_{ms}} \right) \geq T \right] \qquad (16)$$

where T_{Di} is the design strength of the reinforcing elements at the ith layer

L_{ip} is the length of the reinforcement in the resistant zone under consideration.

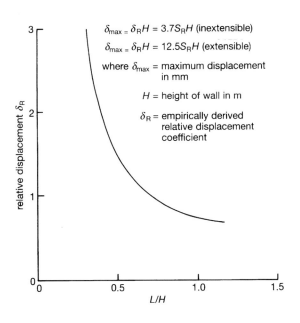

$\delta_{max} = \delta_R H = 3.7 S_R H$ (inextensible)

$\delta_{max} = \delta_R H = 12.5 S_R H$ (extensible)

where δ_{max} = maximum displacement in mm

H = height of wall in m

δ_R = empirically derived relative displacement coefficient

relative displacement δ_R

L/H

Fig. 6.22. Internal displacement of vertical structures

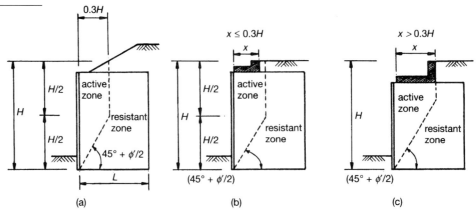

(a)   (b)   (c)

*Fig. 6.23. Coherent gravity
analysis—position of maximum
tension in reinforcement:
(a) wall; (b) wall with
concentrated load; (c) bridge
abutment*

Limit mode 6—deformation

Deformation should be checked at the following stages:

(*i*)   after considering limit modes 1, 2, 3 and 4 to determine if the correct design stress, K_{des}, has been used

(*ii*)   at the end of construction

(*iii*)   at the end of the specified life.

The *post construction* strains should not exceed a predetermined maximum dependent on the serviceability limit, Table 6.2.

An estimate of internal movements can be made from observations of in-service structures. Alternatively, theoretical empirical methods may be used for prediction (see Theory). An empirical method developed by the USA Federal Highway Administration (FHWA) is shown in Fig. 6.22.

6.3.2.3 Rigid structures—coherent gravity analysis

The coherent gravity analysis for the internal stability of reinforced soil structures contains four basic assumptions:

(*a*)   The reinforced mass is divided into two fundamental zones, an active zone and a resisting zone, divided by the line of maximum tension in the reinforcement, Fig. 6.23.

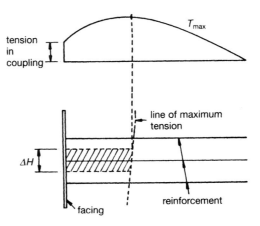

*Fig. 6.24. Distribution of
tensile force in strip
reinforcement*

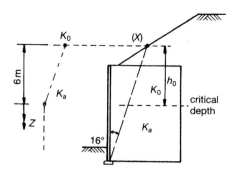

Fig. 6.25. State of stress assumed by the coherent gravity analysis

(b) The state of stress within the reinforced mass varies. Cohesionless fill only is used within the structure.

(c) An apparent coefficient of adherence between the reinforcing elements and the fill based on pullout tests is assumed, Fig. 6.24.

(d) A Meyerhof pressure distribution is assumed to exist beneath and within the reinforced fill. (The assumption of a Meyerhoff pressure distribution implies that the coherent gravity method of analysis is not compatible with anchored earth, as logically there is no vertical loading on the anchor. This anomaly is overcome in some design methods by assuming an additional vertical loading applied to the anchor equivalent to the vertical over-burden pressure. Under these conditions, it is not possible to draw a diagram of forces for the structure.)

Stress state. An empirical coefficiency K_{des}, relating to the state of stress, is adopted. K_{des} varies with depth, Fig. 6.25.

$$K_{des} = K_0 \left(1 - \frac{h}{h_0} \right) + K_a \frac{h}{h_0} \quad \text{when } h \le (h_0 = 6 \text{ m}) \tag{17}$$

$$K_{des} = K_a \quad \text{when } h > (h_0 = 6 \text{ m}) \tag{18}$$

$$\text{and } K_a = \tan^2 \left(\frac{\pi}{4} - \frac{\phi'}{2} \right); \quad K_0 = (1 - \sin \phi')$$

h_0 is measured from the position X defined in Fig. 6.25.

Limit mode 1—sliding
This check is similar to the general case using the tie-back analysis.

Limit mode 2—bearing tilt
The bearing pressure assumed with the coherent gravity analysis is shown in Fig. 6.26.

$$q_r = \frac{R_v}{B}$$

$$= \frac{R_v}{L - 2e}$$

where L is the reinforcement length at base

e is the eccentricity of resultant load R_v about the centre-line of the base

B is defined in Fig. 6.26

q_r is the factored bearing pressure active on the base of the wall.

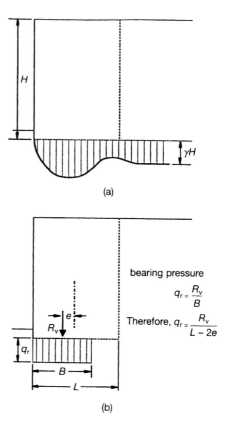

Fig. 6.26. Pressure distribution along base of wall: (a) pressure imposed at base; (b) idealized bearing pressure

The imposed bearing pressure q_r should be less than the ultimate bearing capacity of the foundation soil:

$$q_r \leq \frac{q_{ult}}{\gamma_{bc}} + \gamma D_m \tag{19}$$

where D_m is the embedment depth of the wall

γ_{bc} is the partial factor for bearing capacity of the foundation

q_{ult} is the ultimate bearing capacity of the foundation

R_v is the resultant factored vertical loads.

Limit mode 3—element rupture

The maximum tensile force T_i, to be resisted by the ith layer of reinforcement at depth h_i, is given by:

$$T_i = T_{pi} + T_{si} + T_{fi} - T_{ci} \tag{20}$$

where (*i*) T_{pi} is the vertical load due to self-weight, surcharge and bending moment resulting from an external load

$$= K_{des}\sigma_{vi}V \tag{21}$$

where $\sigma_{vi} = \dfrac{R_{vi}}{L_i - 2e_i}$

where R_{vi} is the resultant factored load excluding strip loads acting on the ith layer of reinforcement

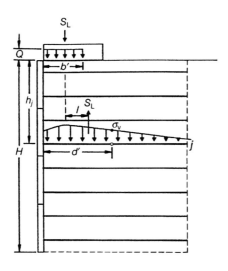

(ii) T_{si} is the vertical strip loading

$$= K_{des}\sigma_v(h_i, d')V \tag{22}$$

where $\sigma_v(h_i, d') = \gamma_{ff}\dfrac{Q}{2}\left[F_B\left(\dfrac{d'+b'}{h_i}\right) - F_B\left(\dfrac{d'-b'}{h_i}\right)\right]$

where $F_B = \dfrac{2}{\pi}\left[\dfrac{x}{1+x^2} + \tan^{-1}(x)\right]$, $\tan^{-1}(x)$ in radians

where $x = (d'+b')h_i$, $(d'-b')\,h_i$ as shown

Q is the pressure beneath the strip footing, Fig. 6.27

(iii) T_{fi} is the horizontal shear F_i applied to a strip contact area of width b, Fig. 6.28

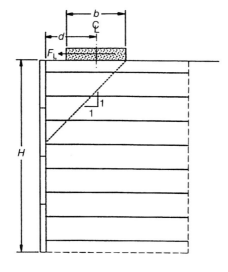

$$= \frac{2\gamma_{\mathrm{ff}} F_i V}{d + (b/2)} \left[1 - \frac{b_i}{d + (b/2)} \right] \tag{23}$$

(*iv*) T_{ci} is the influence of cohesive fill

$$= 2 \frac{c'}{\gamma_{\mathrm{ms}} V} \sqrt{K_{\mathrm{des}}} \tag{24}$$

Rupture. The capacity of the reinforcement elements of *i*th level in the structure should satisfy:

$$\frac{T_{\mathrm{D}}}{\gamma_n} \geq T_i \tag{25}$$

Limit mode 4—element pullout
The coherent gravity method of analysis was developed using metallic strip reinforcement; experimental observations have shown that the coefficient of friction of the reinforcement varies under different conditions (see Chapter 4, Theory). The analysis uses an apparent friction coefficient, μ^*.

Apparent friction coefficient, μ^*. The apparent friction coefficient μ^* is assumed to vary with depth in accordance with Fig. 6.29.

(*a*) $\mu^* = \mu_0 \left(1 - \dfrac{h}{h_0} \right) + \dfrac{h}{h_0} \tan \phi'$ if $h \leq 6\,\mathrm{m}$ $\tag{26}$

where $\mu_0 = 1 \cdot 2 + \log\ C_{\mathrm{u}}$
    (high adherence reinforcement and grids)
    $C_{\mathrm{u}} = $ the uniformity coefficient ($D60/D10$)
    $\mu_0\ = $ the $0 \cdot 4$ (smooth strip reinforcement)

(*b*) $\mu^* = \tan \phi'$, if $h > 6$ m $\tag{27}$
    (high adherence reinforcement and grids)

   $\mu^* = 0 \cdot 4$ (smooth strip reinforcement)

   h_0 is measured from position X defined in Fig. 6.29.

Reinforcement adherence. The maximum adherence capacity T_i of the *i*th layer is given by:

$$T_i \leq \frac{2B\mu}{\gamma_{\mathrm{p}} \gamma_n} \int_{L-L_r}^{L} \gamma_{\mathrm{ff}} \sigma_{\mathrm{v}}(x) \mathrm{d}x \tag{28}$$

Fig. 6.29. Apparent friction coefficient μ^* varies with depth

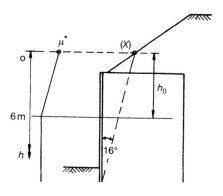

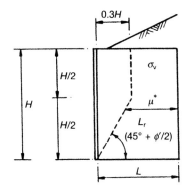

Fig. 6.30. Position of the line of maximum tension with the coherent gravity method of analysis

where γ_p is the partial factor for pullout resistance

B is the width of the reinforcement

L_r is the length of reinforcement beyond the line of maximum tension, Fig. 6.30.

μ is the value of the coefficient of friction or μ^*

$\sigma_v(x)$ is the vertical stress along length (x) of the reinforcement.

Limit mode 5—wedge/slip circle stability
The coherent gravity method of analysis does not normally consider this limit mode of failure. Failures of retaining walls in 1986 designed using this method suggest that this limit condition should be checked (Lee *et al.*, 1993).

Limit mode 6—deflections
This check is the same as the general case using the tie-back method of analysis.

6.3.2.4 Tie-back analysis—geogrids

Although both the tie-back wedge analysis and the coherent gravity analysis may be used with grid reinforcement, simplified analyses based on the superior adhesion capacity or pullout performance have been developed. Analysis is based on an assumption of a Coulomb failure mechanism, Fig. 6.31.

Local tensile stress. For a uniform vertical distribution of horizontal grids, the force T_i exerted in geogrid i at depth h_i:

$$T_i = K_{des} \, \gamma_{ms} \, h_i \, \Delta H \qquad (29)$$

Pullout resistance. The total pullout resistance, F_T, is a combination of the frictional resistance F_F presented by the grid, plus the anchor resistance F_R of the grid:

$$F_T = F_F + F_R \qquad (30)$$

The frictional resistance F_F per unit length of longitudinal wire diameter d:

$$F_F = \mu \, \pi \, d \, \sigma'_v \qquad (31)$$

where μ is the coefficient of friction between the fill and the reinforcing elements

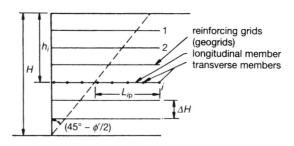

*Fig. 6.31. Position of the
assumed failure plane with the
tie-back method of analysis*

σ'_v is the effective vertical stress on the ith grid

$= \gamma_{ff}\gamma h$

The anchor resistance per transverse member based on the Terzaghi–Buisman bearing capacity expression:

$$F_R/N_W = \gamma_{ms}dc'N_c + \frac{\sigma'_v}{2}d^2N_\gamma + \sigma'_v dN_q \tag{32}$$

where N_w is the number of transverse members outside the Coulomb failure wedge

N_c, N_γ, N_q are the Terzaghi bearing capacity factors.

Since d is small, for a cohesionless fill:

$$\frac{F_R}{N_W} = \sigma_v dN_q \tag{33}$$

Total pullout resistance per unit width:

$$F_T = (\sigma'_v \pi dL\mu M) + (\sigma'_v dN_q N)$$
$$= \sigma'_v d(\pi L\mu M + N_q) \tag{34}$$

where M is the number of longitudinal members in grid per unit width

N is the number of transverse elements outside the Coulomb wedge

μ is the coefficient of friction defined above

σ_v is the vertical stress on the reinforcement defined above.

6.3.2.5 Anchored earth structures

Anchored earth may be analysed using the limit mode procedure of the tie-backwedge analysis, modified to accommodate the different behaviour between strip reinforcement and anchors. The resistance of the anchor elements is the same for both local and wedge stability, with the proviso that the anchors are embedded beyond the failure plane.

When a triangular anchor is proposed (TRRL, 1981), Fig. 6.32, the pullout resistance, F_t may be taken as:

F_t is the shaft resistance, P_s+ anchor resistance P_A $\qquad$ (35)

where P_A is the lesser of $(P_f + P_B)$ or $2P_B$

P_f is the friction of top and bottom of cohesionless soil within

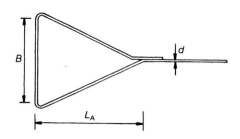

Fig. 6.32. Triangular anchor
(after TRRL, 1981)

the area of the triangular anchor $= L_A B \sigma'_v \tan \phi'_{des}$

P_B is the bearing force in front of the anchor $= 4 K_p \sigma'_v B d$, $\qquad$ (36)
Fig. 6.33

$$K_p = \frac{1 + \sin \phi'}{1 - \sin \phi'}$$

$$P_s = \mu \pi L_i \sigma' \tan \phi'_{des}$$

Analysis is in accordance with Section 6.3.2.2. General design case—tie-back analysis, limit modes 1–6 inclusive, with the following amendments:

(*i*) The expression for T_{hi} equation (8) is replaced by:

$$T_{hi} = K_{des} \gamma_{ms} \gamma h_i V \qquad (37)$$

(*ii*) The expression for T_{si} equation (10) is replaced by:

$$T_{si} = K_{des} \gamma_q \frac{S_i}{D_i} \left(1 + \frac{6e}{D_i} \right) \qquad (38)$$

where D_i is defined as before.

Note 1. As the pullout resistance of each anchor is based on the mobilization of the passive resistance of the fill extending beyond the anchor perimeter, where design requires anchors to be placed at close proximity, interference between the action of anchors may occur. A practical solution may be found in staggering the lengths of adjacent anchors.

Note 2. The assumed distribution of tensile force in the anchor differs from that of a strip reinforcement, Fig. 6.34.

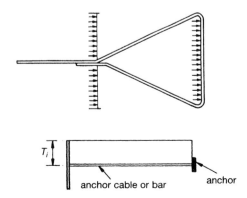

Fig. 6.33. Action of triangular
anchor (after Mair et al., 1983)

anchor cable or bar   anchor

Fig. 6.34. Assumed distribution
of tensile force in an anchor

Apparent earth pressure distribution:

| for wall material | for pole and anchor |

10 kN/m^2 5 kN/m^2

H H

$K_a = 0.4$ $K_a = 0.18$

$p = \gamma H K_a$

Fig. 6.35. Earth pressure distribution used in multi-anchored walls

6.3.2.6 Multi-anchored walls

The retaining system developed by Fukuoka (1980) for the Japanese Ministry of Construction uses a system of rectangular steel anchor plates fixed to a steel tie, which is itself hinged to a facing. The main features of the multi-anchor system are the use of any rational number of rectangular plate anchors and the introduction of a turn-buckle to tension the tie members. Construction details of the system are shown in Chapter 8.

The multi-anchored wall system was developed based on the provision of factors of safety against failure rather than on satisfying limit states. Analysis of a multi-anchored structure can be based on the concept of limit mode, assuming a sliding wedge mode of failure.

The state of stress of the soil within the structure is assumed to vary. In the lower part of the wall, an active stress state is used; at the top, a stress state based on compaction stresses is assumed. Two sets of earth pressures are used for different parts of the system, Fig. 6.35. The lower earth pressure is used to determine the resistance to pullout developed by the anchors.

Wall face

Where a steel facing is used, Fig. 6.36, the plate thickness t may be calculated from:

$$\frac{\sigma_h r}{t} \leq \sigma_a \tag{39}$$

where $\sigma_h = K_a \gamma h (\geq 10 \text{ kN/m}^2)$

σ_a is the permissible tensile stress in facing material

r is the radius of the facing element.

Fig. 6.36. Plan of wall face between anchors in multi-anchored walls with a steel facing

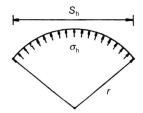

S_h

σ_h

r

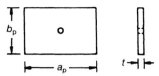

Fig. 6.37. Anchor plate dimensions

Pole

The pole supporting the facing, Fig. 7.9 (Chapter 7), is analysed as a simple beam spanning the vertical distance S_v between successive tie members, a distance S_h apart. The bending moment of the pole M_p may be determined from:

$$M_p = \frac{S_v^2 S_h \sigma_h}{8} \tag{40}$$

where $\sigma_h = K_a \gamma h (\geq 5 \text{ kN/m}^2)$.

For a steel pipe:

$$\frac{M_p}{Z} \leq \sigma_p \tag{41}$$

where σ_p is the permissible tensile stress of the material forming the pole
   Z is the section modulus.

Tie bar

Tension in the tie bar is based on the resistance of the anchor plates to pullout. The frictional resistance of the tie bar is neglected:

$$T_i = \gamma K_a \sigma_{hi} a_p b_p \tag{42}$$

where $\gamma K_a \sigma_{hi} \geq 5 \text{ kN/m}^2$

   a_p and b_p are the dimensions of the anchor plate, Fig. 6.37.

Anchor plate

Anchors are formed from rectangular plates acting as a cantilever, Fig. 6.37. The load in the plate is assumed to be uniform. Maximum bending moments in the plate M_p may be determined from:

$$M_p = \frac{q_p a_p^2}{8} \tag{43}$$

where $q_p = T_i / a_p b_p$

   $a_p b_p$ are the dimensions of the plate

   T_i is the maximum tensile stress in the ith layer tie bar.

Anchor plate thickness t is calculated by:

$$Z = \frac{t^2}{6}$$

where $Z = q_p a_p^2 t^2 / 8 \leq \sigma_a$ \hfill (44)

6.3.2.7 Serviceability related to strain of reinforcing materials

With high modulus reinforcing materials, the strain of the structure due to the extension of the reinforcements is small. In the case of some polymeric

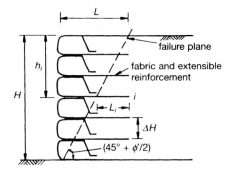

*Fig. 6.38. Geosynthetic
reinforced soil structure*

reinforcements, developed in the form of fabrics or geogrids, relatively large strains may occur, Fig. 6.38. The magnitude of strain of the reinforcing material is governed by the tension in the reinforcement. The tension distribution in the reinforcement may be assumed to be in accordance with Fig. 6.39.

At the base of the structure, $L_i = L$, and the tension distribution may be assumed to be as in Fig. 6.39(a).

The maximum tension in the reinforcement T_{max} at depth $h_i = H$:

$$T_{max} = K_a \gamma H \, \Delta H \tag{45}$$

The tension $T_{(x)}$ at distance x from the facing reduces to:

$$T_{(x)} = K_a \gamma \Delta H \left(1 - \frac{x}{L} \right) \tag{46}$$

Assuming a linear elastic behaviour of the reinforcement, the reinforcement strain ϵ at a distance x from the face may be obtained from:

$$\epsilon = \frac{T_{(x)}}{E_r a_r} \tag{47}$$

The extension of the reinforcement δx over an element of length dx is:

$$\delta x = \epsilon \, dx \tag{48}$$

Extension over length L of the reinforcement L_s is:

$$L_s = \sum (\delta x) = \int_0^L \epsilon \, dx \tag{49}$$

From equation (47),

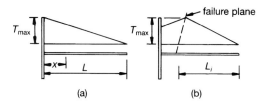

*Fig. 6.39. Tension distribution
in extensible reinforcement*

$$L_s = \int_0^L \frac{T_{(x)}}{E_r a_r} \, dx = \frac{K_a \gamma H \Delta H}{E_r a_r} \int_0^L \left(1 - \frac{x}{L}\right) dx$$

$$L_s = \left(\frac{K_a \gamma H \Delta H L}{2 E_r a_r}\right) \tag{50}$$

Polymeric reinforcements may be susceptible to creep; to overcome long-term serviceability problems, strains may be reduced by controlling working stress levels.

6.3.3 Sloping structures and embankments

Steep slopes, within 20° of the vertical may be designed as vertical structures
The design methods used with sloping structures and embankments are derived from equilibrium methods. The limit state philosophy can be used in which case the soil weight and external loading are increased by partial load factors. The soil properties and the reinforcement base strength are achieved using appropriate partial factors, Table 6.3. Design is satisfied when:

$$\frac{\text{reduced resistances}}{\text{increased disturbances}} \geq 1$$

The most common design methodology in current use is to use lumped factors of safety against appropriate failure conditions.

The external and internal stabilities of an embankment are frequently considered together; this is because failure planes originating within the embankment may intersect the subsoil.

6.3.3.1 Reinforced embankments (cohesive soils)

Embankments may be constructed of cohesive or cohesionless soils. In the case

Table 6.3. Partial factors for sloping structures and embankments

Partial factor		Ultimate limit state	Serviceability limit state
Materials	Soil materials:		
	applied to $\tan \phi'_{des}$	$\gamma_{ms} = 1{\cdot}0$	$\gamma_{ms} = 1{\cdot}0$
	applied to c'	$\gamma_{ms} = 1{\cdot}6$	$\gamma_{ms} = 1{\cdot}6$
	Reinforcement	Varies*	Varies*
Load factors	Soil unit weight	$\gamma_{es} = 1{\cdot}5$	$\gamma_{es} = 1{\cdot}0$
	External dead loads	$\gamma_{ff} = 1{\cdot}2$	$\gamma_{ff} = 1{\cdot}0$
	External live loads	$\gamma_q = 1{\cdot}3$	$\gamma_q = 1{\cdot}0$
Soil/reinforcement interaction	Sliding on surface of reinforcement	$\gamma_s = 1{\cdot}3$	$\gamma_s = 1{\cdot}0$
	Pullout	$\gamma_p = 1{\cdot}13$	$\gamma_p = 1{\cdot}0$
Sliding	Base of structure of soil/soil interface	$\gamma_{ss} = 1{\cdot}2$	$\gamma_{ss} = 1{\cdot}0$

* See Chapter 11, Example 4 for typical treatment.

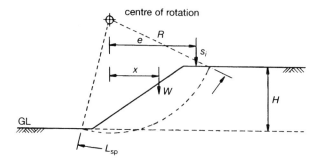

centre of rotation

Fig. 6.40. Slip circle failure in embankment

of cohesive soils, slip failures are normally rotational, the slip surface approximating to an arc of a circle. The equilibrium of the embankments may be determined from a comparison of the disturbing moment of the rotating soil mass and the restraining or resisting moment developed by the materials forming the embankment and subsoil. The critical slip circle is defined as the slip circle producing the lowest factor of safety, FS, where:

$$FS = \frac{\text{restraining moment}}{\text{disturbing moment}}$$

Assuming an arbitrary datum below the embankment and using a matrix of centres of circles and radii, the critical circle may be established, Fig. 6.40

where W is the weight of earth segment

S is the shear strength of soil

$$= c + \sigma \tan \phi$$

L_{SP} is the length of slip plane.

For equilibrium,

disturbing moment = resisting moment

Factor of safety,

$$FS = \frac{\text{resisting moment}}{\text{disturbing moment}}$$

$$= \frac{S L_{SP}}{(Wx + S_i e)}$$

If the calculated factor of safety FS is less than required, the resisting moment can be increased by the addition of reinforcement positioned to cut the shear plane, Fig. 6.41.

Using equation (51), h_0 is the maximum height of embankment which may be constructed without reinforcement, and the reinforced block in an embankment of height H is $(H - h_0)$. Thus, from Figs 6.40 and 6.41:

$$\text{disturbing moment} = Wx + S_i e \tag{51}$$

$$\text{resisting moment} = S L_{SP} + Ty \tag{52}$$

where T is the total tension resistance of the N layers of reinforcement

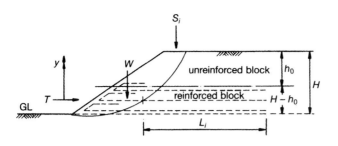

Fig. 6.41. Reinforced embankment

y is the lever arm about the centre of rotation of the reinforcement.

Reinforcement tension. Assuming a factor of safety FS:

$$SL_{SP} + Ty = (Wx + S_ie)(FS) \tag{53}$$

$$T = \frac{(Wx + S_ie)(FS) - SL_{SP}}{y} \tag{54}$$

$$T_{max} = \frac{(Wx + S_ie)(FS) - SL_{SP}}{Ny} \tag{55}$$

Adherence length. For good adherence between the reinforcement and the soil, $\mu \approx 1$. The adherence length of the reinforcement L_i is given by:

$$L_i = \frac{T_{max}(FS)}{S_{av}} \tag{56}$$

where S_{av} is the average shear strength of soil.

Note. Additional consideration to embankment analysis is given later in section 6.4, Computer-aided design and analysis.

6.3.3.2 Sloping structures

Steep embankments may require facing in order to prevent erosion. In these cases, the embankment becomes a sloping structure, Figs 6.42 and 6.43. As a result of their steepened profile, they are often formed from a cohesionless or cohesive frictional fill.

Reinforcement tension. The tensile force T_i resisted by the ith layer of reinforce-

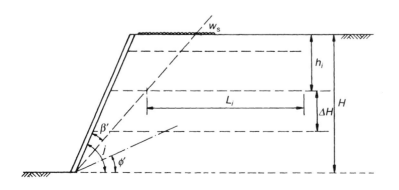

Fig. 6.42. Sloping structures

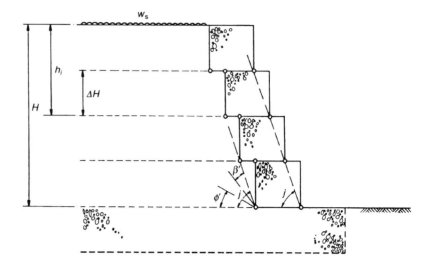

Fig. 6.43. Earth reinforced
gabions

ment per unit width of embankment:

$$T_i = K_a(\gamma h_i + w_s) \sin j(\Delta H) \tag{57}$$

where $\Delta H = H/n$, for uniform vertical distribution of
the reinforcement $\tag{58}$

$$K_a = [\tan(90 - j + \beta') - \tan(90 - j)]$$

$$x \frac{\sin(j - \beta' - \phi')}{\sin(\beta' + \phi')} \tag{59}$$

K_a is a maximum when $\beta' = \dfrac{j - \phi'}{2}$ $\tag{60}$

Adherence length. For good adherence between the reinforcement and the soil,
$\mu = 1$. The adherence length of the reinforcement L_i is given by:

$$L_i = \frac{T_i}{2(\gamma h^* + w_s) \tan \phi'} (FS) \tag{61}$$

where h^* is an effective value taking into account the angle of the slope j.

6.3.4 Foundations

6.3.4.1 Foundation mattresses for embankments

Embankments constructed on weak subsoils are prone to failure if the
embankment loading resulting from the surcharge and the self-weight exceeds
the bearing capacity of the subsoil, Fig. 6.44.

For equilibrium, the bearing capacity

$$q_{ult} > (\text{applied loading intensity} \times FS) \tag{62}$$

From Fig. 6.44,

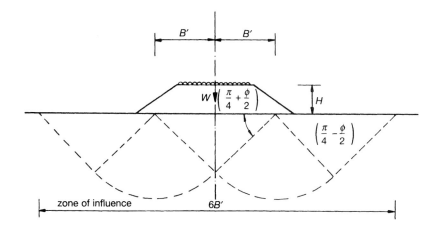

Fig. 6.44. *Failure mechanism of an embankment*

$$q_{ult} = (2 + \pi)c_u \quad \text{where } \phi = 0 \tag{63}$$

$$q_{ult} = 5 \cdot 14 c_u \tag{64}$$

At the base of the embankment,

$$\text{loading intensity} = \frac{W}{2B'} + w_s \tag{65}$$

where $2B^1$ is the effective width of the embankment.

$$\text{Therefore, } FS = \frac{(2 + \pi)c_u}{[(W/2B^1) + w_s]} \tag{66}$$

If, in equation (66), FS is less than permissible, the effective distribution width $2B^1$ may be increased by the use of a reinforcing mattress at the base of the embankment, Fig. 6.45.

6.3.4.2 Geogrid mattresses or geocells

In the case of very soft subsoils, the construction of embankments may result in significant settlements and the use of additional fill material, Fig. 6.46. A geogrid mattress may improve the rigidity or stiffness of the embankment and produce relief to the underlying soil by reducing the loading intensity, equation (65).

One possible settlement profile of a soil embankment founded upon a yielding or weak subsoil is shown in Fig. 6.46. The associated strain in the mattress resulting from the settlement suggests that a stiff or semi-stiff

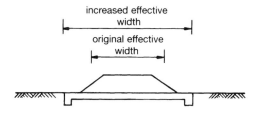

increased effective
width

original effective
width

Fig. 6.45. *Reinforcing mattress used at base of embankment to increase the effective distribution width*

173

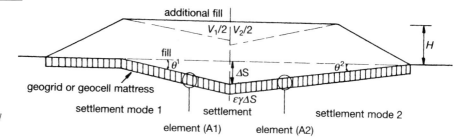

Fig. 6.46. Geocell reinforced embankment on soft ground

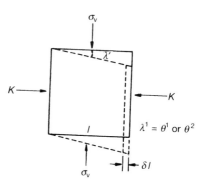

Fig. 6.47. (after Hay, 1982)

reinforcing material is required for the construction of the mattress and geocell.

Considering an element of the geogrid mattress, Figs 6.46 and 6.47, the distortion is θ^1 or θ^2 depending on settlement mode. For no spread of the embankment, there is no volume change in the mattress; as the mattress distorts, the horizontal elements of the grid strain, Fig. 6.48. For no volume change, when any cell has distorted, N, the horizontal elements of the geogrid mattress, increase in length:

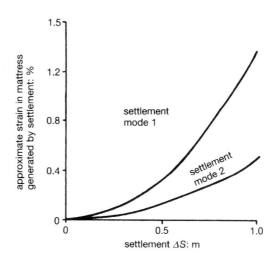

Fig. 6.48. Geogrid element A1 or A2

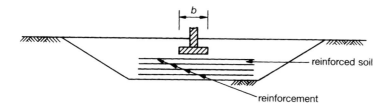

Fig. 6.49. Reinforced foundation

$$l + \delta l \approx l' = \frac{1}{\cos \lambda'} \tag{67}$$

$$\delta l = l\left(\frac{1}{\cos \lambda'} - 1\right) \tag{68}$$

For a pin jointed geogrid. Total shear resistance provided by mattress, S_{rm} = shear resistance of mattress fill + distortial resistance.

$$S_{rm} = K\sigma_v \tan \phi + \delta l/E_{ar}(\text{mattress}) \tag{69}$$

(Geogrids formed from plane materials may display additional resistance to distortion at joints or node points.)

6.3.4.3 Structural footings

The bearing capacity and settlement characteristics of the subsoil beneath a foundation or footing may be increased by a layer of reinforced soil placed beneath the footing, Fig. 6.49. Binquet and Lee (1975) have defined the benefit of a reinforced soil foundation in terms of the bearing capacity ratio q_r as:

$$q_r = q/q_0 \tag{70}$$

where q_0 is the average contact pressure of footing on unreinforced subsoil

q is the average contact pressure of footing on reinforced soil and the settlement is constant.

Failure modes. The following failure modes may be considered, Fig. 6.50:

(a) reinforcement tension failure
(b) reinforcement adherence or pullout failure
(c) soil failure above the reinforced soil layer.

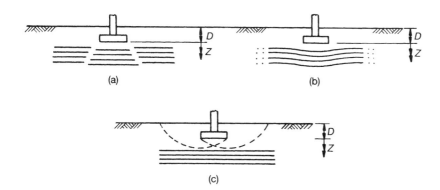

Fig. 6.50. (a) Tension failure; (b) adherence failure; (c) failure above reinforced layer

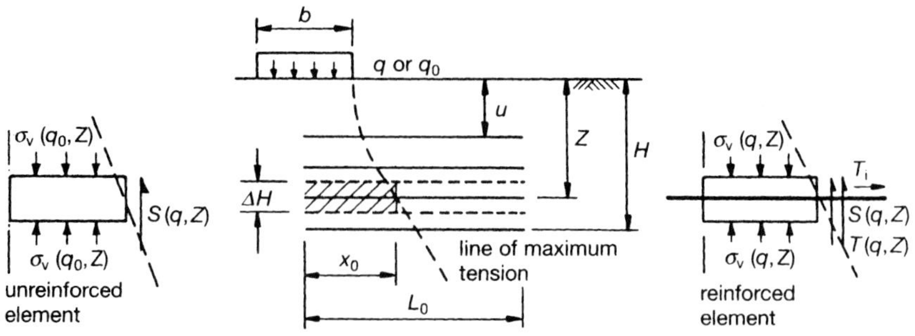

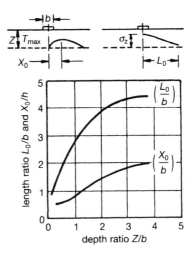

Fig. 6.51. Dimension components
for reinforced foundation bearing
capacity theory

unreinforced
element

reinforced
element

Tension failure. Assuming a Bousinesq distribution of stress beneath the footing, Binquet and Lee (1975) have defined the location of the maximum stress in the reinforcement, Figs 6.51 and 6.52. The length L_0 is defined as the point at which the vertical stress σ_v is one per cent of the applied pressure. L_0 may be determined by elastic theory, Fig. 6.52.

It is assumed that reinforcement tension force T varies inversely with the total volume of reinforcement. In terms of strip or grid reinforcement place in (n) horizontal layers:

$$T(Z, n) = T\frac{(Z, n = 1)}{n} \tag{71}$$

From Fig. 6.51, the difference in bearing pressure (q_0 and q) when $q > q_0$, is defined by Binquet and Lee (1975) as:

$$\sigma_v(q, Z) - \sigma_v(q_0, Z) = S(q, Z) - S(q_0, Z) + T(Z, n) \tag{72}$$

$$T(Z, n) = \sigma_v(q, Z) - \sigma_v(q_0, Z) - S(q, Z) + S(q_0, Z) \tag{73}$$

where

$$\sigma_v(q, Z) = \int_0^{X_0} \sigma_z(q, x, Z)\mathrm{d}x$$

$$S(q, Z) = \tau_{xz}(X_0, Z)\Delta H$$

where τ_{xz} is the maximum shear stress at every depth Z.

Fig. 6.52. Dimensionless lengths
of reinforced soil slab (after
Binquet and Lee, 1975)

Defining the shear stress and the normal stress in dimensionless forms:

$$\sigma_v(q, Z) = J\left(\frac{Z}{b}\right) qb \qquad (74)$$

where

$$J\left(\frac{Z}{b}\right) = \int_0^{x_x} \frac{\sigma_z(Z/b)\,\mathrm{d}x}{qb}$$

$$S(q, Z) = I\left(\frac{Z}{b}\right) q\,\Delta H \qquad (75)$$

where

$$I\left(\frac{Z}{b}\right) = \frac{T_{\max}(Z/b)}{q}$$

and

$$T(Z, n) = \frac{1}{n}\left[J\left(\frac{Z}{b}\right)b = 1\left(\frac{Z}{b}\right)\Delta H\right]q_0\left(\frac{1}{q_0} - 1\right)$$

The values of $J(Z/b)$ and $I(Z/b)$, the normal and shear stresses beneath the footing, may be determined by conventional means, Binquet and Lee's derivations are given in Fig. 6.53.

Adhesion resistance. The adhesion resistance of the reinforcing members is a function of the normal pressure acting over the adhesion length ($L_0 = X_0$).
   The total normal force acting on the adhesion length:

$$\sigma_v A(q, Z) = A \int_{X_0}^{L} \sigma_z(q, x, Z)\,\mathrm{d}x \qquad (76)$$

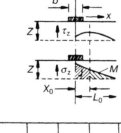

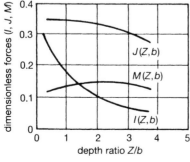

Fig. 6.53. Dimensionless forces relating to the bearing capacity of reinforced soil (after Binquet and Lee, 1975)

177

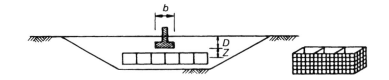

*Fig. 6.54. Geocell reinforced
foundation under a footing*

where A is the plan area of strip reinforcement, or the area of grid per unit width.

$$\sigma_{\rm v}(q, Z) = AbM\left(\frac{Z}{b}\right)q \tag{77}$$

and

$$M\left(\frac{Z}{b}\right) = \frac{\int_{0}^{L_0} \sigma_{z}\left(\frac{Z}{b}\right){\rm d}x}{bq} \tag{78}$$

The dimensionless parameter $M(Z/b)$ is derived in Fig. 6.53. The total normal force σ_N on the reinforcement at depth Z:

$$\sigma_N = \sigma_{\rm v}(q, Z) + A\gamma(L_0 - X_0)(Z + D) \tag{79}$$

where D is defined in Fig. 6.50.

When the coefficient of friction between the soil and the reinforcing element is defined as μ, the reinforcement frictional resistance $T_{\rm f}$ per unit length of footing at depth Z in terms of the bearing capacity ratio may be taken as:

for strip reinforcement

$$T_{\rm f} = 2\mu A_{\rm strip}\left[M\left(\frac{Z}{b}\right)bq_0\left(\frac{q}{q_0}\right) + \gamma(L_0 - X_0)(Z + D)\right] \tag{80}$$

for grid reinforcement

$$T_{\rm f} = \mu A_{\rm grid}\left[M\left(\frac{Z}{b}\right)bq_0\left(\frac{q}{q_0}\right) + \gamma(L_0 - X_0)(Z + D)\right] \tag{81}$$

Geogrid cell footings
An alternative to the use of the strip or grid reinforcement in a reinforcement soil foundation is a geogrid mattress or geocell. The principal advantage of a cell construction, Fig. 6.54, is that the adhesion resistance is not an issue, as the form of construction provides its own anchor. For analysis purposes, equations (71)–(76) hold where Z is redefined as the depth beneath the footing to the mid-height of the cell.

*Fig. 6.55. (a) reinforced cutting;
(b) reinforced block*

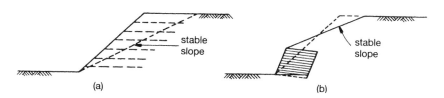

6.3.5 Cuttings
6.3.5.1 Slopes and slope failures

The use of reinforcement to increase slope stability occurs usually after slope failure. Accordingly, slope reinforcement is normally a remedial technique, Fig. 6.55.

The reinforced soil block in Fig. 6.55(b) may be analysed in a conventional manner using, from Section 6.3.2.2, 6.3.2.3 and 6.3.2.4.

(*a*) tie-back analysis
(*b*) coherent gravity analysis
(*c*) tie-back analysis—geogrids.

The reinforced cutting in Fig. 6.55 may be analysed in a similar manner to a reinforced embankment, in accordance with Fig. 6.41 and equations (51)–(55). Alternatively, the cutting may be assumed to have failed in a bilinear mode, Fig. 6.56, and the equilibrium of the repaired section equates to the needs of the bilinear slip plane.

6.3.6 Soil nailing

A number of standards/recommendations exist for the analysis of soil nailing, including the Japanese Design Code (JHPC, 1987), Recommendations Clouterre (FHWA, 1993) and the British Standard (BS 8006: 1995). The general consideration in Europe is that, for near-vertical walls' reinforcement with horizontal nails, the tensile component of the reinforcement is the dominant action and the contribution of shear/bending is of a second order of magnitude. At the serviceability condition, the contribution of shear/bending is negligible. Direct comparison between the different codes used in Europe is not possible as the UK code is a limit state code considering both the serviceability limit state and the ultimate limit state. The French analyses are based mainly on the consideration of circular arc or log-spirals, while the German preference is for a bilinear wedge. The UK Code accepts any form of failure mode (circular, log-spiral or wedge). A summary of the load factors in common use in the UK and parts of Europe is given in Tables 6.4 and 6.5.

The Japanese practice of soil nailing appears different from that used in Europe and North America, being based on a more empirical approach. In Japan, the situations in which soil nailing can be used are identified in respect of the geological conditions of the site, Fig. 6.57. In addition, the analytical models suitable for different conditions are suggested, Fig. 6.58. Interestingly, and somewhat controversially, the Japanese Guidelines do not appear to

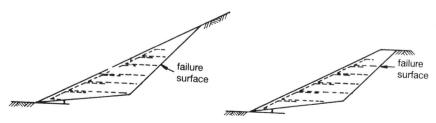

Fig. 6.56. Bilinear slip in cutting

Table 6.4. Load and safety factors for soil nail slope design (UK)

Partial factor		Ultimate limit state	Serviceability limit state
Load factors	Soil unit weight, e.g. embankment fill	$\gamma_{es} = 1\cdot3$	$\gamma_{es} = 1\cdot0$
	External dead loads, e.g. line or point loads	$\gamma_{ff} = 1\cdot2$	$\gamma_{ff} = 1\cdot0$
	External live loads, e.g. traffic loading	$\gamma_q = 1\cdot3$	$\gamma_q = 1\cdot0$
Soil material factors	to be applied to tan ϕ'_{cv}	$\gamma_{ms} = 1\cdot0$	$\gamma_{ms} = 1\cdot0$
	to be applied to tan ϕ'_p	$\gamma_{ms} = 1\cdot25$	$\gamma_{ms} = 1\cdot0$
	to be applied to c'	$\gamma_{ms} = 1\cdot6$	$\gamma_{ms} = 1\cdot0$
Reinforcement material factor	to be applied to the reinforcement base strength	The value of γ_m should be consistent with the type of reinforcement to be used and the design life over which the reinforcement is required	
Soil/reinforcement interaction factors	Sliding across surface of reinforcement	$\gamma_s = 1\cdot3$	$\gamma_s = 1\cdot0$
	Pullout resistance of reinforcement	$\gamma_p = 1\cdot3$	$\gamma_p = 1\cdot3$

consider corrosion of the soil nails to be of particular significance; furthermore, the Japanese Guide provides no teaching on durability, other than to conclude that: 'special maintenance and management for slopes constructed by the method is not needed'.

The design of soil nailing is based on equilibrium methods. Most design is based on the use of computer methods, see Section 6.4. Bruce and Jewell (1987) have developed expressions which can be used for preliminary design based on four dimensionless parameters, Table 6.6.

Table 6.5. Load and safety factors for permanent soil nail wall design

Global	Yield	Pullout	Load static	Load imposed	Soil friction	Soil cohesion	
1·0	1·7	2·0	1·0	1·0	1·0	1·0	Germany[1]
1·5	1·5	2·0	1·0	1·0	1·0	1·0	France[2]
1·5	1·5	1·5	1·0	1·0	1·0	1·0	France[3]
1·125	1·15	1·4	0·95–1·05	0·9–1·2	1·2	1·5	Clouterre recommendations[4]

(1) Practice by Bauer Spezialtiefbau GmbH in 1993.
(2) Practice by CLC SA/Solrenfor in 1993.
(3) Practice by Seetauroute in 1993.
(4) Recommendations Clouterre (1991).

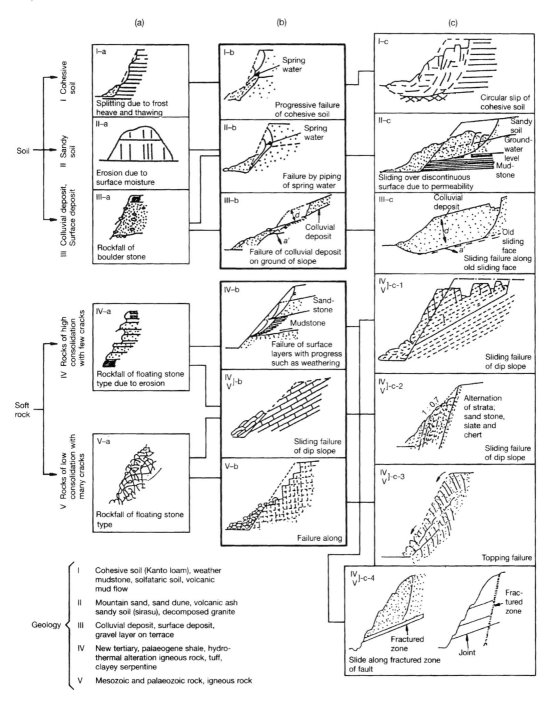

Fig. 6.57. Geological conditions suitable for soil nailing (after JHPC, 1987): (a) rockfall erosion, splitting; steep slope without unstable factors; (b) surface failure; slope with unstable factors such as soil, rock (physical properties), or ground water; (c) slope with unstable factors such as failure in a large scale, sliding failure or geological structure. (The inside of thick line shows applicable slope failure types of reinforced earthwork with steel bars)

Length ratio: $\dfrac{\text{Maximum nail length}}{\text{Excavation height}}$

Bond ratio: $\dfrac{\text{Hole diameter} \times \text{length}}{\text{Nail spacing}}$

Strength ratio: $\dfrac{(\text{Nail diameter})^2}{\text{Nail spacing}} \times 10^{-3}$

Performance ratio: $\dfrac{\text{Outward movement (top of structure)}}{\text{Excavation height}}$

The analysis of soil nailing by kinematical methods has been developed by Gudehus (1972) and Stocker *et al.* (1979).

Gässler and Gudehus (1981, 1983) have produced a tabular form of analysis for the simple case illustrated in Fig. 6.59.

A double wedge failure mechanism is assumed in cohesionless soil ($c_u = 0$). If

nail length $= L_n$

ratio $\lambda = L_n/H = 0{\cdot}6$ or $0{\cdot}7$

nail orientation to horizontal $= 10°$

face slope $= 10°$

Fig. 6.58. Relationship between the failure mode and analytical method (after JHPC, 1987)

Calculation technique	1 Study by circular slip failure	2 Study by wedge type slip failure	3 Study by earth-pressure
Model example	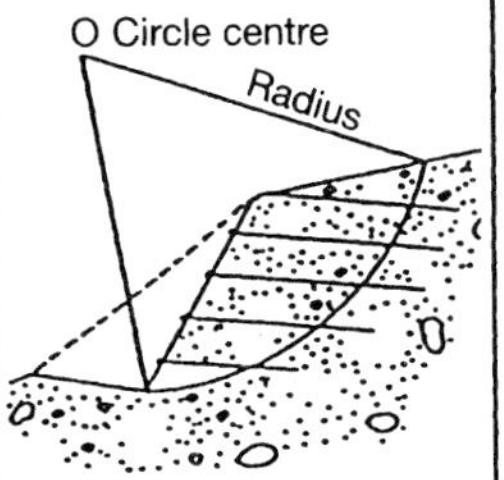 O Circle centre / Radius / O	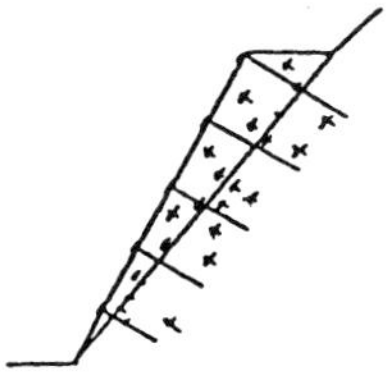	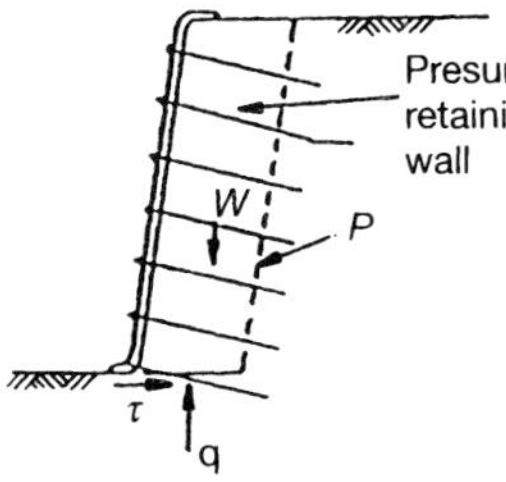 Presumed retaining wall / W / P / τ / q
Features of slopes	In the case where soil or the colluvial deposit is thick	Soft rock with developed bedding or joint, and dip slope	Usual slopes constructed by concrete block retaining wall or retaining wall
	Compounded slip In the case where sliding surface can almost be presumed by conditions of geological structure and known deformation.		

Table 6.6. *Comparison of drilled, grouted and driven nails—case histories (after Bruce and Jewell, 1987)*

	Drilled and grouted	Driven
Length ratio	0·5–0·8	0·5–0·6
Bond ratio	0·3–0·6	0·6–1·1
Strength ratio	0·4–0·8	1·3–1·9
Performance ratio	0·001–0·003	No data

then

$$\text{nail force } NF = \left(\frac{T_m}{\gamma S_v S_h} \right) \tag{82}$$

where T_m is the mobilized shear force per unit of nail length, measured from pullout tests *in situ*

ϕ is the frictional angle for *in situ* soil

FS is the global factor of safety

S_v is the vertical spacing of nails
S_h is the horizontal spacing of nails.

Using design charts, Figs 6.60(a) and (b), a satisfactory array of nails may be obtained, together with the slip plane angle ϕ_n relevant to the minimum value of FS.

6.4 Computer-aided design and analysis

Computer-aided design is an established practice in many civil engineering fields. The analysis of earth reinforcement and soil structures may benefit also from these techniques, and some problems are susceptible to the more sophisticated computer systems associated with computer-aided design.

It is convenient to compare the various application areas with the possible or preferred analytical procedures, Table 6.7. In all, except for the centrifuge methods, computer-aided design and analysis are possible; all the empirical and limit analysis methods described previously may be developed into computer systems. Computer-aided analysis approaches computer-aided design when the power or facilities of the computer are used to explore analytical techniques

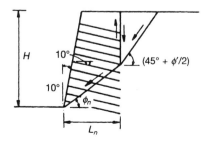

Fig. 6.59. *Failure mechanism for soil nailing*

183

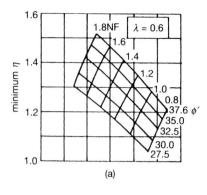

(a)

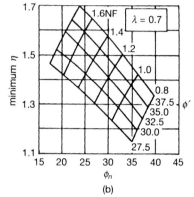

(b)

Fig. 6.60. Design charts for soil nailing (after Gässler and Gudehus, 1982)

Table 6.7. Possible or preferred analytical procedures

Application	Internal analysis	External/global analysis	Parametric studies
Retaining walls			
good subsoil	EM	EM/LA	FE/FD
weak subsoil	EM	LA/FE*	FE(CF)
Bridge abutments			
good subsoil	EM	EM/LA	FE/FD
weak subsoil	EM	LA/FE*	
Embankment			
good subsoil	EM/LA	EM/LA	FE/CF
weak subsoil	EM/LA	FE/LA	FD(CF)
Foundations	LA	LA/FE	LA/FE
Industrial structures	EM	?	LA/FE
Dams	EM	LA/FE	FE(CF)

* Consider settlement limitations.
EM Empirical analysis.
CF Centrifuge test.
LA Limit analysis.
FE/FD Finite element analysis/Finite difference.

or procedures which otherwise would be impossible or impractical. The finite element method of analysis is an example of the former, while the iterative search for the least stable slip circle, considering every circle within a practical range, is an example of the latter.

6.4.1 Finite element analysis/finite difference

The internal analysis of reinforced soil structures using the finite element method is possible, but the limitations posed by the known behaviour mechanism of reinforcement in soil cannot be ignored (see Chapter 4, Theory). Accordingly, finite element analysis of internal stability may not offer benefits over the empirical and limit equilibrium techniques.

The external design of reinforced soil structures, including retaining walls, bridge abutments and embankments, supported on good subsoil, rests within the compass of conventional analytical methods. On weak subsoils, however, the use of simple empirical design rules may produce design difficulties. Finite element techniques may be used to consider the overall displacement of the reinforced soil structure on the subsoil and the influence of the reinforced embankment or foundation mattress on the subsoil stability.

Application of the finite element method is through the use of mathematical models. Reinforced soil problems demand the use of critical state models or nonlinear elastic models capable of mimicking the construction sequence, Fig. 6.61, producing vectors of horizontal and vertical strains at the base and within the structure, and determining the residual strength of the subsoil.

In practice, there are two methods of approach to this form of mathematical modelling. One method is to use an elasto-plastic stress/strain model based on the concept of critical-state soil mechanics which includes realistic volumetric and shear behaviour for the soil. This will conduct analyses in terms of effective stress (it can model pore pressures) and covers the effects of stage-by-stage construction. The other method is to use a nonlinear elastic model such as that proposed by Duncan and Chang (1970). This uses a hyperbolic curve to represent the results of triaxial tests of representative samples of soil. The tangent of the hyperbola is used to provide an expression for an incremental deformation modulus, the increments giving an excellent representation of step-by-step construction. The hyperbolic model does not allow for dilatancy and is sensitive to criticism regarding the rate of volume change. However, a tangent modulus can be obtained to simulate changes in Poisson's ratio. Since the action of correctly orientated tensile reinforcement is to suppress the

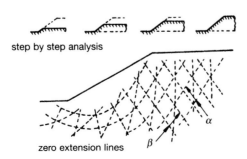

step by step analysis

zero extension lines

Fig. 6.61. Analysis modelling construction sequence

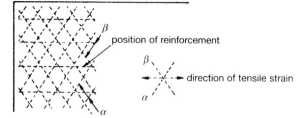

Fig. 6.62. *Continuum analysis of reinforced soil*

dilation of the soil, one of the main objections to the use of the hyperbolic model is removed.

The form of the analysis is analogous to the finite element analysis used with many bridge decks in which the analysis is used to determine shears, bending moments and reactions from which the necessary reinforcement is derived using Wood–Armer equations. In the case of reinforced embankments, the analysis has to be taken to a second iteration.

(a) An initial finite element analysis using incremental procedures (step-by-step construction) is used to derive the direction of principal total stress for each increment, together with the α and β zero-extension lines, Fig. 6.61.

(b) Additional elements are added to the idealization to represent the reinforcement and the analysis is repeated, Fig. 6.62. In view of the uncertainty in the use of any soil model, it seems prudent to restrict the

Table 6.8. Assumptions and requirements in a finite element model

Variable	Composite material (unit cell)	Special elements	'Full' model
Foundation	✓	✓	✓
Stage construction	✓	✓	✓
Fill properties	×	(✓)	✓
Reinforcement properties	×	(✓)	✓
Reinforcement prestress	×	×	✓
Stress distribution in reinforcement	×	✓	✓
Facing/reinforcement connection stresses	×	✓	✓
Compaction stresses in fill	×	×	✓
Construction technique (incremental—full height)	×	×	✓
Compressible backfill layers	×	✓	✓

(✓) Assumes part composite material.
✓ 2D analysis sheet or grid reinforcement.
3D analysis strip reinforcement.

position of the reinforcement to the middle third of the tensile strain arc, practical considerations permitting.

The second analysis is required to check that the reinforcement stresses do not exceed limiting values based on stress levels or adhesion characteristics derived from laboratory tests. The second analysis can also be used to demonstrate the realignment of the zero extension characteristics.

In a cohesionless soil the zero extension characteristic cannot be plotted directly, since the hyperbolic model analysis ignores volumetric strains. However, Roscoe (1970) has shown that a value for volumetric strain ($\nu = 20°$) for dense sands holds over a wide stress range (covering the internal stresses generated in embankments up to a height of 30 m). Using this, the zero extension lines can be derived from:

$$\frac{dy}{dx} = \tan \left[\xi \pm \left(\frac{\pi}{4} - \frac{\nu}{2} \right) \right] \tag{83}$$

where ξ is the direction of the major principal strain rate ϵ.

In the case of reinforcement beneath an embankment, cohesive soils are usually present and an analysis based on undrained conditions is appropriate. Full idealization of a reinforced soil structure is complex. Table 6.8 illustrates the assumptions and requirements for different levels of sophistication in the finite element model.

6.4.2 Iterative techniques

Vertically sided structures
Iterative methods are common during analysis for limit modes 3 and 4. As the number of possible solutions may be limited so the number of iterations may be low, and cycling through the conventional analytical procedure will normally suffice.

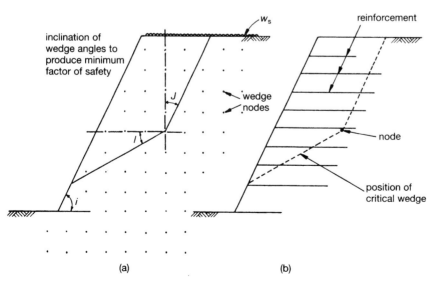

Fig. 6.63. Two-part wedge analysis (after Jewell, 1982)

Checking against wedge and slip circle stability, limit mode 5, is frequently undertaken using computer techniques. In these cases the number of potential failure planes can be very large, and iterative techniques to determine limiting conditions may be the logical method.

Sloping structures

With sloping structures the number of possible solutions is magnified, and iterative techniques based on simple controlling parameters may be useful. An example of iterative design for reinforced slopes has been described by Jewell (1982), Fig. 6.63.

(a) A systematic search over a defined grid is made of a series of failure wedges. For simplicity, a two-part failure is assumed and the most critical combination of the two wedge angles (I, J) is determined, Fig. 6.63.

(b) The results of (a) produce a measure of the stability or instability of the embankment.

(c) From (b), an assumed arrangement of reinforcement may be defined and the two-part wedge analysis repeated, but with a value of mobilized reinforcement force included in the equilibrium system wherever a trial wedge surface intersects a reinforcing member. The analysis is iterative, the slope angle being varied if required, and various reinforcements being investigated.

(d) Stable conditions exist when the mobilized reinforcement forces and the mobilized soil strength resisting failure are less than the limiting or maximum values developed at failure.

A similar approach is possible for cuttings, in particular the analysis of cutting failures, which may be analysed using two-part wedges.

Fig. 6.64. Analysis of a tied embankment (after Jewell, 1982)

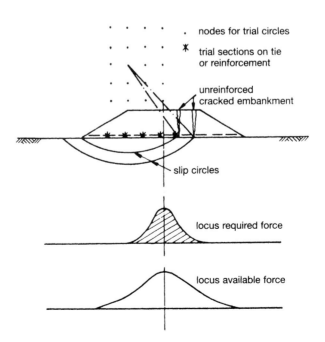

nodes for trial circles

trial sections on tie or reinforcement

unreinforced cracked embankment

slip circles

locus required force

locus available force

Table 6.9. Computer codes for the analysis of soil nailed structures

Name	Origin	Analysis considered
CLOUDIN	France (Solrenfor)	Shear, bending and tension (log spiral)
CRESOL	UK (Cardiff University)	Shear, bending and tension (log spiral)
Nail-Solver	UK (Oxford University)	Tension only
Soil-Nailer	USA (Caltrans)	Tension only
STARS	France (Ecole Polytechnique)	Tension only
TALREN	France (Terrasol)	Shear, bending and tension

A similar technique may be applied to embankments tied at the base (Jewell, 1982). The iterative method illustrated in Fig. 6.64 assumes a circular slip failure mechanism, based, as above, on a defined grid in which the circle cuts the reinforcing element at the base of the embankment; the force in the tie or reinforcing member required to ensure stability is determined for each circle. In addition, consideration of the potential failure circles, the available resisting force from the ties or reinforcing members is determined, considering all the factors which affect the behaviour of reinforcement (see Chapter 4, Theory). Provided that the locus of maximum required force is contained within the locus of available force, stability is ensured.

Soil nailing
A number of computer codes have been developed as an aid to the design of soil walls, Table 6.9.

6.5 Seismic design
6.5.1 General

In the normal course of events, a reinforced soil structure is acted on by a static thrust P from the retained fill. During an earthquake, the retained fill exerts an additional dynamic horizontal thrust P_{AE} on the reinforced soil wall or abutment. The dynamic horizontal thrust may be evaluated using the pseudo-static Mononabe–Okabe analysis method and added to the static forces acting on the structure, Fig. 6.65.

In addition to the external forces, the reinforced soil mass is subject to an horizontal inertial force P_{IR}. This force is a function of the active effective mass of the structure M and the maximum horizontal acceleration a_m, to which the structure is subjected (Segrestin *et al.*, 1988; Seed and Whitman, 1970):

$$P_{IR} = Ma_m P_{AE} \tag{84}$$

$$a_m = \alpha_m g \tag{85}$$

$$\alpha_m = (1\cdot45 - \alpha)\alpha \tag{86}$$

where g is the acceleration due to gravity

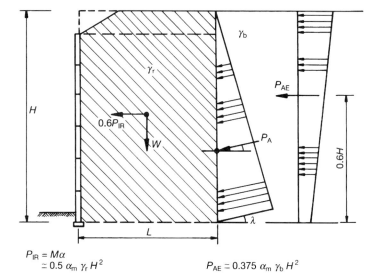

Fig. 6.65. Seismic external stability of a reinforced soil wall (after US Department of Transportation, 1989)

$P_{IR} = M\alpha$
$\simeq 0.5\,\alpha_m\,\gamma_r\,H^2$

$P_{AE} \simeq 0.375\,\alpha_m\,\gamma_b\,H^2$

Minimum allowable safety factors for seismic = 75% static

α is the maximum ground acceleration coefficient

α_m is the maximum structure acceleration coefficient at the centriod.

It has been suggested that:

$$P_{AE} = 0.375\,\alpha_m\gamma H^2 \tag{87}$$

$$P_{IR} = 0.5\,\alpha_m\gamma H^2 \tan\beta \tag{88}$$

where β is the angle subtended by an active wedge of soil which is subjected to seismic acceleration.

It is assumed that the horizontal forces associated with P_{AE} and P_{IR} will not peak simultaneously, and a factored value for P_{IR} of 60% may be accepted (FHWA, 1989).

Thus, the total horizontal force P_{HS}, resulting from seismic conditions:

$$P_{HS} = P + P_q + P_{AE} + 0.6\,P_{IR} \tag{89}$$

where P is the backfill thrust on the reinforced soil block, per metre 'run'
P_q is the resultant of active earth pressure on the reinforced soil block per metre 'run' due to a uniform surcharge.

6.5.2 External stability

The external stability of a reinforced soil structure is evaluated by summing the static forces acting on the structure, P and P_q, with the external seismic thrust, P_{AE}, and the inertial force, P_{IR}, and comparing this with the available resisting forces.

Overturning
Factor of safety against overturning

$$= \frac{\Sigma \text{ restoring moments}}{\Sigma \text{ overturning moments}}$$

$$= \frac{\left(W\left(\dfrac{L}{2}\right) + w_s\right)L}{H\left(\dfrac{P}{3} + \dfrac{P_q}{2} + 0.6 P_{AE} + 0.6 \dfrac{P_{IR}}{2}\right)} \tag{90}$$

Sliding
Factor of safety against sliding

$$= \frac{\Sigma \text{ resisting forces}}{\Sigma \text{ horizontal forces}}$$

$$= \frac{(W + w_s)L\mu_f}{P + P_q + P_{AE} + 0.6\ P_{IR}} \tag{91}$$

Note: In seismic conditions, reduced factors of safety are usually accepted. The US Federal Highway Administration (FHWA) design guidelines for reinforced soil structures adopts the following factors in respect of overturning and sliding, FHWA (1989):

overturning—dynamic factor of safety $= 0.75$ static factor of safety
$$= 0.75 \times 2.0 = \underline{1.5}$$
sliding—dynamic factor of safety $\qquad = 0.75$ static factor of safety
$$= 0.75 \times 1.5 = \underline{1.1}$$

6.5.3 Internal stability

A seismic load induces an internal inertial force P_I, acting horizontally on the structure in addition to the existing static forces. This will result in an incremental dynamic increase in the maximum tensile force in the reinforcement. It is assumed that the location of the maximum tensile force does not change during seismic loading. This assumption is conservative in respect of reinforcement rupture, limit mode 3, and considered acceptable in respect of pullout resistances, limit mode 4, Fig. 6.14.

6.5.3.1 General case—tie-back analysis

Where extensible reinforcement is used and the $K_{des} = K_a$. The tie-back method of analysis can be adapted to accommodate seismic conditions, by adding the tensile force developed by the internal inertial force P_I, to the static forces.

Limit mode 3—element rupture
The maximum ultimate limit state tensile force T_i, a depth of h_i below the top of the structure, is given by:

$$T_i = T_{hi} + T_{wi} + T_{fi} + T_{mi} + T_{si} \tag{7 Section 6.3.2.2.}$$

The maximum ultimate limit state tensile force in seismic T_{is} conditions is given by:

$$T_{is} = T_i + P_{Ii} \tag{92}$$

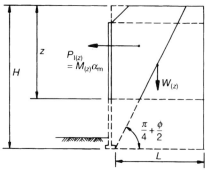

$$\sigma_h = K\sigma_v = KW = K\gamma H$$

$$T_{m1} = \frac{KWS_v}{n} \quad \text{per unit width of wall face}$$

Dynamic increment

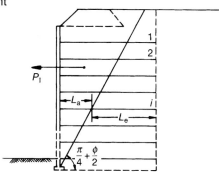

$$T_{m2} = P_I \frac{nbL_e}{\Sigma n_i b_i L_{ei}} = R_i$$

$$T = T_{m1} + T_{m2}$$

$$T_0 = 85\text{–}100\% \ T_m$$

Fig. 6.66. Internal seismic stability of a reinforced soil wall (extensible reinforcement) (after US Department of Transportation, 1989)

where $P_{Ii} = \gamma_{ms} M_i \alpha_m g$

$\qquad = (\gamma_{ms} \gamma b_i \alpha_m g)\nu$

and $\alpha_m = (1{\cdot}45 - \alpha)\alpha$

α is the maximum ground acceleration coefficient

α_m is the maximum structure acceleration coefficient at centroid

M_i is the mass of the active zone of the reinforced fill at a depth of b_i below the top of the structure, Fig. 6.66.

Rupture. The tensile strength of the ith layer of reinforcing elements required to satisfy load and stability in a seismic event is:

$$\frac{T_{DS}}{\gamma_n} \geq T_{is} \tag{93}$$

where T_{DS} is the design strength of the reinforcement in seismic conditions

γ_n is the partial factor for ramification of failures.

Note: During seismic conditions an increase in the characteristic strength of polymeric reinforcements is often accepted. The design method with respect to seismic loading detailed above was developed for inextensible reinforcements.

Any extensibility of the reinforcement affects the overall stiffness of the reinforced soil mass. As the use of extensible reinforcements reduces the overall stiffness, it is expected to have an influence on the lateral earth pressures induced by the seismic loading. As the stiffness of the structure decreases, damping should increase. In addition, there is a factor of safety in the design tension for the potential creep of extensible reinforcement under long-term static loading. This provides an additional factor of safety against dynamic overload for polymeric reinforcements.

Limit mode 4—element pullout
The perimeter P_i of the ith layer of reinforcing elements required to satisfy local stability requirements is:

$$P_i = \frac{T_{is}}{\mu L_{ei}\left(\dfrac{\gamma_{es}\gamma_{hi} + \gamma_{ef}w_s}{\gamma_n\gamma_p} + \dfrac{\alpha_{sr}c'L_{ei}}{\gamma_{ms}\gamma_n\gamma_p}\right)} \tag{94}$$

where T_{is} is the design strength of the reinforcement in seismic conditions (equation (92)).

Limit mode 5–wedge stability
Wedge stability needs to be checked for earthquake conditions (see 6.3.2.2).
   The total resistance of the layers of elements T_s, anchoring the wedge during a seismic event, is obtained from:

$$\sum_{i=1}^{m}\left[\frac{T_{is}}{\gamma_n} \text{ or } \frac{P_iL_{ip}}{\gamma_p\gamma_n}\left(\mu\gamma_{es}\gamma_{ni} + \gamma_{ms}\gamma_{ni}\,\alpha_m g + \mu\gamma_{ef}w_s + \frac{a_{sr}c'}{\gamma_{ms}}\right) \geq T_s\right] \tag{95}$$

6.6 Tension membranes

6.6.1 Embankments over voids

Voids can occur under an existing highway embankment or other earthworks owing to a variety of causes, including cavities resulting from solution activity, the collapse of underground mine workings, differential foundation settlements, or thawing of subsurface ice lenses (thermokasts), Fig. 6.67.
   Tensile reinforcing elements may be used to support highway embankments and other soil structures over voids, Fig. 6.68. Bonaparte and Berg (1987) have subdivided these applications into two categories:

(*a*) design to resist complete collapse into the void, while accepting loss of serviceability
(*b*) design to limit deformation so as to maintain serviceability of the structure over the void.

In the first category the reinforcement is temporary and is acting as a safety net; in the second, the reinforcement action is permanent and the reinforced soil structure behaves as a beam or slab with sufficient bending stiffness to limit deformations to an acceptable level.
   The two categories are dependent on the size of the potential void. In the case of very large voids (≥ 10 m), the reinforcement can be considered to be

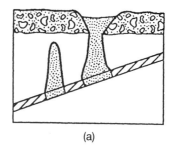

(a)

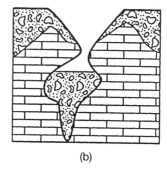

(b)

Fig. 6.67. (a) Crown hole; (b) solution cavities

providing a safety role, and the reinforcement function is to prevent a catastrophic situation from occurring with a potential risk to life. In this case, the support of the road would be required for a matter of hours until the void had been backfilled or the structure repaired.

The second category of design occurs when the void is relatively small (typically 2–5 m). In this condition, the reinforcement material has to have sufficient strength and stiffness to support the structure within the required serviceability limits for the life of the facility. In the case of a highway, the life of the facility is typically assumed to be 60 years.

The ability of any reinforcing material to fulfil the requirement function depends on the size of the void to be bridged, the nature and height of the earth

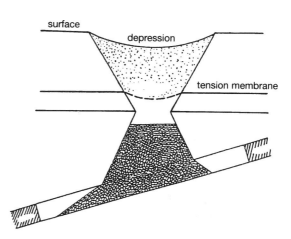

Fig. 6.68. Tension membrane supporting embankment over void

fill supported by the reinforcement together with any surcharge loading and the serviceability limits specified. Current practice in the UK is to limit differential surface deformations to 1 per cent for motorways and trunk roads and to 2 per cent for lower class roads (BS 8006: 1995).

6.6.1.1 Analytical methods

Current methods of design of tension membranes adopt a conservative approach, owing to the uncertainties involved and the simplicity of the analytical techniques employed. The soil and reinforcement are assumed to be resting initially on a firm foundation. With the development of a void under the reinforcement the overlying soil deflects into the void. The deflection of the soil layer generates arching within the soil above the reinforcement, and the load in the reinforcement over the void is less than the theoretical weight of the soil above the void. Deflection of the reinforcement into the void mobilizes part of the reinforcement strength and the material will act as a tension membrane supporting loads normal to the plane, Fig. 6.69. As a result of the reinforcement straining, three cases can be considered:

(a) the soil–reinforcement system fails, Fig. 6.69(a)
(b) the soil–reinforcement system exhibits limited deflection and the system bridges the void, Fig. 6.69(b)
(c) the soil reinforcement deflects until the reinforcement comes in contact with the bottom of the void; in this case, part of the load is transmitted to the bottom of the void and the tension in the reinforcement is reduced, Fig. 6.69(c).

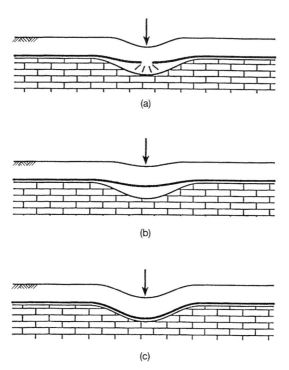

(a)

(b)

(c)

Fig. 6.69. Action of reinforcement spanning a void: (a) soil reinforcement system fails; (b) soil reinforcement deflects but bridges void; (c) soil reinforcement deflects to bottom of void

The case illustrated in Fig. 6.69(a) represents a deficient design. The case illustrated in Fig. 6.69(b) represents the classical design case for tension membranes. The case illustrated in Fig. 6.69(c) is typically that associated with the use of reinforcement to spread the load of embankments or soil structures on support piles. Although it is usual to assume that no support is provided by the subsoil beneath the void, analytical studies using the finite element method have shown significant subsoil support is provided in these circumstances (Jones et al., 1990).

Tension membrane theory

The length of reinforcement required for embedment beyond the edges of the voids is determined by the soil reinforcement interface shear behaviour.

The pullout capacity of membrane reinforcement may be estimated from:

$$P_{\mathrm{u}} = \sigma_{\mathrm{v}} L \mu \tan \phi' \tag{96}$$

where L is the embedded length of reinforcement.

Assuming that the reinforcement spanning a void transmits tensile stresses but not shear stresses and that the soil reinforcement interface above the void is frictionless, the applied stress to the membrane is normal. For plane strain conditions, the shape of the membrane is circular. The tensile resistance T_{max} in a layer of reinforcement spanning an infinitely long void of width b has been determined by Giroud (1981) as:

$$T_{\mathrm{max}} = k \left[\frac{2f(\epsilon)}{180/\pi} \sin -1 \left(\frac{1}{2f(\epsilon)} \right) - 1 \right] \tag{97}$$

$$f(\epsilon) = \frac{1}{4} \left(\frac{2y}{b} + \frac{b}{2y} \right) \tag{98}$$

where k is the secant tensile stiffness of reinforcement (kN/m)

y is the maximum vertical reinforcement deflection (m).

Strain in the reinforcement ϵ is obtained by dividing the reinforcement tensile resistance by the secant tensile stiffness.

Arching

Arching above the void may cause a reduction in vertical stress. In the case of a void under an embankment, arching reduces the stress and the tension in the reinforcement spanning the void by transferring part of the stress to adjacent stable portions of the embankment. Terzaghi (1943) has presented an approximate method for calculating the vertical stress on a horizontal plane at the base of a soil mass due to yielding of part of the base. It is assumed that yielding produces vertical shear surfaces from the edge of the void to the surface. An expression for the vertical stress under plane strain yielding conditions is given by:

$$\sigma_{\mathrm{v}} = \frac{by}{2K \tan \phi} \{ 1 - \exp[-(K \tan \phi \, 2b/b)] \} \tag{99}$$

where K is the ratio of horizontal to vertical earth pressure along vertical shear surfaces.

Similarly, Kezdi (1975) has developed an equivalent expansion for vertical stress developed over a circular yielding surface of diameter D

$$\sigma_v = \frac{Dy}{4K \tan \phi}\{1 - \exp[-(K \tan \phi \, 4h/D)]\} \tag{100}$$

Handy (1985) has suggested the following value for K in equation (99):

$$K = 1 \cdot 06(\cos^2 \theta + K_a \sin^2 \theta) \tag{101}$$

where $K_a = \tan^2(45° - \phi/2)$,

$\theta = 45° + \phi/2$.

Bonaparte and Berg (1987) using equations (97–99) have developed design charts shown in Figs 6.70 and 6.71.

Soil arching can be destroyed by vibrations, percolating groundwater and other external factors. In such a case, the tension membrane theory provides a lower bound conservative estimate of soil–reinforcement behaviour.

Combined arching and tension membrane theory
Giroud *et al.* (1988) have developed an analytical method for the design of reinforcement spanning a void, which combines both arching and tension membrane theory. The problem is acknowledged to be one of complex soil reinforcement interaction, and the solution provided involves uncoupling the soil response due to arching from the reinforcement response associated with the tension membrane theory. As a result, a two-step approach is used. Firstly, the behaviour of the soil fill is analysed using classical arch theory, this provides pressure at the base of the soil layer on that portion of the reinforcement located above the void. Secondly, tension membrane theory is used to establish a relationship between the pressure on the reinforcement, the tensile stress and strain in the reinforcement and the deflection.

An inherent assumption in this uncoupled two-step approach is that the

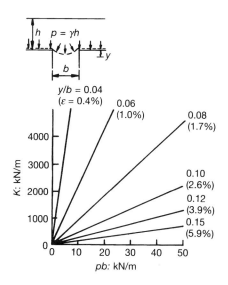

Fig. 6.70. Required secant tensile stiffness for a tensioned membrane spanning an infinitely long void of width b(m) (chart from Bonaparte and Berg, 1987; based on equations from Giroud, 1981)

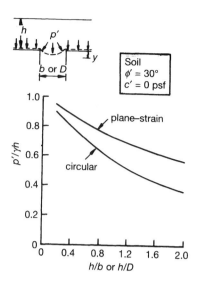

Fig. 6.71. Theoretical reduction in vertical stress due to cohesionless soil arching over an infinitely long void of width b(m) or a circular void of diameter D(m) (after Bonaparte and Berg, 1987)

deformation required to generate a soil arch is compatible with a tensile strain required to mobilize the reinforcement tension. This assumption has not been verified.

BS 8006: 1995 method

The tension membrane theory assumes that the deflected shape of the reinforcement is circular; in reality, the weight of the fill acting on the reinforcement over the unsupported void causes the membrane to deform into the shape of catenary. However, to simplify the analysis, BS 8006: 1995 assumes that the load is actually distributed along the horizontal span of the reinforcement rather than along the actual deflected length. In this case, the shape of the deflected reinforcement is parabolic. The equation governing the extension of an unsupported membrane which assumes the shape of a parabola is given by Fig. 6.72.

$$\epsilon = \frac{8d^2}{3D^2} \tag{102}$$

where ϵ is the extension of the reinforcement

d is the maximum deflection of the reinforcement

D is the cavity diameter, Fig. 6.72.

Equation (102) cannot be solved directly as it has two unknowns (ϵ and d). In addition, what has to be determined is the maximum allowable reinforcement extension which satisfies the maximum acceptable differential deformation criterion at the surface of the pavement or embankment. By utilizing the geometry of the subsided soil mass and assuming a constant volume of soil in the subsided soil mass ($v_1 = v_2$), a relationship can be derived for the maximum allowable extension $\epsilon_{\max}$, in the unsupported reinforcement, given a maximum acceptable differential deformation, d_s/D_s:

$$\epsilon_{\max} = \frac{8 \left(d_s/D_s\right)^2 (D + 2H/\tan\theta_d)^6}{3D^6} \tag{103}$$

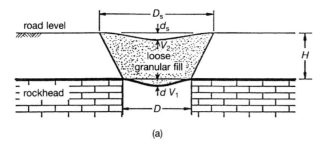

(a)

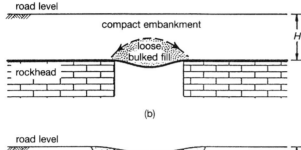

(b)

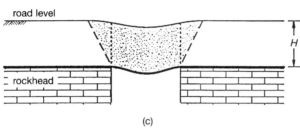

(c)

Fig. 6.72. Tension membrane theory (BS 8006): (a) assumed working condition (arching action assumed in determining tension in reinforcement); (b) actual working condition (arching action assumed); (c) ultimate condition (no arching action)

where d_s/D_s is the differential deformation criteria of the surface of the embankment

D is the cavity diameter

H is the height of the fill of the embankment

θ_d is the angle of draw of the embankment.

The angle of draw θ_d is usually equated to the peak angle of friction of the fill ϕ'_{peak}. If the value of d_s/D_s is set at the maximum acceptable limit required to maintain serviceability at the surface of the embankment or highway, the resulting value of ϵ_{max} derived from equation (103) will be the maximum allowable extension in the reinforcement over the period of time in which the material is required to act.

In the development of the value of ϵ_{max} of equation (103) it is assumed, frequently, that no soil arching occurs in the embankment material above the unsupported reinforcement. This is because it is thought that vibrations caused by traffic can result in the breakdown of the arching action. In this situation, the reinforcement is required to support the full weight of the fill material together with any surcharge for the life of the structure.

Validity of current methods

The BS 8006: 1995 method is based on tension membrane theory and is acknowledged as being conservative. This is accepted as a consequence of the simplicity of the method which ignores many of the soil parameters involved in the analytical problem. An advance on the BS 8006 method would be to couple arching theory with tension membrane theory as advised by Giroud *et al.* (1988); however, this requires additional assumptions which have not been verified.

Experimental tests indicate that the tension membrane theory can be considered to be a lower bound conservative estimate of soil–reinforcement behaviour, Fluet *et al.* (1986). The theory appears to describe accurately the condition where a void exists *before* construction. Where a void is seen to occur *after* construction, the existing theories for analysing soil–reinforcement supporting an embankment are seen to be inaccurate and over-conservative. It can be concluded that the accurate analysis of any reinforcement system supporting an embankment over a void needs to consider both the geometry of the problem and also to include a proper evaluation of the material properties, both in respect of the reinforcement, the embankment fill and the subsoil support conditions and their combined behaviour. This can best be achieved using continuum methods.

6.6.1.2 Modelling the design problem

Void geometry

The successful modelling of any reinforced soil structure spanning over a void can be achieved only if the parameters used in the analysis are accurately described. A fundamental requirement is a knowledge of the size of any potential void. The selection of unrepresentative void dimensions will result automatically in the analysis of an impossible or improbable problem. The majority of surface discontinuities produce circular depressions at the surface. The size of the void can vary enormously from 1 m to 40 m. The latter was the size of the sinkhole which occurred under the Vera Cruz road in Pennsylvania in 1983 and which caused the collapse of a bridge (Bonaparte and Berg, 1987). The majority of occurrences are smaller than this. Many sinkholes or mining depressions produce voids of a limited depth, others result in extensive voids. The collapse of shafts usually results in deep voids.

Surface geology

The surface geology can have a significant effect on both the development of a void and the behaviour of any reinforcement used to support an embankment over a void. Two basic geometries can be identified, the first being where rock supports the embankment directly, and the second where a soil layer exists beneath the reinforcement, Fig. 6.73. In both cases, the analytical model is required to describe accurately the material properties of the supporting soil or rock.

Parameter values

The height and material properties of the embankment supported by a reinforcing member will influence the reinforcement/soil performance, and, in any modelling exercise, realistic material properties of any fill material are required. Use of layered fills may influence the problem and, accordingly, care has to be taken to describe the geometry of any embankment. However, unlike

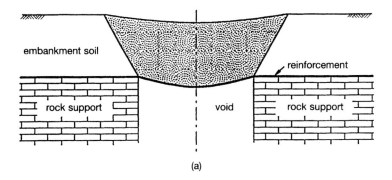

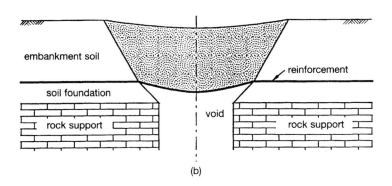

Fig. 6.73. Embankment
supported: (a) directly on soil;
(b) on soil foundation overlying
rock

many reinforced soil modelling problems, modelling the actual construction process is not necessary in this case and a realistic solution to the problem can be obtained by 'switching on' gravity.

6.6.1.3 Reinforcement orientation

Reinforcement over a void may be required in either a single direction, as in the case of a long slender void such as a trench, or in two directions, when the void is either circular or an irregular shape. In the first case, reinforcement with anisotropic strength characteristics are efficient; in the latter, strength in two dimensions is required. This may be achieved using grid reinforcement, individual fabric reinforcement, or at least two layers of strip reinforcement laid orthogonally.

6.6.2 Reinforcement on embankment piles

Embankment piles are used extensively in Scandinavia and south-east Asia to support fills, lightweight structures and bridge abutments. Loads generated by the fill or structures are carried by friction along the sides of the piles or through piled caps at the top of the piles, Fig. 6.74. The piles and pile caps are normally designed to carry the total load of the embankment (Swedish Road Board, 1974).

A development described by Broms and Wong (1985) and Reid and Buchanan (1984) is to provide a polymeric reinforcing membrane spanning

201

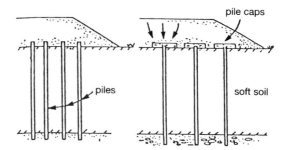

Fig. 6.74. Embankment piles reinforced by pile caps

in catenary between the pile caps. The tension membrane reduces the horizontal loads on the piles caused by the placement and compaction of fill; this reduces the possibility of tilting the pile caps and the punching through of the fill between the piles. Importantly, the provision of the tension membrane permits the spacing of the piles to be increased and the size of the pile caps to be reduced, Fig. 6.75. The design of tension membranes of piles is given comprehensive treatment in BS 8006: 1995.

6.6.3 Land reclamation

The requirement to reclaim land covered by very soft alluvial or clay deposits is growing. Yano *et al.* (1982) describe the problem associated with coastal

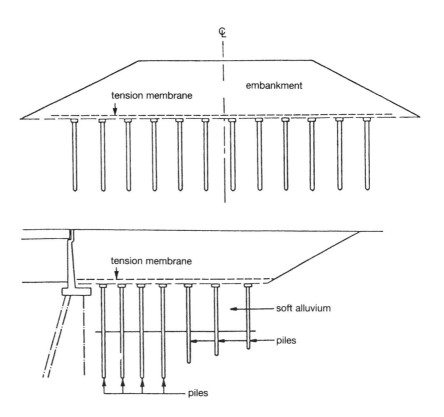

Fig. 6.75. Bridge abutment support piling (after Reid and Buchanan, 1984)

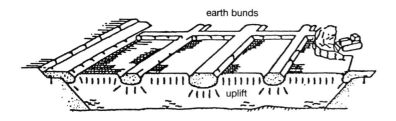

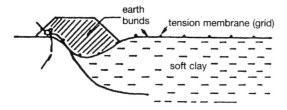

Fig. 6.76. Land reclamation using tension membranes (after Yano et al., 1982)

areas including the bays of Tokyo and Osaka, where soft marine clay has previously been deposited over wide areas. Usually this material has little bearing capacity but may be in a potentially prime location. A number of techniques have been developed to reclaim these areas, including solidification with cement. An alternative method is the use of tension membranes in the form of a grid of high-strength geosynthetic materials. The technique used is to create a large lattice of earth bunds surrounding and covering the area to be reclaimed. Once these bunds have been placed, tension membranes are laid covering the open soft material and anchored to the earth bunds. Reclamation is achieved by controlled filling of the areas between the bunds, the fill being placed on top of the tension membrane. The earth bunding settles, balancing the tension in the tension membranes. Uplift is maintained by the tension membrane, and the fabric acts as both a tension member and a separator, preventing loss of soft soil. Fig. 6.76 illustrates the technique.

Design for the system is empirical; Yano *et al.* (1982) reported that tension membranes formed from 12 mm dia. polypropylene ropes laid on a 0·5 m grid and having a breaking load of 1600–2000 kgf are successful.

6.7 References

BINQUET J. and LEE K. L. (1975). Bearing capacity tests on reinforced earth slabs. *J. Geot. Engng Div. Proc. ASCE*, **101**, GT12, 1241–1276.

BOLTON M. D. and PANG P. L. R. (1982). Collapse limit states of reinforced earth retaining walls. *Géotechnique*, **32**, No. 4, 349–367.

BOLTON M. D. (1991). McGown, Yeo and Andrawes (eds). Reinforced soil: laboratory testing and modelling. *Performance of reinforced soil*. Thomas Telford, London, 287–298.

BONAPARTE R. and BERG R. R. (1987). The use of geosynthetics to support roadways over sink-hole prone areas. *Proc. 2nd Multidisciplinary Conf. on Sink-holes and Environmental Impact of Karst*, Orlando, 437–445.

BROMS B. B. and WONG I. H. (1985). Embankment piles. *Proc. 3rd Int. Geot. Sem. on Soil Improvement Methods*, Singapore.

BRITISH STANDARDS INSTITUTION (1995). *Code of practice for strengthened/reinforced soils and other fills.* HMSO, London, BS 8006, 156.

BROWN R. E. W. and ROCHESTER T. A. (1979). Reinforced Earth—technical guidance provided by the Department of Transport in England. *C.R. Coll. Int. Reinforcement des Sols*, Paris.

BRUCE D. A. and JEWELL R. A. (1986–87). Soil nailing applications and practice. *Ground Engng*, Part I, **19**, no.8, 10–15; Part 2, **20**, no.1, 21–23.

COOK D. S. (1989). D. A. Shercliff (ed.). Design of road embankments over mineral workings using high strength geotextile membranes. *Reinforced embankments.* Thomas Telford, London, 157–167.

DEPARTMENT OF TRANSPORT (1978). *Reinforced earth retaining walls and bridge abutments for embankments.* Tech. Memo BE 3/78.

DUNCAN J. M. and CHANG C. (1970). Non-linear analysis of stress and strain in soils. *J. S.M. and Found. Engng, ASCE*, **96**, 1629–1653.

FEDERAL HIGHWAY ADMINISTRATION (FHWA) (1989). *Reinforced soil structures. Volume 1, Design and construction guidelines.* US Department of Transportation, Washington DC.

FEDERAL HIGHWAY ADMINISTRATION (FHWA) (1993). *Recommendations Clouterre 1991.* (English Translation), Soil Nailing Recommendations 1991. Report FHWA-SU-026, Washington DC, 302.

FLUET J. E., CHRISTOPHER B. R. and SLATER A. R. (1986). Geosynthetic stress–strain response under embankment loading conditions. *Proc. 3rd Int. Conf. on Geotextiles*, Vienna, **1**, 175–180.

FUKUOKA M. (1980). Static and dynamic earth pressure on retaining walls. *Proc. 3rd Australian–New Zealand Conf. on Geomechanics*, Wellington, **3**, 3–37.

GÄSSLER G. and GUDEHUS G. (1981). Soil nailing—some soil mechanical aspects of *in situ* Reinforced Earths'. *Proc., ICSMFE*, Stockholm, **3**.

GÄSSLER G. and GUDEHUS G. (1983). Soil nailing—statistical design. *Proc. VIII ECSMFE*, Helsinki.

GIROUD J. P. (1981). Designing with geotextiles. *Matériaux et Constructions*, **14**, no. 82, July–Aug., 257–272.

GIROUD J. P., BONAPARTE R., BEECH J. F. and GROSS B. A. (1988). Load carrying capacity of a soil layer supported by a geosynthetic overlying a void. *Theory and practice of earth reinforcement*, Balkema, 185–190.

GUDEHUS G. (1972). Lower and upper bounds for stability of earth-retaining structures. *Proc. 5th Eur. Conf. SMFE*, Madrid, **1**.

HANDY R. L. (1985). The arch in soil arching. *J. Geotech. Engng Div., ASCE*, **111,** no. 3, Mar., 302–318.

HAY C. (1982). Private communication. Henderson and Busby Partnership.

IWASAKI K. and WATANABE S. (1978). Reinforcement on railway embankments in Japan. *ASCE Spring Conv.*, Pittsburgh, Apr.

JABER M. B. (1989). *Behaviour of reinforced soil walls and centrifuge model tests*, University of California, Berkeley, PhD Diss. 239.

JAPAN HIGHWAY PUBLIC CORPORATION (1987). *Guide for design and construction on reinforced slope with steel bars.* Tokyo.

JEWELL R. A. (1982). A limit equilibrium design method for reinforced embankments on soft foundations. *Proc. 2nd Int. Conf. Geotextiles*, Las Vegas, **3**.

JEWELL R. A. (1985). Limit equilibrium analysis of reinforced soil walls. *Proc. 11th Int. Conf. on Soil Mechs and Foundation Engng*, San Francisco. Balkema, The Hague, **3**, 1705–1708.

JONES C. J. F. P. (1989). *Review of the effects of mining subsidence on reinforced earth.* Transport and Road Research Laboratory, TRRL Contractor Report 123.

JONES C. J. F. P. (1994). Tatsuoka and Leshchinsky (eds). The economic construction of reinforced soil structure. *Recent case histories of permanent geosynthetic–reinforced soil retaining walls*. Balkema, 103–116.

JONES C. J. F. P., LAWSON C. R. and AYRES D. J. (1990). Geotextile reinforced piled embankments. *4th Int. Conf. on Geotextiles and Geomembranes and Related Products*. Balkema, The Hague, **1**, 155–161.

JONES C. J. F. P. and BELLAMY J. B. (1973). Computer prediction of ground movements due to mining subsidence. *Géotechnique*, **23**, no.4, 515–30.

JONES C. J. F. P. and EDWARDS L. W. (1975). *Finite element analysis of M180 Trent Embankment*. Report to North Eastern Road Construction Unit, West Yorkshire MCC.

KEMPTON G. T. (1992). The use of reinforcement geotextiles to support road embankments over areas subjected to mining subsidence. *Highways & Transportation*, Dec., 21–31.

KEZDI A. (1975). Winterkon and Fang (eds). Lateral earth pressure. *Foundation engineering handbook*. Van Nostrand Reinhold Company, New York, 157–220.

KINNEY T. C. (1986). *Tensile reinforcement of road embankments on polyzonal ground by geotextiles and related materials*. Interim Report to State of Alaska, Department of Transportation and Public Facilities, Fairbanks Alaska, Mar.

LEE K. L., JONES C. J. F. P., SULLIVAN W. R. and TROLINGER W. (1993). Failure and deformation of four reinforced earth walls in Eastern Tennessee, USA. *Géotechnique*, **44**, no. 3, 397–426.

MAIR R. and HIGHT D. (1983). Private communication. Geotechnical Consulting Group, London.

MINISTERES DES TRANSPORTS (1979). *Direction generale des transports interieurs—les ouvrages en Terre Armee–recommendations et regles de l'art*, Paris.

MITCHELL J. K. and VILLET W. C. B. (1987). *Reinforcement of earth slopes and embankments*. National Cooperative Highway Research Program Report 290, Transportation Research Board, National Research Council, Washington DC.

MURRAY R. T. (1980). Fabric reinforcement earth walls: development of design equations. *Ground Engng.*, **13**, no. 7, Oct.

MURRAY R. T. (1982). Fabric reinforcement of embankments and cuttings. *Proc. 2nd Int. Conf. Geotextiles*, Las Vegas.

MURRAY R. T. and IRWIN M. J. (1981). *A preliminary study of TRRL anchored earth*. TRRL Supplementary Report 674.

NETLON LTD (1982). *Designing with Tensar*. Netlon Ltd., Blackburn, July.

PETERSON L. M. (1980). *Pullout resistance of welded wire mesh embedded in soil*. Utah State University, MSc Thesis.

RECOMMENDATIONS CLOUTERRE (1991). *Soil nailing recommendations—1991*. French National Research Project Clouterre, US Department of Transport, Federal Highway Administration Report FHWA-SA-93-026 (English Translation 1993).

REID W. M. and BUCHANAN N. W. (1984). Bridge approach support piling. *Proc. Conf. on Piling and Ground Treatment*, Thomas Telford, London, 267–274.

ROSCOE K. H. (1970). The influence of strains in soil mechanics (10th Rankine Lecture). *Géotechnique*, **20**, no. 2, 129–170.

SEED H. B. and WHITMAN R. V. (1970). Design of earth retaining structures for dynamic loads. *ASCE Speciality Conf. on Lateral Stresses and Earth Retaining Structures*, Cornell University, 103–147.

SEGRESTIN P. and BASTIC M. J. (1988). Seismic design of reinforced earth retaining walls: the contribution of finite element analysis. *Proc. Int. Symp. on Theory and Practice of Earth Reinforcement*, Kyushu, Japan.

SIMS F. A. and BRIDLE R. J. (1966). Bridge design in areas of mining subsidence. *J. Inst. Highway Engng.*, Nov., 19–34.

STOCKER M. F., KORBER G. W., GASSLER G. and GUDEHUS G. (1979). Soil nailing. *C.R. Coll. Int. Reinforcement des Sols*, Paris.

SUTHERLAND R. J. M. (1973). Typical cost multipliers for urban road design. *Highway and Road Const.*, Dec.

SWEDISH ROAD BOARD (1974). *Bankpalming (embankment piles)*. Report TV121.

SWEETLAND D. (1982). Private communication. Netlon Ltd., Blackburn, Lancs.

TATSUOKA F. (1992). Ochiai, Hayashi and Otaiu (eds). Roles of facing rigidity in soil reinforcement. *Earth reinforcement practice*. Balkema, 831–870.

TATEYAMA M., TATSUOKA F., KOSEKI J. and HARII K. (1995). Damage to soil retaining walls for railway embankments during the Great Hanskin-Awaji Earthquake, January 17, 1995. *1st Int. Conf. on Earthquake Geotechnical Engineering, IS-Tokyo '95*, Tokyo.

TERZAGHI K. (1943). *Theoretical soil mechanics*. John Wiley & Sons, New York, 510.

US DEPARTMENT OF TRANSPORTATION (1989). *Reinforced soil structures, Vol. 1—Design and construction guidelines*. National Technical Information Service, Report FHWA-RD-89-043, 287.

WALKINSHAW J. L. (1975). *Reinforced earth construction*. US Department of Transportation, Arlington, Virginia, Report FHWA-DP-18.

YANO K., WATARI Y. and YAMANOUCHI T. (1982). Earthworks on soft clay using rope-netted fabrics. *Proc. Symp. on Recent Developments in Ground Techniques*, Bangkok, 225–237.

7

Construction

7.1 Introduction

Construction of reinforced soil structures must be of a form determined by the theory and in keeping with the assumed idealization and analysis. The theoretical form of the structure may be quite different from an economic prototype, and attention should be paid to the method of construction throughout the design process.

Speed of construction is usually essential to achieve economy, this may be achieved by the simplicity of the construction technique. Hambley (1979) has detailed those aspects of simple construction relevant to earth retaining structures; these may be amended in the case of reinforced earth structures to be:

(a) Use materials which are readily available and easy to use.
(b) Construct as much as possible with plant at existing ground level to ease access and use natural crust and drainage.
(c) Design excavations with a level base.
(d) Use foundations with simple shapes and details.
(e) Form all surfaces horizontal or vertical.
(f) If required anticipate re-use of formwork.
(g) Where possible design structures to be stable at all stages of construction.
(h) Fix reinforcement and place soil in one plane at a time.
(i) Use medium size reinforcement, avoid both small sections and also heavy large elements which can be difficult to transport and fix without lifting equipment.
(j) Make structures wide enough for a man to get inside reinforcement.

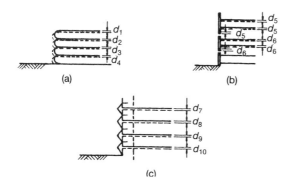

(a)

(b)

(c)

Fig. 7.1. Three methods used for constructing reinforced soil structures: (a) concertina method; (b) telescope method; (c) sliding method

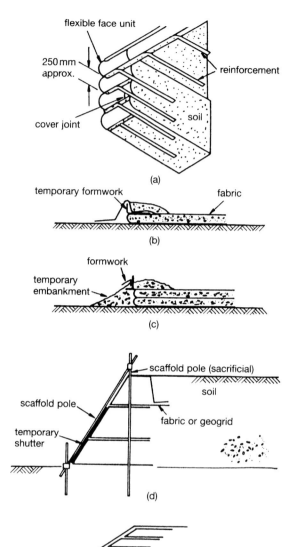

Fig. 7.2. (a) Reinforced soil with metallic face units; (b) method of construction, fabric wall; (c) method of construction, fabric wall; (d) method of construction of sloping traverse; (e) method of construction, fabric or geogrid or cutting

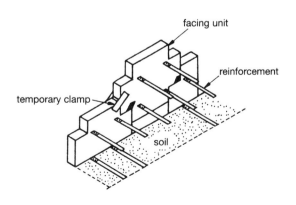

Fig. 7.3. Telescope method of construction

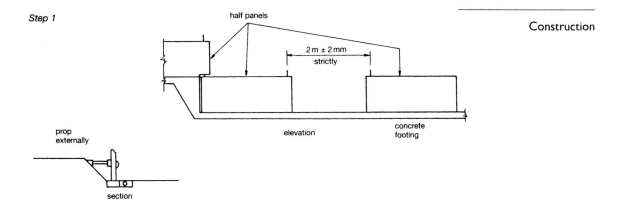

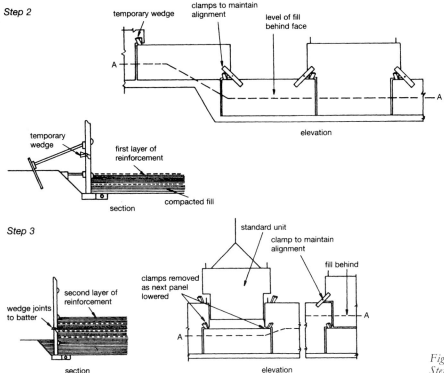

Step 1

half panels

2m ± 2mm
strictly

prop
externally

elevation

concrete
footing

section

Step 2

temporary wedge

clamps to maintain
alignment

level of fill
behind face

A

A

elevation

temporary
wedge

first layer of
reinforcement

compacted fill

section

Step 3

standard unit

clamp to maintain
alignment

fill behind

clamps removed
as next panel
lowered

second layer of
reinforcement

wedge joints
to batter

A

A

section

elevation

*Fig. 7.4. Construction sequence:
Steps 1–3*

Step 1
*Cast footing and drainage (approximately 150 × 300 mm) with top surface level. Erect half panels. Erect full size panels and fix
temporary wedges to create horizontal gap between units. Clamp adjacent units together and prop first rows from front.*

Step 2
*Place fill and compact to level of first row of reinforcement A–A compaction within 2 m of the face must be undertaken with care so as
not to create excessive distortion of the facing. Place reinforcement and attach to facing, note any form of reinforcement may be used, grid,
strip, chain or plank. Continue filling.*

Step 3
*When filling has reached level (A–A) remove clamps. Place another row of panels and wedges. Replace clamps at higher level and
continue cycle. As erection proceeds, remove temporary wedges to permit vertical settlement of soil mass and facing.*
Note: When extensible reinforcement is used, a degree of prestressing is required before the fill is placed.

209

7.2 Construction methods

In keeping with the general rules detailed above, constructional techniques compatible with the use of soil as a constructional material are required. The use of soil, deposited in layers to form the structure, results in settlements within the soil mass caused by gravitational forces. These settlements within the soil result in the reinforcing elements positioned on discrete planes moving together as the layers of soil separating the planes of reinforcement are compressed. Construction techniques capable of accommodating this internal compaction within the soil fill are required. Failure to accommodate the differential movement may result in loss of serviceability or worse.

The three constructional techniques which can accommodate differential vertical settlements ($d_1 - d_{10}$) within the soil mass are shown in Fig. 7.1. Except for some special circumstances, every reinforced soil structure constructed above ground uses one or other of these forms of construction.

7.2.1 Concertina method

The constructional arrangement of the concertina method developed by Vidal (1966), is shown in Fig. 7.1(a). Differential settlement within the mass ($d_1 - d_4$) is achieved by the front or face of the structure concertinaing. The largest modern reinforced soil structures have been built using this approach, and it is the form of construction frequently used with fabrics and geogrid reinforcing materials in both embankments and cuttings, Figs 7.2 (a, b, c, d, e). Since the facing must be capable of deforming, a flexible hoop-shaped unit made from steel or aluminium is normally used. Fabrics and geogrids usually provide their own facing. The method is often used with temporary structures when the distortion of the facing is accepted.

7.2.2 Telescope method

In the telescope method of construction, developed by Vidal (1978), the settlement within the soil mass (d_5, d_6) is achieved by the facing panels closing up an equivalent amount to the internal settlement. This is made possible by supporting the facing panels by the reinforcing elements and leaving a discrete horizontal gap between facing panel, i.e. the facing panels hang from the reinforcing elements. The horizontal gap between the facing panels may be produced by the use of compressible gaskets employed to hold the panels apart during the placing of the soil fill, Fig. 7.3. Failure to provide a large enough gap between facing elements can result in crushing and spalling of the facing as the soil fill is compressed under the action of gravity.

The closure between panels will vary from application to application depending upon the geometry of the structure, quality of fill material, size of the facing panels and the degree of compaction achieved during construction. Typical movements, reported by Findlay (1978), show vertical closures of 5–15 mm for facing panels 1·5 m high. The shape and form of the facing panel must be compatible with the procedure adopted, and reinforced concrete cruciform or tee-shaped panels covering 1–4 m^2 and 150–250 mm thick are typical.

The construction sequence for the telescope method is shown in Fig. 7.4.

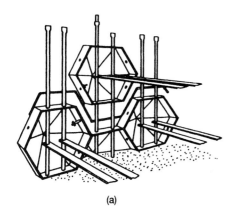

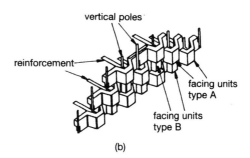

(a)

(b)

7.2.3 Sliding method

In the sliding method of construction developed by Jones (1978), differential settlement within the fill forming a reinforced soil structure can be accommodated by permitting the reinforcing members to slide relative to the facing. Slideable attachments can be provided by the use of grooves, slots, vertical poles, lugs or bolts. If vertical poles are used these may form the structural elements of the facing and the facing may become non-structural providing only a covering, whose role is to protect the completed structure from the elements and prevent erosion. With this arrangement, the type and form of the non-structural part of the facing can be chosen to suit the particular application or environment, Figs 7.5(a), (b). The erection sequence for a non-structural facing is shown in Fig. 7.6.

If a structural facing is used, the connecting element, the vertical pole, may be reduced in size to an appropriate form. The facing may be rigid or semi-rigid; up to a height of approximately 10 m a rigid facing may be used; for heights above 10 m an elemental form of facing is appropriate. Where a full-height rigid facing is used, this is erected and propped before filling starts. The erection sequence is shown in Fig. 7.7. A notable difference between the use of a flexible or a rigid facing is that uniform compaction can be applied to the fill behind a rigid facing without fear of distortion.

Fig. 7.5. (a) Sliding method of construction. (b) perspective illustrating assembly of components

7.3 Reinforcing systems

7.3.1 Anchored earth

Anchored earth systems have developed from a combination of the techniques used in reinforced soil and soil anchoring. Figs 7.8 and 7.9 show anchored earth methods originating from Austrian and Japanese practitioners. The Austrian application involves polymeric strips connecting concrete wall blocks and semi-circular anchors, while the Japanese application exploits the local passive resistance of small rectangular anchor plates. A further scheme developed in the UK is illustrated in Fig. 7.10. This system employs reinforcing steel bent into triangular anchors, pullout resistance is mobilized by friction along the straight portion of the steel element and by passive pressure mobilized at the triangular anchor. Anchored earth concepts have been

211

Construction

Step 1

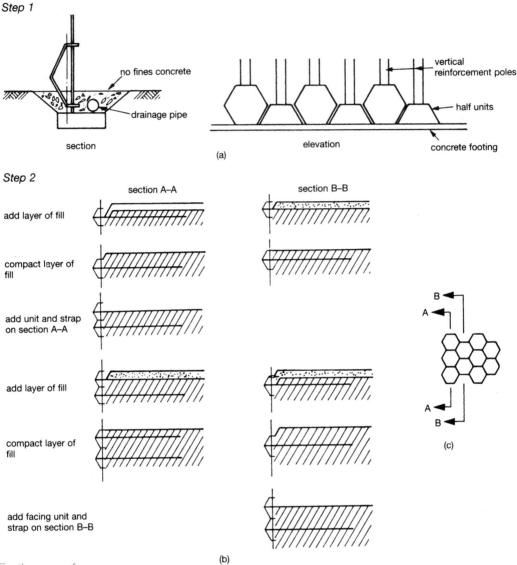

section

elevation

(a)

Step 2

section A–A   section B–B

add layer of fill

compact layer of fill

add unit and strap on section A–A

add layer of fill

compact layer of fill

add facing unit and strap on section B–B

(b)

(c)

Fig. 7.6. Erection sequence for a non-structural facing (see text example): Steps 1 and 2

Step 1
Cast footing (approximately 100 × 150 mm) with top surface level. Erect half unit. Erect first full units. Erect vertical reinforcement poles. Place porous drainage pipe. Place no fines concrete.

Step 2
Note: Reinforcement position is at mid-height of facing unit. Any form of reinforcement may be used, grid, strip, chain, plank. Compaction of the fill close to the face is restricted to small plant so as not to distort the facing. Speed of construction is dependent upon the speed of placing the fill. A bold facing is usually used to disguise any inconsistencies of distortions caused during construction. Construction can be stopped at any level or position without any fear for the safety of the workers. No propping is used.

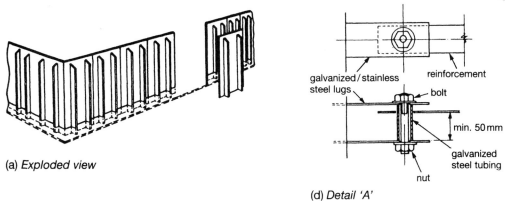

(a) Exploded view

(d) Detail 'A'

galvanized/stainless steel lugs

reinforcement

bolt

min. 50 mm

galvanized steel tubing

nut

temporary prop

concrete footing

temporary folding wedge

(b) Staged

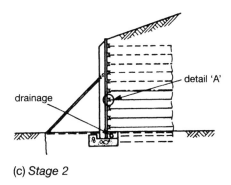

drainage

detail 'A'

(c) Stage 2

Fig. 7.7. Sliding method of construction: rigid facing

Step 1
Cast footing (approximately 150 × 300 mm) plus upstand. Erect and prop facing panels.

Step 2
Construct drainage. Place fill layers and compact. When fill level with top of first pair of connecting lugs attach first level of reinforcement. Continue filling. When filling is complete, or when sufficient fill has been placed to stabilize facing remove props and folding wedges.

213

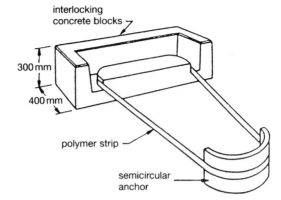

interlocking
concrete blocks

300 mm

400 mm

polymer strip

semicircular
anchor

*Fig. 7.8. Wall system with
concrete blocks, polymer strips
and anchors*

extended to the use of waste automobile tyres as illustrated in Fig. 7.11 which
uses polymer strips to connect the anchors together, or Fig. 7.12 where the side
walls of used tyres are linked on to reinforcing elements to form anchors.

7.3.2 Soil nailing

Soil nailing is a method of reinforcing natural ground in order to increase the
shear strength of the soil and provide an element of tensile strength. The
construction method is illustrated in Fig. 7.13 and includes the following steps:

(a) Excavate soil layer of 1·0–1·5 m depth.
(b) Protect exposed face of soil, normally using spray concrete reinforced with
wire mesh.
(c) Drive in nail or nails at a predetermined angle of inclination and spacing,
using percussion, rotary drilling, flushing, vibration or pneumatic.
(d) Grout area between the nails and the soil to ensure bond between the soil
and the reinforcing member.

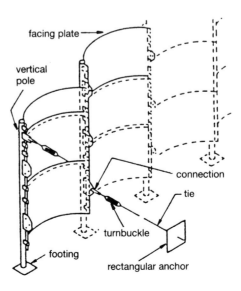

facing plate

vertical
pole

connection

tie

turnbuckle

footing

rectangular anchor

*Fig. 7.9. Wall system with
facing plates and rectangular
anchors*

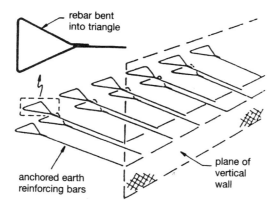

Fig. 7.10. *Anchored earth with triangular rebar reinforcement*

(*e*)   Cover the exposed ends of the nail with spray concrete.
(*f*)   Repeat processes (*a*)–(*e*).

Soil nailing is a process in which excavation walls and slopes are stabilized in situ by the installation of relatively short, fully bonded steel bars or other reinforcing materials according to a regular and relatively closely spaced pattern. The development of the soil nailing technique has been described by a number of authors including Bruce and Jewell (1986–1987) and Mitchell and Villet (1987).

The early forms of soil nailing were adapted from procedures following the New Austrian Tunnelling Method (NATM) and its application is often cited as a development of this technique, Fig. 7.14. In 1972, soil nailing was employed first to stabilize a railway cutting in heavily cemented sand above the water-table near Versailles, France (Rabejac and Toudic, 1974). This construction resulted eventually in a 18–22 m high reinforced soil slope inclined at 70°. Benches, typically 100 m long, were excavated from the top down at regular height intervals of 1·4 m. The soil nails varied in length from 4 m to 6 m. Each

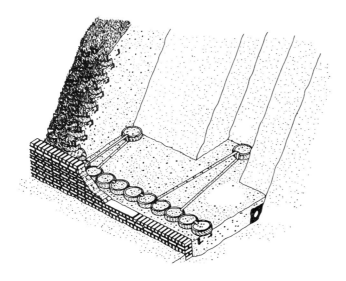

Fig. 7.11. *Earth retention with waste tyres and geotextiles*

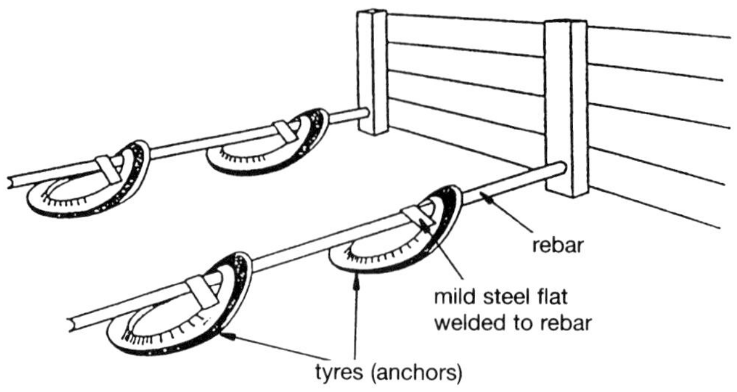

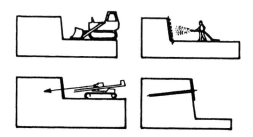

Fig. 7.12. Tie anchored timber wall (TAT), California

rebar

mild steel flat welded to rebar

tyres (anchors)

Fig. 7.13. Soil nailing construction method

nail consisted of two 10 mm diameter bars grouted in 100 m diameter drilled holes. Facing along the slope was provided by a 50–80 mm thick mat of reinforced shotcrete.

The first North American application of reinforced soil nailing occurred in 1976 to provide temporary excavation support for basement construction at a hospital at Portland, Oregon (Shen *et al.*, 1981). The soils at the site were cohesive dense silty sands above the water-table. Excavation proceeded from

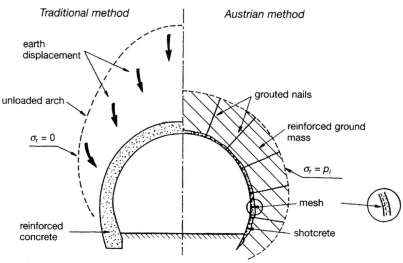

Fig. 7.14. Traditional and Austrian tunnelling methods (after FHWA, 1993)

σ_r confining pressure
p_i initial pressure

Table 7.1. *Types of nail in use in Europe*

Formed	Nail	Soil type	Remarks
Driven/ Percussion (Hupinoise)	Steel tubes (45 mm diameter) Steel angles (50 mm × 50 mm × 5 mm, 60 mm × 60 mm × 6 mm)	Cohesionless soils	Used for temporary works Care should be used with 'loose' soil Tubes can be driven with upward inclination and act as drains as well as nails
Driven (simultaneous grouting)	Steel bar	Dense medium sand	A sacrificial enlarged tip (55 mm) is driven in front of the anchor to produce an open anchor
Predrilled (without casing) (a) Rotary boring displacing soil	Closed ended tube (70 mm) driven into soil by simultaneous pushing and rotation Steel nail (22 mm) grouted into tube	Medium dense sand	Soil around the nail is displaced and compacted, increasing shear bond between nail and soil
Predrilled (b) Dry auger	Conventional nail grouted into hole	Clay soils Soft rocks Cemented sand (Residual soil)	Not applicable where boulders or layers of hard rock are present
Predrilled (c) Cased			Soil nailing becomes less economic when casing is required
Jet bolting	Special steel profile driven into ground by vibro-percussive hammer with grout jetted down nail to the head (200 bar pressure)	Medium dense sand	Grout under pressure assists installation of the nail
Ballistic	Nails up to 6 m long 'fired' into the ground by air launcher	Cohesiveless soil Cohesion soil Made ground	Air launch pressure is varied depending upon nature of the ground Very fast installation technique Can be installed in upward direction

the top down at height intervals of 1·5 m to a maximum depth of 13·7 m. The soil nails consisted of 25–38 mm diameter steel bars grouted in augered holes, approximately 7–8·5 m long. The face of the excavation was covered with mesh reinforced shotcrete, approximately 25–50 mm thick.

Most nails are installed either by track mounted drill rigs or by percussion methods. The types of nail most in use are identified in Table 7.1. A relatively novel development is to install the nail by ballistic penetration. The key to this new approach is the acceleration of the nail from the front, thereby putting the nail in *tension* during the period of time it is penetrating the soil, Fig. 7.15. During installation the nail displaces the soil, and the pullout resistance is related to a smaller surface area than those resulting from grouted nails. The cross-section of the nails used with the ballistic method are usually larger than the grouted nails, and the rate of installation of the nails can be significantly higher than with other methods.

Soil nailing challenges conventional thinking in that it demonstrates that in-situ walls can be constructed from the top down, without pre-existing embedment. At every stage during the construction, excavation beneath the toe of the stabilized soil is required. This is only possible if the soil is able to provide self-supporting stability. Over excavation can lead to failure, and the depth of excavation for each step of the wall depends on different types of soil, Fig. 7.16. Table 7.2 provides typical excavation depths in current use in Europe (Gässler, 1990).

Fig. 7.15. The ballistic method of installing a soil nail (after Myles, 1993): (a) nail being presented for loading; (b) nail loaded in launcher; (c) nail firing by air pressure release; (d) nail impact with collet release; (e) nail fully installed and arrested

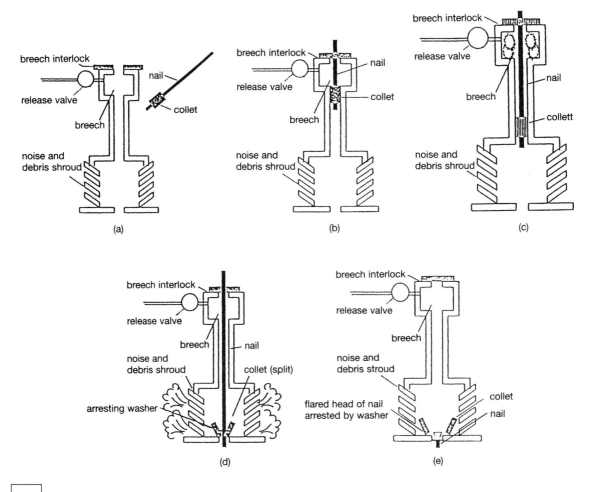

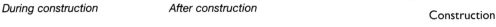

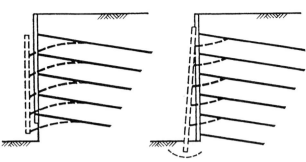

During construction   After construction

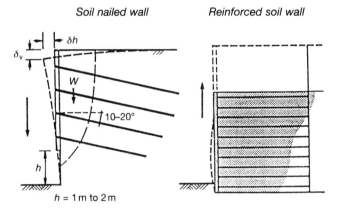

Soil nailed wall   Reinforced soil wall

$h = 1\,m\ to\ 2\,m$

Fig. 7.16. Potential movements of soil nailed walls (after Schlosser, 1990)

7.3.3 Lateral earth support systems

The construction system for the lateral earth support system is essentially the same as the methods used for soil nailing, but with one important difference. After the reinforcement and grouting has been placed in position, a prestressing force equal to 59–80 per cent of the capacity of the reinforcing anchors is applied to each reinforcing element.

7.3.4 Polymer impregnation

The development of earth retention systems can be viewed as an evolutionary process, in which methods for supporting soil have involved progressively more alterations and insertions of reinforcing elements. One of the goals has been to transform soil into an engineered medium, with enhanced mechanical properties.

Recent innovations include polymer impregnation of soils formed by mixing soil with a small continuous filament of polymer. The reinforcing element described by Leflaivre *et al.* (1983) is a polyester filament of 0·1 mm diameter and a tensile strength of 10 N. Construction requires the simultaneous projection of sand, water and filament, Fig. 7.17. Approximately 0·1–0·2

Table 7.2. *Excavation depths of near vertical cuts (after Gässler, 1990)*

Type of soil	Cut depth: m average	Cut depth: m maximum
Gravel (sandy)	0·5 (with capillary cohesion)	1·5 (cemented)
Sand	1·2 (middle dense with capillary cohesion)	up to 2·0 (cemented)
	1·5 (dense with capillary cohesion)	
Silt	1·2 (depends upon structure/ stability of grain skeleton and water content)	2·0
Clay	1·5 (normally consolidated)	2·5 (overconsolidated)

per cent of the composite material is filament, producing a total length of reinforcement of 200 000 m per m³. Cohesive strengths between 100 kPa to 200 kPa have been reported for this material which is known as Texsol, Fig. 7.18. Similar mechanical behaviour has been developed by the inclusion of small 60 mm × 40 mm polymeric grids into sand. As reported by Mercer *et al.* (1984), the inclusion of 0·2 per cent by weight of grids can increase the shear strength of sandy soil by 25–60 per cent, Table 7.3.

7.4 Labour and plant

Labour and plant requirements for the construction of reinforced soil structures are minimal, and no specialist equipment or skills are required. Erection of a normal vertically faced structure of 500–1000 m exposed area is undertaken, usually by a small construction team of three to four men deployed to cover the main construction elements, namely erecting the face, placing and compacting the fill, and placing and fixing the reinforcement. A comparison in labour requirements for different forms of retaining walls has been given by Leece (1979), Table 7.4.

The plant requirements during construction normally include aids to the placing and compaction of fill, and some form of small crane or lifting device, although the latter is not required when a non-structural facing is used, Fig. 7.6. Where a method specification is employed, as with the Department of

Fig. 7.17. Texsol machine

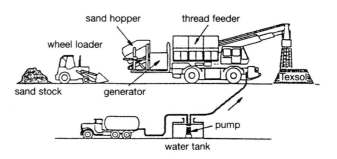

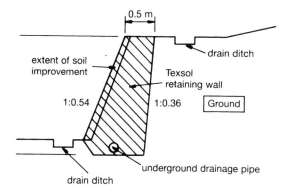

Fig. 7.18. Formation of reinforced soil block

Transport's *Specification for highway works* (1991), the compaction plant used within 2 m of the facing consists normally of the following forms:

(*a*) vibro tampers
(*b*) vibrating plate compactors with a mass < 1000 kg
(*c*) vibrating rollers with a mass/metre width < 1300 kg and a mass < 1000 kg.

The equipment needs when constructing reinforced embankments are the same as for conventional construction.

Construction equipment used for soil nailing or lateral earth support systems may be specialized, although conventional equipment employed for the installation of ground anchors may be used.

7.5 Rate of construction

Construction of reinforced soil structures is normally rapid. Construction rates for vertically faced structures of $40-200$ m^2 per day may be expected and usually the speed of erection is determined by the rate of placing and compacting the fill. However, in some cases, the economic production of facing units may determine the construction rate, particularly if an original or unique facing is required. Construction is normally unaffected by weather except in extreme situations.

Table 7.3. *Values of cohesion and soil-inclusion interaction types in sand reinforced with synthetic inclusions*

	Inclusions			
	Fibres 50 mm	Fibres 150–200 mm	Grid-type plates 60 × 40 mm	Continuous filaments
Cohesion: kN/m^2	10	100	50	200
Types of interaction	Friction Extensibility	Friction Extensibility	Friction Extensibility Entanglement Interlocking	Friction Extensibility Entanglement Curvature effect

Table 7.4. Comparison in labour requirements for different forms of retaining walls (Leece, 1979)

Type of wall	Labour content: manhours/m^2
Reinforced soil—without traffic barrier	4·1
Reinforced soil—with traffic barrier	4·7
Mass concrete	11·2
Reinforced concrete	11·5
Crib walling	13·3

7.6 Damage and corrosion

Care must be taken that facing elements and reinforcing members are not damaged during construction. Vehicles and tracked plant must not run on top of reinforcement; a depth of fill of 150 mm above the reinforcement is frequently specified before plant can be used. Reinforcement should be stored in a safe dry environment, and non-metallics should be stored away from ultraviolet light. With soil nailing and lateral earth support systems, the precautions usually adopted to cater for the corrosion problems of anchor systems are relevant.

7.7 Distortion

Reinforced soil structures are prone to distortion, particularly during construction. Many of the construction details adopted in practice are chosen to minimize distortion or the effects of distortion.

7.7.1 Concertina construction

Structures built from fabrics and geogrids or constructed as temporary structures and using the concertina method of construction are particularly prone to distortion of the face. The degree of distortion cannot be predicted. An accepted method of overcoming the problem is to cover the resulting structure either with soil or with some form of facing. An alternative is to provide a rolling block against which the compaction plant can act.

7.7.2 Telescope construction

An estimate of the internal movements and distortion of the facing can be made from observations of prototype structures. Typical vertical movements within the soil mass which are transmitted to the facing are illustrated in Fig. 7.19(a). Horizontal movements of the facing are made up of two components:

(*a*) horizontal movements at the joints
(*b*) tilt of the facing units.

Joint movement during construction is not normally significant and is likely to be 2–5 mm depending on construction details. Tilt of the facing panels can be significant and may have a marked effect on the facial appearance of the structure, although other elements of serviceability are unlikely to be affected by tilt of the facing. All facing panels in this form of construction tilt, the pivot point depending on the geometry of the facing, Fig. 7.20. Panel tilt of a typical cruciform structure, constructed with facing panels approximately 1·5 m square is illustrated in Fig. 7.19(b). The overall tilt of this structure is 2·5 per cent. To accommodate the forward tilt of the facing panels, caused mainly by compaction of the fill, it is normal practice to incline the panels backwards between 1 in 20 and 1 in 40.

7.7.3 Sliding method construction

When a non-structural facing is used, distortion of the facing is likely to occur, the degree being dependent on compaction. The distortion is accommodated by:

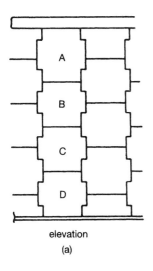

elevation

(a)

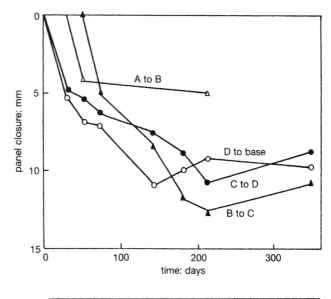

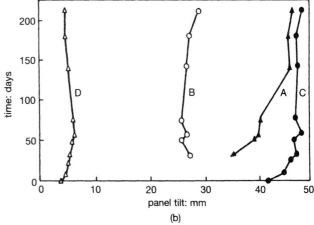

Fig. 7.19. (a) Vertical movement of panels; (b) variations in panel tilt (after Findlay, 1978)

223

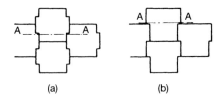

Fig. 7.20. *Pivot point of facing panels*

(a)   (b)

(*a*) using light plant in the 2 m zone adjacent to the facing (section 7.4)
(*b*) using bold architectural features to mask the distortion, such as in Fig. 7.5(a); these can disguise forward rotations and major bulges by creating a face without a natural sight line.

When a structural facing is used and the construction method is as in Fig. 7.7, the horizontal movement of the facing will be limited to the joint movement capacity provided by the reinforcement/facing connection. Typical movements are 2–4 mm.

7.8 Logistics

The speed of construction must be catered for if the full potential of the use of reinforced soil structures is to be realized. Normally, this will cause little or no problems with the reinforcing materials, but the production and delivery rate of the facing units may cause problems, particularly if multiple use of a limited number of shutters is expected for economy.

Transport may cause difficulties and the choice of structural form and construction technique may depend ultimately on the ease and economy of moving constructional materials. As an example, the lightweight of the fabrics and geogrid materials, with their ability to be transported in rolls, makes them suitable for air freight.

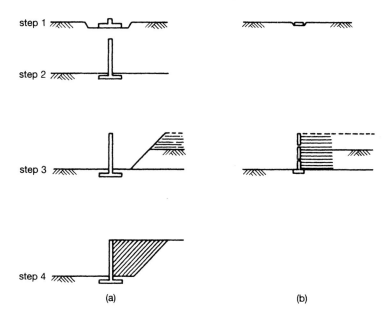

Fig. 7.21. Construction sequence for: (a) conventional structure; (b) reinforced soil structure

(a)   (b)

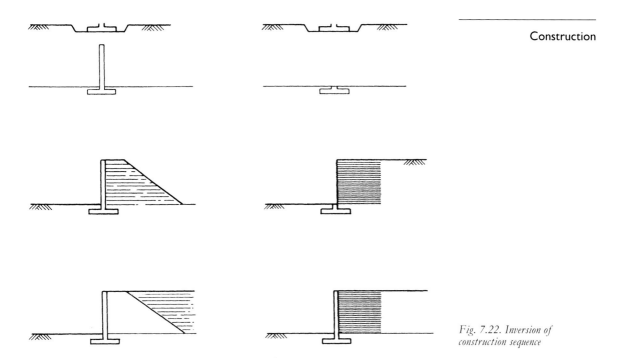

Fig. 7.22. Inversion of construction sequence

7.9 Contractor's construction sequence

Reinforced soil structures encourage the use of non-conventional construction sequences. It is possible to streamline the construction sequence and eliminate some steps as illustrated in Fig. 7.21.

Alternatively, it is possible to invert a construction sequence as in Fig. 7.22, where the backfill of a structure is placed before the structure itself. This is achieved by forming the backfill into a temporary reinforced soil structure and using the face of this structure as the back shutter for the permanent structure. This technique has been used successfully in bridgeworks construction.

7.10 Construction tolerances and serviceability limits

Reinforced soil structures deform during construction. Consideration should be given to providing the necessary tolerances to permit a structure to attain a stable configuration, and also to ensure that construction and post-construction movements are within acceptable limits.

Construction tolerances
The construction tolerances detailed in Table 7.5. have been found to be acceptable.

Serviceability limits
Post-construction movements of reinforced soil structures result from:

(*a*) internal creep strain of polymeric reinforcement
(*b*) foundation settlements
(*c*) internal settlements within the fill

Table 7.5. Construction tolerances

Concrete panel, full height units and semi-elliptical steel facings	
Overall height	± 50 m
Horizontal alignment	± 50 mm
Horizontal level of panels	± 10 mm
Differential alignment	⊅25 mm over 3 m straight edge

Note: Geotextile wraparound facings cannot usually be constructed to close tolerances.

Table 7.6. Serviceability limits on post-construction internal strains

Structure	Overall limiting strain, per cent
Bridge abutments	0·5
Walls	1·0
Steel slopes	1·5
Embankments	5·0
Note: Polymeric reinforcement	⊅$(f) \times$ performance limit strain during life of structure at working temperature

Table 7.7. Maximum differential settlement along line of facing

Maximum differential settlement	
1 in 1000	Not significant
1 in 200	Normally safe limit for concrete facings without special precautions
1 in 50	Normal safe limit for semi-elliptical steel facings or geotextile facings
1 in < 50	Distortion may affect retaining ability of geotextile facings

Table 7.8. Vertical internal settlement of retained fill

Structural form	Required movement relative to height of structure
Discrete panels	Closure of 1 in 50
Full-height of panels	Movement capacity of connections 1 in 50
Semi-elliptical steel facings and geotextile wraparound facings	Vertical distortion 1 in 50

(d) uniform or differential settlements resulting from mining or other voids beneath the structure.

Post-construction internal strains which have been found tolerable are detailed in Table 7.6.

7.10.1 Settlements

Reinforced soil structures can tolerate settlements greater than those acceptable with conventional structures. Reinforced soil is an acceptable construction technique in areas of mining subsidence. Uniform settlement of a reinforced mass presents no problem; however, checks must be made to ensure that drainage systems, surfaces and supported structures can accept the movements.

Differential settlement
The effects of differential settlement must be considered in respect of:

(a) disruption of the facing
(b) additional internal strains imposed on the reinforced soil mass
(c) differential movements imposed on bridge decks or other structures supported by the reinforced soil structure.

The tolerance of reinforced soil structures to differential settlements along the line of the facing is shown in Table 7.7. Differential settlements normal to the face of the structure will result in rotation of the reinforced soil block. Backward rotations (into the fill) of 1:50 have been experienced in reinforced soil structures without any distress being experienced. However, consideration should be given to a differential settlement producing additional strain in the reinforcement.

Internal settlement
Reinforced soil structures settle internally, so the construction system and the construction tolerances must be able to accommodate these movements. Table 7.8 provides typical values of internal strains.

7.11 Propping forces

The construction of reinforced soil structures formed using full height facing units requires the use of temporary props. The horizontal loads supported by the props (P_L) may be determined from:

$$P_L = \frac{K_a \gamma h_t^2}{6} \tag{1}$$

where h_t is the height of fill above the toe.

If a recognized construction sequence is used, the prop loading developed is less than that derived by equation (1). A proven sequence of releasing the facing is as follows:

(a) fill to height below the prop height, but higher than half prop height
(b) remove wedges holding the toe of the footing
(c) remove prop.

With this sequence, the horizontal propping force P_L may be reduced to:

$$P_L = \frac{K_a \gamma h_t^3}{6 h_p} \tag{2}$$

where h_p is the height of prop.

7.12 Pressure relieving systems

The design of reinforced soil structures is a problem of soil structure interaction which involves satisfying two soil performance requirements. The first is to develop sufficient stress in the soil to mobilize fully its shearing resistance either through friction of adhesion at the interface between the soil and the reinforcement. The second is to limit strains in the soil so that lateral strains within the structure, in particular post-construction strains, comply with serviceability criteria, Table 7.6. The advantage of permitting a degree of horizontal yielding to reduce the lateral stresses within a reinforced soil structure has been demonstrated by Naylor (1978); see Chapter 4, Theory. Practical application of this technique has been addressed by Jones (1979) and McGown *et al.* (1987). At present, the development of boundary yielding appears best achieved when using a full height facing, when the removal of the

Fig. 7.23. Wall construction using a compressible fill layer (after McGown et al., 1987): (a) unconnected reinforcement; (b) reinforcement connected to facing; (c) after removal of prop, reinforcement connection adjusted

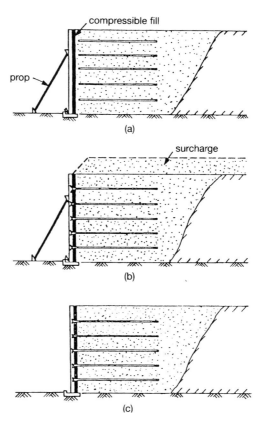

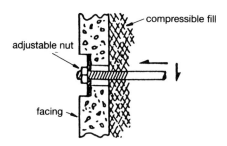

Fig. 7.24. Adjustable reinforcement connection

temporary props, used to support the structure during construction, can be the trigger for lateral yielding. Yielding can be achieved either by the use of a compressible layer of fill or material next to the facing or by the adoption of yielding connections between the reinforcement and the facing. Use of a compressible layer is illustrated in Fig. 7.23, while Figs 7.24 and 7.25 give details of adjustable or yielding reinforcement connection.

7.13 Influence of construction practice on reinforced soil

Construction practice is critical to reinforced soil behaviour. It is difficult to determine the degree of importance of separate elements on the construction system adopted. Table 7.9 provides an indication of the potential weightings which might be relevant based on a consideration of the end use of the structure.

The use of the proposed structure probably has the major influence on the choice of structure and construction adopted. As an example, some materials may be used as both facing and reinforcement, although the appearance of the resulting structure may not be acceptable other than for temporary, industrial or military applications. Similarly, the influence of fill properties on the

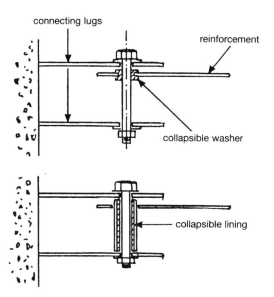

Fig. 7.25. Yielding reinforcement connection

Table 7.9. Relative importance of factors influencing the behaviour of reinforced soil structures

Factor	Application					
	Temporary structure	Short life	Permanent structure	Industrial structure	Marine structure	Military structure
Drainage	***	***	***	***	***	***
Distortion	*	*	***	*	*	*
Subsoil conditions	**	**	**	**	***	***
Fill properties	*	**	***	**	***	**
Reinforcement properties	*	**	***	**	***	**
Facing	*	*	***	*	**	**
Aesthetics	*	*	***	*	**	*
Durability	*	*	***	*	***	(***)
Rate of construction	***	**	**	**	**	***

* Secondary importance; ** Important; *** Very important; (***) May require self-destruct material.

durability of reinforcing materials may not be important with temporary constructions but are critical with permanent structures.

Subsoil conditions are important in all reinforced constructions and are often the primary technical reason for the choice of this type of structure. Some constructional details are better than others at accommodating significant differential settlements. In particular, the wrap around or gabion style facings are able to accept major distortions.

7.13.1 Reinforcement

Reinforcing elements may take a variety of forms. The form of the reinforcement affects the soil reinforcement interaction and influences the method of construction. As an example of the latter, flexible reinforcements formed from polymers need to be tensioned to restrict deformation of the facing. Polymeric reinforcements may also be extensible and susceptible to creep.

The general location of the reinforcement is fixed during the design process; however, some choice may be exercised over the size of the reinforcement used and it has been shown that larger reinforcing elements placed at greater spacing may be more efficient than smaller elements placed close together. The orientation of the reinforcement is usually fixed during design. Where reinforcement is designed to provide high pullout resistance, it is best orientated along the line of principal tensile strain. Change in the orientation of the reinforcement with respect to the principal strains reduces its effectiveness as a tensile member. Reinforcement placed in fill is often laid horizontal owing to construction constraints. In the case of vertical structures, this orientation may be near the optimum. However, with sloping structures, the optimum orientation of the reinforcement close to the base of the structure may be at an angle of between 20° to 30° to the horizontal which reflects the rotation of the strain field. In the case of soil nailing, construction methods capable of installing inclined reinforcing elements are usually preferred. The orientation of the nails determines the subsequent level of displacement of the structure.

The method used for installing the reinforcing elements in soil nailing has an influence on the performance of the structure and significantly affects the pullout capacity of the nails. Plumelle *et al.* (1990) have indicated that bored nails used with a high pressure backfill grout provide the highest pullout capacity compared with other installation techniques. Pullout capacity of nails also depends on the shape of the borehole developed during installation. Smooth cylindrical holes in stable soil can produce an initial stress on the reinforcement equal to zero, and the resultant initial pullout capacity of the nail is low. With regular boring, a ribbed effect is created which results in dilatancy and an increase in normal stress during pullout: the effect of the initial state of stress on driven nails or nails installed by the new method of firing into the ground by an air launcher.

7.13.2 Facings

Facings, with the exception of full height panels, are particularly influenced by the method used to compact the backfill. Although distortion does not influence the basic assumption used in the analysis, distorted facings have a major effect on aesthetics and need to be limited to acceptable construction tolerances, Table 7.5. Full height facings are propped during part of the construction procedure. The effect of the propping is to ensure the development of the strain field assumed in the analysis, while at the same time limiting distortion.

The use of rigid facings for reinforced soil walls can provide additional structural stability when compared with structures built using elemental facing systems (Tatsuoka, 1992). The structural advantage obtained from the use of rigid facings is illustrated clearly by consideration of the limit mode method of analysis (Chapter 6, Design and analysis). Limit mode 5 covers mechanisms of failure which in the usual reinforced soil structure can pass through any elevation in the structure. In the case of structures erected with rigid facings, the potential failure planes are reduced, and must pass below the toe of the structure. Experience shows that retaining walls, including reinforced soil walls, seldom fail on a plane passing through the toe. The usual critical failure plane passes through the face of the structure at a point one-third above the toe and, in the case of structures formed from element facing units and masonry structures, occurs when the facing distorts to a point where mechanical stability is lost at which point failure is inevitable and usually very rapid (Lee *et al.*, 1993).

For a number of years the non-proprietary system (the York method) used to construct vertical reinforced soil structures in the UK has been based on the use of a rigid facing formed from steel H sections and concrete planks (king post and panel construction) (Jones *et al.*, 1990). This form of construction can be used with any form of reinforcement and any type of fill. It has proved to be competitive for use in the construction of retaining walls and bridge abutments.

The system presents two advantages. Firstly, the use of the H section and concrete planks produces a facing with *flexural rigidity*, thereby providing the most stable structure form identified by Tatsuoka (1992), but with the added advantage of providing this rigidity/stability at all stages, including during construction. Secondly, the facing can be left as a king post and panel structure or can be provided with an additional face treatment to enhance the aesthetics of the construction. In industrial conditions, in a contemporary environment or with temporary structures, the appearance of a king post and panel structure may itself be acceptable. In the case of permanent structures in sensitive locations, an architectural treatment is usually appropriate.

As the H pile and panel system is non-proprietary, any form of reinforcement can be selected by the designer. Connection of the reinforcement to the H pile and panel facing system can follow a range of methods. The most appropriate has been found to be the sliding connection which allows for settlement of the soil fill without causing additional stresses to the face/reinforcement connections, while providing a full strength structural connection. This results in the most stable structure while retaining all the structural benefits associated with flexural rigidity.

7.13.3 Backwall friction

Backwall friction is a function of compactive effort, with large compaction plant tending to have a greater influence. The use of heavy compaction is restricted to the rear of reinforced soil structures or to an area within the body of the fill remote from the wall face. The implication of backfill friction on the rear of the reinforced soil block can be profound. Ignoring the effect can add 30–70% to the vertical stress assumed in the design. Using backfill friction in the analysis would simplify the design, by allowing a uniform pressure distribution under the structure, and would reduce the reinforcement requirements.

Backwall friction has been advocated as the explanation of reduced vertical and lateral pressures adjacent to the facing (Murray and Farrar, 1990). The effect of settlement on the vertical pressure σ_w, adjacent to the facing, may be determined from the assumed settlement profile shown in Fig. 7.26.

$$\frac{d\sigma_w}{dz} = \gamma - \frac{2K\sigma_w \tan \phi_w}{x} \tag{3}$$

where K is the ratio of the horizontal to vertical earth pressure, ϕ_w is the friction angle at the wall face, and x and z are defined in Fig. 7.26. Murray and Farrah (1990) suggest that mobilized friction on the rear of a facing to be $2/3 \phi_w'$.

It is possible to arrange for the effective wall friction at the back of the facing to be very high by the deliberate selection of specific facing shapes, Fig. 7.27. The practical benefit of this is that the facing/reinforcement connection tension is reduced to zero and may be eliminated. This, in turn, moves the locus of maximum tension further into the structure away from the facing.

Should the facing settle relative to the reinforced fill, the backfill wall

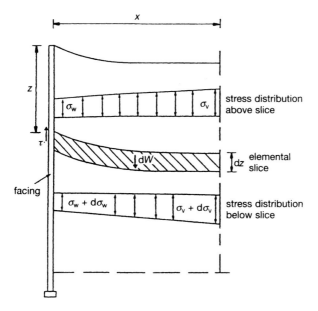

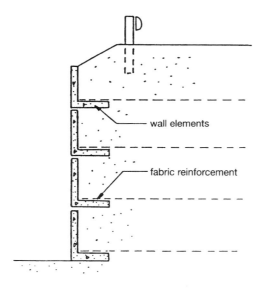

Fig. 7.26. Settlement profile behind reinforced soil wall (after Murray and Farrar, 1990)

friction would be reversed, thereby increasing the overturning force and potentially increasing the loads in the connections. The consequence of wall settlement relative to the backfill has been considered by O'Rourke (1987) and, in the case of conventional structures, failure may occur. In the case of reinforced soil, studies indicate that the pressures under the toe are often less than those under the bulk of the structure, although this is at variance with the accepted trapezoidal or Meyerhoff distributions assumed in analysis. Under these conditions, settlement of the facing relative to the fill is unlikely. However, support for the toe of the wall to resist local settlement and to ensure downward backwall friction may be provided by piling or the use of

Fig. 7.27. 'L' shaped facing (after Broms, 1978)

233

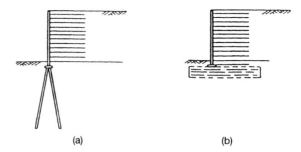

Fig. 7.28. Support methods
under the facing

(a)   (b)

geocell foundations supporting the toe, Fig. 7.28. The construction of conical foundation pads to support the facing as employed in Japan fulfils the same function, Fig. 7.29.

7.13.4 Three-dimensional structures

As a matter of convenience, plane strain conditions are assumed for the analysis of reinforced soil structures, and in the majority of cases the assumption of a two-dimensional analysis is adequate. However, the recent failure of structures in Tennessee indicates that there are conditions where three-dimensional geometry influences structural behaviour.

Figure 7.30 illustrates the case of a reinforced soil wall constructed across a ravine in a mountainous region, which supports a sloping embankment. Three-dimensional finite element analyses were performed to determine the pattern of fill movement and corresponding deformation of the wall. Vector displacements, determined by analysis, are shown for two points on the crest of the wall, either side of the tallest part of the structure. It can be seen that the movements in the ZX plane (i.e. in the plane of the face of the wall) are both towards the centre of the structure. In some conditions, such as with large walls which are curved or articulated in plan or which cross deep or steep-sided ravines, these in-plane movements are significant and must be considered in design.

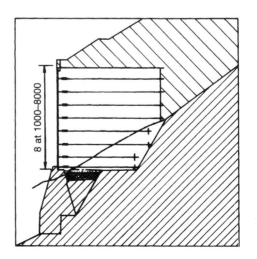

8 at 1000–8000

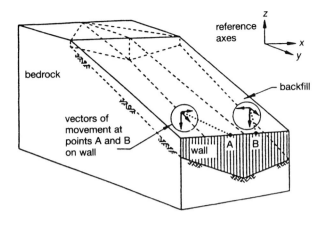

Fig. 7.30. Isometric view of a reinforced soil wall with in-plane movements conveyed to the wall (after O'Rourke and Jones, 1990)

7.14 Trends in construction practice

7.14.1 Proliferation of systems

The market for reinforced soil is showing signs of maturity, identified by the increase in number of construction systems. Early innovations were introduced by the California Highways Department (CALTRANS), one of the most important of which was the introduction of the concept of grid reinforcement. The introduction of grids led directly to the related methods associated with the Hilficker Co. and VSL, both of which use welded wire reinforcements. The CALTRANS initiative culminated in the development of the Georgia Department of Transportation Stabilized Earth system, GASE, in 1981. Other Californian developments have been the tie-anchored timber walls (TAT) and the salvaged guard-rail system (SGR), Fig. 7.31. In Europe, development examples are illustrated by the hybrid systems typified in the Norwegian Tronderblock method and the Austrian anchored earth system, both of which are particularly suited to the scale of construction associated with those countries. Another example of tailoring structures to suit local needs is the LIMI system which has evolved in Finland. This system uses conventional mild steel rebar as the reinforcing element bent into the shape of a hairpin, Fig. 7.32.

Similarly, in Japan new developments applicable to that country's condi-

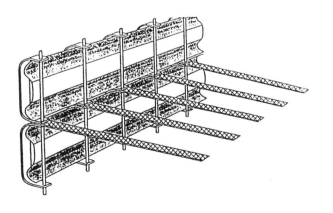

Fig. 7.31. Salvaged guard-rail wall (SGR), California DoT

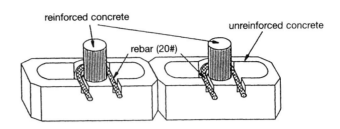

Fig. 7.32. LIMI reinforced soil wall system, Finland

tions and requirements are in evidence based on the need to use fine-grained soil as fill.

7.14.2 Improved forms of reinforcement anchorage

The initial concept of soil reinforcement being solely by friction has been superseded by the introduction of grids and anchors. The use of improved methods of developing soil reinforcement interaction has implications with respect to the nature and quality of the fill which can be used, and these developments are important, particularly where good quality frictional fill is difficult to obtain. Recent developments in this area are the innovations associated with anchored earth systems and the use of composite reinforcements formed from high strength grids used in association with drainage geotextiles.

7.14.3 Use of cohesive fills

The use of nonwoven geotextiles as reinforcement for cohesive fill is a feature of the Swiss Texomural method used to construct steep slopes. Control of the face profile of these structures is achieved with the aid of sacrificial pre-formed metal welded formers lined with a carefully selected geotextile used to encourage vegetation. The development of suitable ground cover as a permanent facing is a feature of the system.

The use of cohesive fill to form steep slopes is practised in Japan. The method relies on geotextile reinforcement providing suction forces in the clay fill. Long-term stability is seen to rely on provision of a continuous rigid facing placed after the backfill of the structure is complete, Fig. 7.33.

7.14.4 Ease of construction

Reinforced soil is attractive because of the economic benefits it offers. The economic equation can become unbalanced and the benefits lost if the

Fig. 7.33. Geotextile reinforced clay embankment with rigid facing (after Tatsuoka et al., 1990)

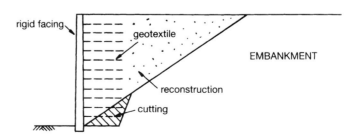

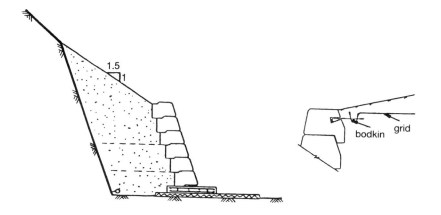

Fig. 7.34. Tronderblock precast concrete facings used for construction of flow height gravity retaining walls

particular construction system adopted is inefficient. The need for buildability remains a major element in the successful implementation of reinforced soil structures. The importance of ease of construction and the associated logistical parameters associated with construction of forest roads in mountainous terrain have been described by Keller (1990). The growth of reinforced soil techniques for temporary works suggests that the criteria which determine the economical construction of reinforced soil are better understood and practised.

7.14.5 Aesthetics

Another method of identifying maturity in reinforced soil technology is the emphasis now being placed on the aesthetic quality of construction. The quality of finishes and construction materials and the appreciation of environmental factors conducive to the sustained growth of vegetation associated with reinforced soil is now as important as the base technology, and receives equal treatment in the trade literature. The range of finishes and architectural treatments appears unlimited.

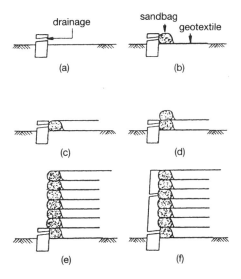

Fig. 7.35. Standard construction procedure of GRS–RW system: (a) base concrete; (b) laying geotextile and sandbag; (c) backfill and compaction; (d) second layer; (e) laying completed; (f) concrete facing erected

237

7.14.6 Hybrid structures

The use of reinforced soil techniques in conjunction with conventional construction is a logical development. The use of tailed gabions is illustrated in Fig. 3.22(b). A similar technique has been developed using the Norwegian Tronderblock system. Tronderblock precast concrete facings are used for the construction of flow height gravity retaining walls, Fig. 7.34. The versatility of the method can be improved by introducing horizontal layers of geogrid reinforcement connected to separate facing elements. The use of reinforcement extends the range for the construction technique from 3 m to 5·5 m.

In Japan, a geosynthetic-reinforced soil retaining wall system (GRS–RW system) with a continuous rigid cast-in-place concrete panel has been used to construct permanent retaining wall structures supporting railway embankments (Tatsuoka *et al.*, 1992). Fig. 7.35 shows the standard construction method of the GRS–RW system.

7.14.7 Vertical reinforcement system

A method of constructing soil structures using corrugated steel elements orientated in a vertical plane is shown in Fig. 7.36. The system, which has been developed in Canada, is rigid in the vertical plane. The behaviour is described by Schlosser and Delage (1987) as that of a rigid beam; indeed, the structural effect illustrated in Fig. 7.36 is that of a Howe beam used by Coyne (1945).

Fig. 7.36. Vertical reinforcing system (after Schlosser and Delage, 1987)

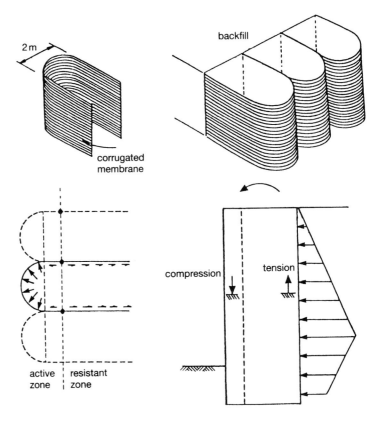

7.15 Education

In the construction of conventional earth retaining structures, the importance of good site practice is recognized as essential to the production of an acceptable durable product. With repeated practice over a long timescale, good construction technique, associated primarily with the production of good quality concrete, is an established broad-based skill within the construction industry. On account of the relatively short life of the reinforced soil technique, wide knowledge of construction know-how is confined generally to specialist practitioners. With the major growth in interest in the concept of internally stabilized structures, there is an identified need for construction training, both among designers but more importantly among contractors and site staff.

The proper education of site staff in the production of good concrete is accepted as essential to quality assurance in the construction industry. Similarly, education of those responsible for construction of reinforced soil structures is required. It is possible to conclude that the construction techniques associated with reinforced soil are not as complex as those associated with the production of concrete structures, but they cannot be ignored. The basic concepts of reinforced soil and the sensitivity of the design assumptions to construction practices must be understood by those responsible for the construction. Without this knowledge the technical and financial benefits offered by the use of reinforced soil are in jeopardy, and use of the technique will not reach its full potential.

Because reinforced soil is essentially an elegantly simple concept, the basics can be identified and transmitted quickly to those not familiar with the technique. The most productive and simple method used by some practitioners is through training videos, giving detailed construction systems, supported by the presence of experienced personnel seconded to site during the critical early part of new structures, or new reinforcing material products. Demonstration projects aimed at identifying construction technology have become accepted procedure.

7.16 References

BRITISH STANDARDS INSTITUTION (1995). *Code of Practice for strengthened/reinforced soils and other fills*. BSI, London, BS 8006.

BROMS B. B. (1978). Design of fabric-reinforced retaining structures. *Symp. on Earth Reinforcement*. ASCE Special Convention, Pittsburg, 282–304.

BRUCE D. A. and JEWELL R. A. (1986 and 1987). Soil nailing: applications and practice. *Ground Engng*, Part I, **19**, No. 8, 10–15; Part 2, **20**, No. 1, 21–23.

DEPARTMENT OF TRANSPORT (1978). *Reinforcing earth retaining walls and bridge abutments*. Technical Memorandum (Bridges) BE 3/78, London.

DEPARTMENT OF TRANSPORT (1991). Manual of contract documents for highway works, *Specification for highway works*, **1**, HMSO, London.

FEDERAL HIGHWAY ADMINISTRATION (1993a). *Recommendations Clouterre 1991* (English Translation), Soil Nailing Recommendations 1991, Report No. FHWA-SA-93-026, Washington DC, 302.

FEDERAL HIGHWAY ADMINISTRATION (1993b). *FWHA international scanning tour on soil nailing*. US Department of Transportation, Washington DC, 45.

FINDLAY T. W. (1978). Performance of a reinforced earth structure at Granton. *Ground Engng*, **2**, No. 7, 42–44.

GÄSSLER G. (1990). McGowan, Andrawes and Yeo (eds). *In-situ* techniques of reinforced soil. *Performance of reinforced soil structures*. Thomas Telford, London, 185–196.

HAMBLEY E. C. (1979). *Bridge foundations and substructures*. Building Research Establishment Report, Department of the Environment.

JAPAN HIGHWAY PUBLIC CORPORATION (1987). *Guide for design and construction on reinforced slope with steel bars*. Tokyo, 33.

JEWELL R. A. (1990). Strengths and deformations in reinforced soil design. *4th Int. Conf. on Geotextiles and Related Products*, The Hague, 77.

JONES C. J. F. P. (1978). The York method of reinforced earth construction. *ASCE Symp. on Earth Reinforcement*, Pittsburgh.

JONES C. J. F. P. (1979). Lateral earth pressures acting on the facing units of reinforced earth. *Proc. Int. Conf. on Soil Reinforcement*, Paris, **2**, 445–449.

JONES C. J. F. P. (1988). Yamanouchi, Minura and Ochai (eds). Predicting the behaviour of reinforced soil structures. *Theory and Practice of Earth Reinforcement*. Balkema, 535–540.

JONES C. J. F. P., CRIPWELL J. B. and BUSH D. I. (1990). Reinforced earth trial structure for Dewsbury Road. *Proc. Instn Civ. Engrs*, Part 1, April, **88**, 321–345.

JONES C. P. D. (1990). McGowan, Andrawes and Yeo (eds). *In-situ* techniques for reinforced soil. *Performance of reinforced soil structures*. Thomas Telford, London, 277–282.

KELLER G. R. (1990). Lambe and Hansen (eds). Alternative wall and reinforced fill experiences on forest roads. *Design and performance of earth retaining structures*. ASCE Geotech. Special Publication 25, 155–169.

LEE K., JONES C. J. F. P., SULLIVAN W. R. and TROLLINGER W. (1995). Failure and deformation of four reinforced soil walls in eastern Tennessee. *Géotechnique*, **45**, No. 4, 749–752.

LEECE R. B. (1979). Reinforced earth highway retaining walls in South Wales. *C.R. Coll. Int. Renforcement des Sols*, Paris.

LEFLAIVRE E., KHAY M. and BLIUET J. C. (1983). Un nouveau material: le Texsol. *Bulletin de Liaison du Laboratoire des Ponts et Chaussees*, No. 125, 105–114.

MERCER F. B., ANDRAWES K. Z., McGOWN A. and HYTIRIS N. (1984). A new method of soil stabilization. *Polymer grid reinforcement*. Thomas Telford, London, 244–249.

McGOWN A., MURRAY R. T. and ANDRAWES K. Z. (1987). *Influence of wall yielding on lateral stresses in unreinforced and reinforced fills*. Transport and Road Research Laboratory, Research Report 113, Crowthorne, 14.

MITCHELL J. K. and VILLET W. C. B. (1987). *Reinforcement of earth slopes and embankments*. NCHRP Report 290, Transport Research Board, Washington DC, 323.

MURRAY R. T. and FARRAR D. M. (1990). Reinforced earth wall on the M25 motorway at Waltham Cross. *Proc. Instn Civ. Engrs*, Part 1, **88**, April, 261–281.

MYLES B. (1993). *Practical aspects of soil nailing*. Amerad Jord, Svenska Geotekniska Foreningen, Uppsala, Sept.

NAYLOR D. J. (1978). A study of Reinforced Earth walls allowing slip strip. *Symp. on Earth Reinforcement*, American Society of Civil Engineers Speciality Conference, Pittsburgh, 618–643.

OKASAN KOGYO (1989). *Civil engineering works catalogue.* Technical Guide, Tokyo.

O'ROURKE T. D. (1987). Lateral stability of compressible walls. *Géotechnique*, **37**, No. 2, 145–149.

O'ROURKE T. D. and JONES C. J. F. P. (1990). Lambe and Hansen (eds.) 1970–90 overview of earth retention systems. *Design and performance of earth retaining structures.* ASCE Geotech Special Publication 25, 22–51.

PLUMELLE C., SCHLOSSER F., DELAGE P. and KNOCHENMUS G. (1990). *French National Research Project on Soil Nailing: Clouterre.* ASCE, Cornell, 660–675.

POWELL G. E. and WATKINS A. T. (1990). McGowan, Andrawes and Yeo (eds). Improvement of marginally stable existing slopes by soil nailing in Hong Kong. *Performance of reinforced soil structures.* Thomas Telford, London, 241–247.

RABEJAC S. and TOUDIC P. (1974). Construction d'un mur de soutènement entre Versailles-Chantiers et Versailles-Matelots. *Revue Générale des Chemins de Fer*, **93** ème année, 232–237.

SCHLOSSER F. S. (1990). McGowan, Andrawes and Yeo (eds). *In-situ* techniques for reinforced soil—discussion. *Performance of reinforced soil structures.* Thomas Telford, London, 279–280.

SCHLOSSER F. S. and DELAGE P. (1987). Reinforced soil retaining structures and polymeric materials. *Proc. NATO Advanced Research Workshop on Application of Polymeric Reinforcement in Soil Retaining Structures*, Royal Military College of Canada, Kingston, Ontario, NATO ASI Series, 3–68.

SHEN C., BANG S., RONSTAD K. M., KILCHIN L. and DE NATALE J. S. (1981). Field measurements of an earth support system. *Geotech. Engng Div., ASCE*, **107**, No. GT12, Dec. 1625–1642.

TATSUOKA F. (1992). Roles of facing rigidity in soil reinforcing. *Proc. Int. Symp. on Earth Reinforcement Practice*, Kyushu, Japan, **2**, 831–870.

TATSUOKA F., MURATA O., TATEYAMA M., NAKAMURA K., TAMURA Y., LING H. I., IWASAKI K. and YAMAUCHI H. (1990). McGowan, Andrawes and Yeo (eds). Reinforcing steep clay slopes with a non-woven geotextile. *Performance of reinforced soil structures.* Thomas Telford, London, 141–146.

VIDAL H. (1966). Diffusion Restpeinte de la Terre Armée. *La Terre Armée, Annales de L'Institut Technique du Bâtiment et des Travaux Publics*, **19**, No. 223–4 (July/August), 888–939.

VIDAL H. (1978). The development and future of reinforced earth. *ASCE Symp. on Earth Reinforcement*, Pittsburgh.

8

Construction details

8.1 Introduction

Although the form of any soil structure may vary, the constructional elements required to produce different structures are often similar. In keeping with other forms of construction, poor constructional details will produce an inadequate structure, while good detailing will ensure success. The difference between a good and a poor detail can be subtle, and weaknesses and deficiencies may become apparent only during the construction phase or later during the life of the structure.

Many situations produce common constructional problems. The constructional details shown below, although not necessarily the best possible, have been shown to be efficient and effective in some conditions. In some cases the details represent a compromise between constructional efficiency and durability criteria, in others a compromise has had to be made between the structural requirements and aesthetics.

8.2 Foundations

8.2.1 Geocell mattress

One solution to overcome inadequate bearing capacity beneath the footing of a soil structure is to use a foundation mattress. Typical construction details of a 1 m thick mattress are shown in Figs 8.1–8.10. Greater thicknesses may be attained by constructing successive mattresses on top of one another.

8.2.1.1 Assembly of geocell mattress

(a) Lay out the base and fix the side, two ends, the centre diaphragm and the side diaphragms to the base.
(b) Lift ends of diaphragms through 90° so that they stand vertically, Fig. 8.1.
(c) Form joint between panels by inserting the securing pin through the interlocked panels, Figs 8.2 and 8.7.
(d) Position the adjacent mattress. The joints may be formed in the same way except that the ends and the side diaphragms from the first mattress are fastened into the joint along the side panel of the second, Fig. 8.3.

Construction details

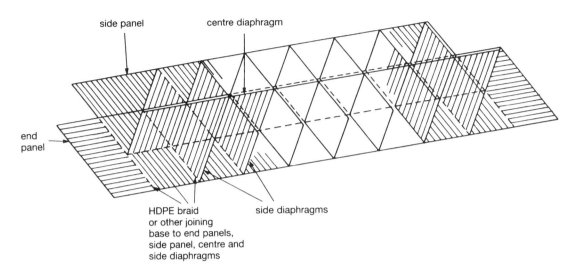

side panel

centre diaphragm

end panel

HDPE braid
or other joining
base to end panels,
side panel, centre and
side diaphragms

side diaphragms

*Fig. 8.1. 8 m × 2 m × 1 m high
geocell matress*

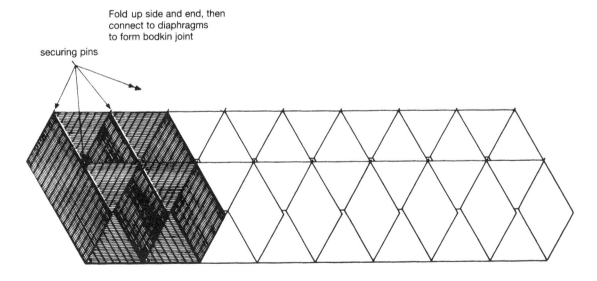

Fold up side and end, then
connect to diaphragms
to form bodkin joint

securing pins

Fig. 8.2. First stage assembly

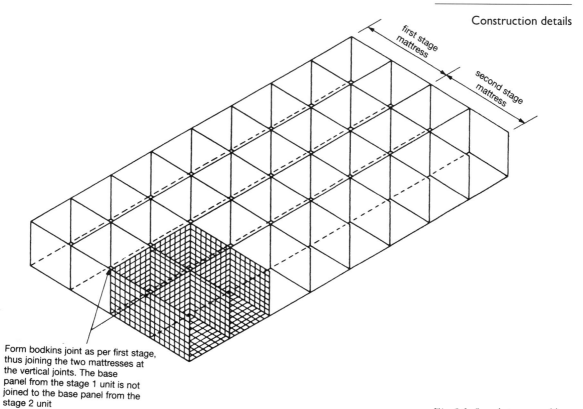

Form bodkins joint as per first stage, thus joining the two mattresses at the vertical joints. The base panel from the stage 1 unit is not joined to the base panel from the stage 2 unit

Fig. 8.3. Second stage assembly

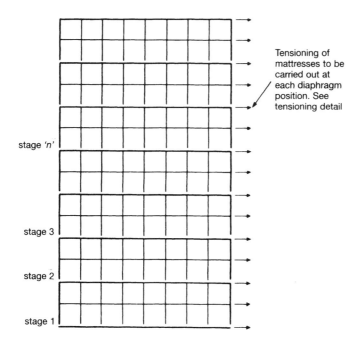

stage 'n'

stage 3

stage 2

stage 1

Tensioning of mattresses to be carried out at each diaphragm position. See tensioning detail

Fig. 8.4. Plan on foundation raft

245

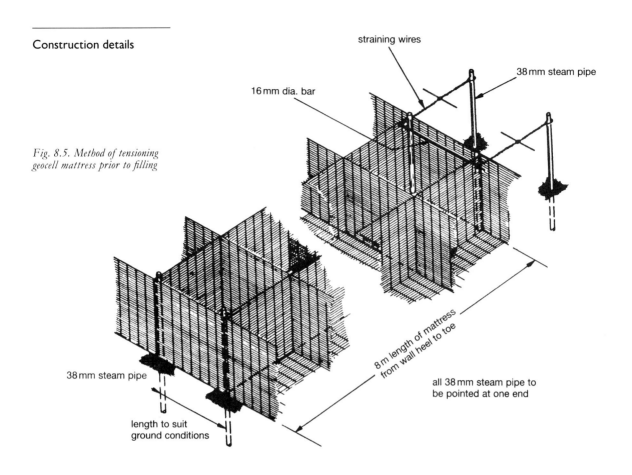

Fig. 8.5. Method of tensioning geocell mattress prior to filling

straining wires

38 mm steam pipe

16 mm dia. bar

8 m length of mattress from wall heel to toe

all 38 mm steam pipe to be pointed at one end

38 mm steam pipe

length to suit ground conditions

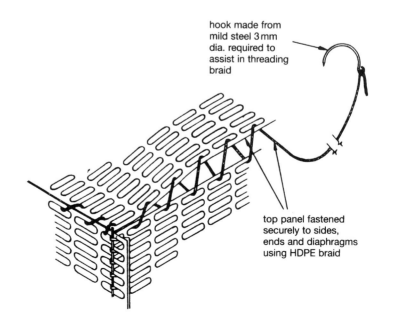

hook made from mild steel 3 mm dia. required to assist in threading braid

top panel fastened securely to sides, ends and diaphragms using HDPE braid

Fig. 8.6. Top panel fixing

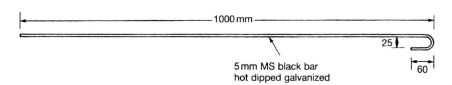

1000 mm

25

5 mm MS black bar
hot dipped galvanized

60

Fig. 8.7. Details of securing pin

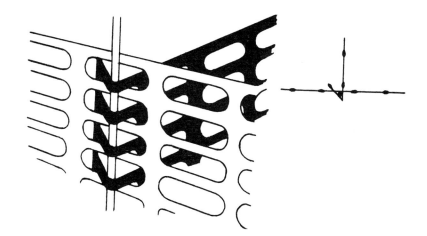

*Fig. 8.8. Joint for diaphragm to
side or end panel*

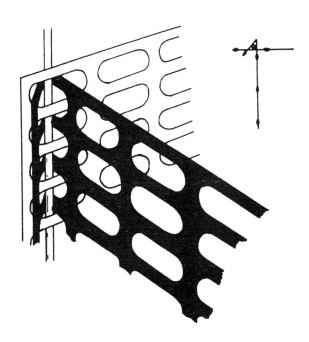

Fig. 8.9. Joint for side to end

247

Construction details

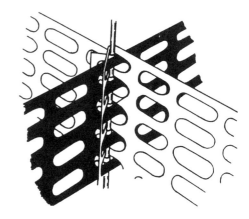

Fig. 8.10. Joint for two diaphragms to centre panel

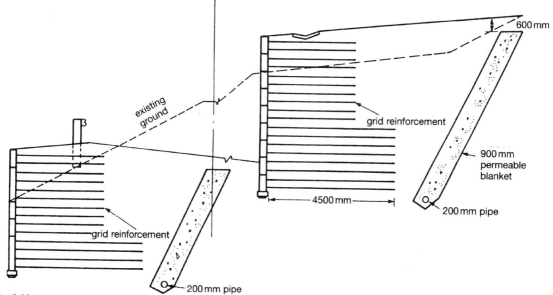

existing ground

grid reinforcement

600 mm

900 mm permeable blanket

4500 mm

200 mm pipe

grid reinforcement

200 mm pipe

Fig. 8.11

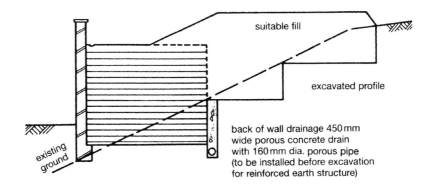

suitable fill

excavated profile

back of wall drainage 450 mm wide porous concrete drain with 160 mm dia. porous pipe (to be installed before excavation for reinforced earth structure)

existing ground

Fig. 8.12

(*e*) The two longitudinal vertical diaphragms of the first mattress (centre diaphragms and side) may be tensioned as shown in Figs 8.4 and 8.5. The process of erection and tensioning is repeated for all subsequent mattresses.

(*f*) The fill material may be placed in the compartments by hand or mechanically. Care is required to prevent the collapse of the mattress cells. When the geocell mattress is acting as a drainage layer as well as a foundation element, the fill used should be angular, hard durable stone, well-graded with size varying between 50 mm and 150 mm. In other conditions the fill should be selected in accordance with the criteria covered in Chapter 5, Materials.

(*g*) Filling of the geocell mattresses may be accomplished in three lifts of one-third height each lift.

(*h*) Once the geocell mattresses have been filled and compacted, a lid to the mattress may be attached using braid, Fig. 8.6.

8.3 Drainage

Proper drainage of reinforced soil and structures is essential. The drainage for each structure or condition should be considered separately on its merits. Typical drainage details are shown in Figs 8.11–8.21, as follows:

8.3.1 Drainage behind the soil structure, Figs 8.11 and 8.12.

8.3.2 Drainage beneath the structure, Fig. 8.13.

8.3.3 Combined drainage and geocell mattress, Fig. 8.14.

8.3.4 Wall drainage, Figs 8.15–8.17

8.3.5 Drainage above a reinforced soil structure, Figs 8.18–8.21.

8.4 Facings

Facings may take a variety of forms dependent upon the design requirements.

8.4.1 Full-height facing

8.4.1.1 Double-tee concrete beam
Prestressed double-tee concrete beam used as a full-height facing, Figs 8.22–8.27. Typical notes relating to the manufacture and erection of this facing unit include:

Manufacture
(*a*) Concrete to be class 45/20 in prestressed units.
(*b*) Minimum cube strength at transfer to be 30 N/mm^2.
(*c*) Concrete to be placed in one continuous operation.
(*d*) Strand and wire to be low relaxation and in accordance with BS 3617: 1971.
(*e*) Characteristic load of strand to be 184 kN and initial force to be 128·8 kN, characteristic strength of wire to be 1570 N/mm^2 and initial prestress to be 1099 N/mm^2 giving an initial force of 15·1 kN per wire.

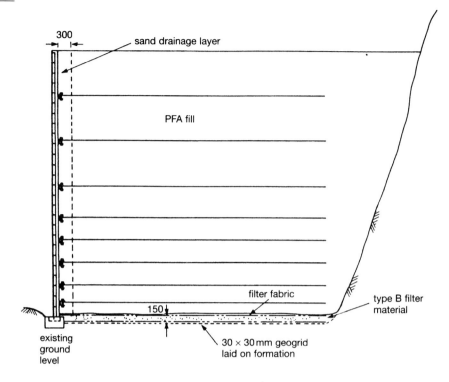

300

sand drainage layer

PFA fill

filter fabric

150

type B filter
material

existing
ground
level

30 × 30 mm geogrid
laid on formation

*Fig. 8.13. Drainage beneath
reinforced structure*

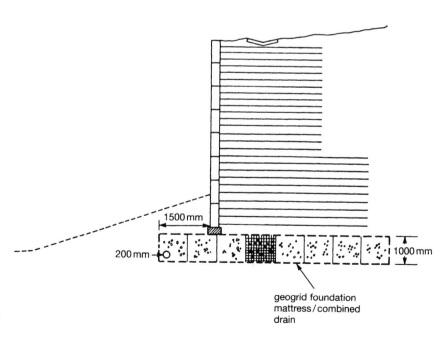

1500 mm

200 mm

1000 mm

geogrid foundation
mattress / combined
drain

*Fig. 8.14. Combined geogrid
foundation and drainage blanket*

250

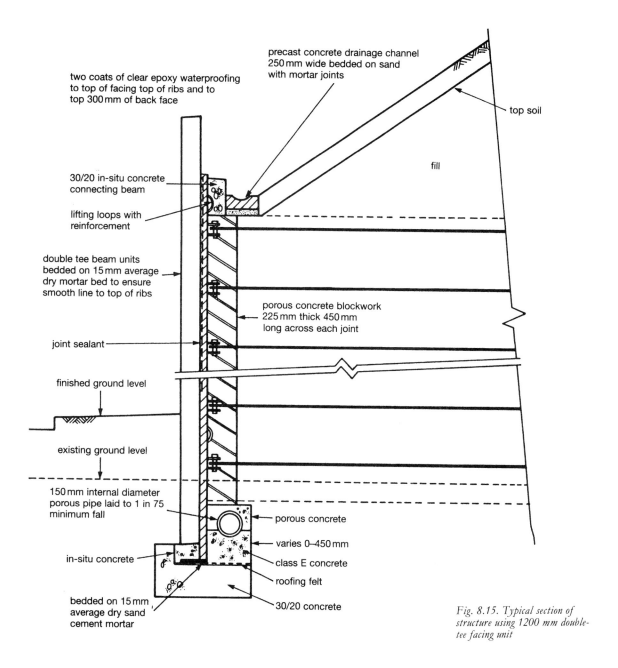

two coats of clear epoxy waterproofing to top of facing top of ribs and to top 300 mm of back face

precast concrete drainage channel 250 mm wide bedded on sand with mortar joints

top soil

fill

30/20 in-situ concrete connecting beam

lifting loops with reinforcement

double tee beam units bedded on 15 mm average dry mortar bed to ensure smooth line to top of ribs

porous concrete blockwork 225 mm thick 450 mm long across each joint

joint sealant

finished ground level

existing ground level

150 mm internal diameter porous pipe laid to 1 in 75 minimum fall

porous concrete

varies 0–450 mm

in-situ concrete

class E concrete

roofing felt

30/20 concrete

bedded on 15 mm average dry sand cement mortar

Fig. 8.15. Typical section of structure using 1200 mm double-tee facing unit

251

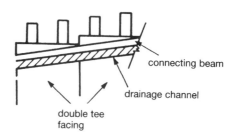

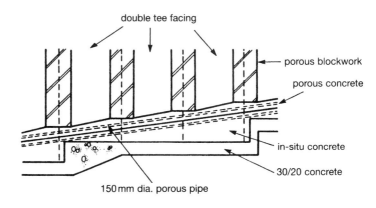

Fig. 8.16. Position of drainage channel

Fig. 8.17. Position of drainage pipe

Fig. 8.18. Full-height structure

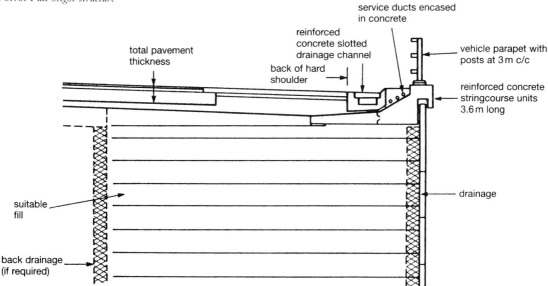

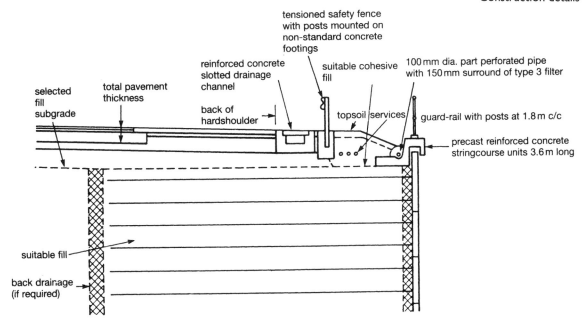

selected
fill
subgrade

total pavement
thickness

reinforced concrete
slotted drainage
channel

back of
hardshoulder

tensioned safety fence
with posts mounted on
non-standard concrete
footings

suitable cohesive
fill

topsoil services

100 mm dia. part perforated pipe
with 150 mm surround of type 3 filter

guard-rail with posts at 1.8 m c/c

precast reinforced concrete
stringcourse units 3.6 m long

suitable fill

back drainage
(if required)

Fig. 8.19. Part height structure

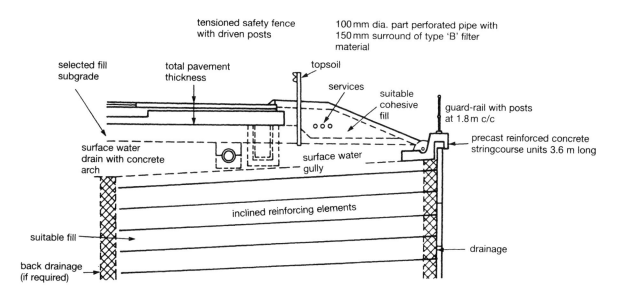

selected fill
subgrade

total pavement
thickness

tensioned safety fence
with driven posts

topsoil

services

suitable
cohesive
fill

100 mm dia. part perforated pipe with
150 mm surround of type 'B' filter
material

guard-rail with posts
at 1.8 m c/c

precast reinforced concrete
stringcourse units 3.6 m long

surface water
drain with concrete
arch

surface water
gully

inclined reinforcing elements

suitable fill

drainage

back drainage
(if required)

Fig. 8.20. Part height structure

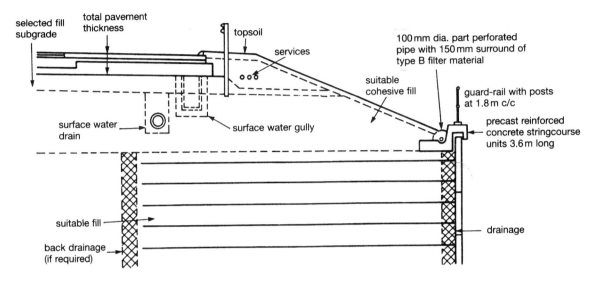

Fig. 8.21. Part height structure

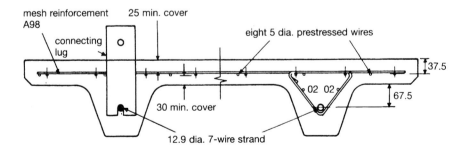

Fig. 8.22. Double-tee facing unit

(*f*) Ends of strand and wire to be cut flush with ends of panels.

(*g*) Reinforcement to be mild steel to BS 4449.

(*h*) Mesh reinforcement to be lapped at least one pitch such that concrete cover is maintained.

(*i*) Concrete cover to reinforcement to be 25 mm (min.) unless otherwise stated.

(*j*) Stainless tying wire to be used throughout.

(*k*) Panels may only be lifted and supported with ribs down and by means and at positions approved by the Engineer.

(*l*) Tolerances are to be in accordance with CP 116, clause 407, except with regard to:

Length ± 6 mm; Cross-section (each direction) ± 6 mm;
Variation in camber 6 mm.

(*m*) Concrete finishes to be F3 – all formed surfaces; U2 – all unformed surfaces.

(*n*) Two coats of approved clear waterproofing to be applied to top face of each unit.

(*o*) The maximum length of beams produced from this design to be 6·25 m (in this example).

(*p*) Two coats of pitch epoxy waterproofing are to be applied to the rear face of each unit.

Construction requirements

(*a*) During the entire operation of placing the reinforced fill material, the facing units shall be supported only at the top of the unit. The support system shall be capable of providing a horizontal force of 10 kN per unit with zero deflection. Differential movement between units shall also be prevented by means of a rigid waling which shall only be removed three days after placing concrete to the connecting beam. Loading forces are to be taken directly against the ribs (other support conditions may be required in other circumstances).

(*b*) Folding wedges shall be placed at the bottom of the facing units in the position shown to prevent movement of the facing units during initial placing of the fill, Fig. 8.27. When the fill reaches the level of the third row of reinforcing elements from the bottom the wedges shall be removed. The *in situ* concrete in front of the units shall not be placed until the support system referred to in (a) above has been removed.

8.4.1.2 Prestressed concrete unit

Full-height prestressed facing unit, faced with masonry, Fig. 8.28.

8.4.1.3 King post and panel

Full-height facing formed from post and infill planking: for utilitarian structures the basic elements are adequate; alternatively, a variety of architectural finishes may be attached, Fig. 8.29.

Full-height units may be used conveniently only to a height of 10 m; above this height, elemental facing units will usually be required. In order to erect a full-height facing unit, temporary propping during part of the backfilling is required; typical propping details are shown in Fig. 8.30.

255

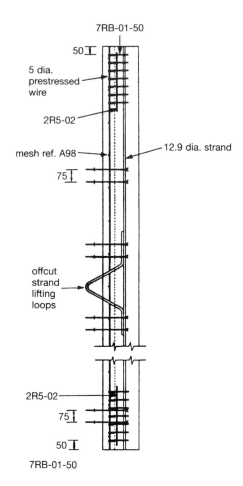

Fig. 8.23. Section

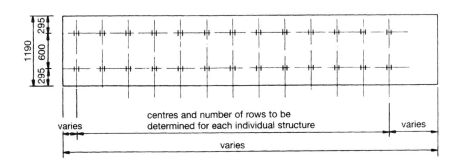

Fig. 8.24

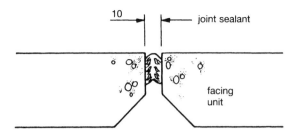

10 — joint sealant

facing unit

Fig. 8.25. Joint between units

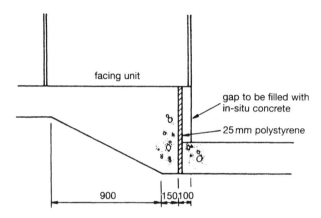

facing unit

gap to be filled with in-situ concrete

25 mm polystyrene

900   150 100

Fig. 8.26. Expansion joint in foundation

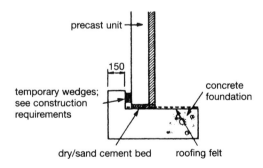

precast unit

150

temporary wedges; see construction requirements

concrete foundation

dry/sand cement bed   roofing felt

Fig. 8.27. Wedging detail

257

Construction details

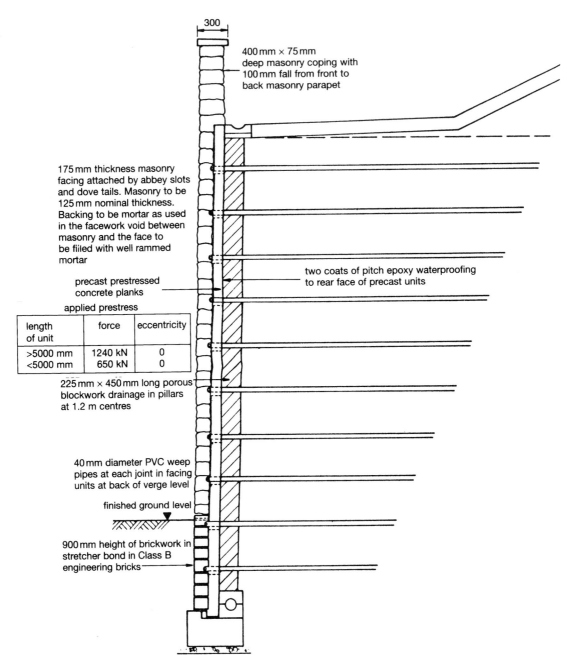

300

400 mm × 75 mm
deep masonry coping with
100 mm fall from front to
back masonry parapet

175 mm thickness masonry
facing attached by abbey slots
and dove tails. Masonry to be
125 mm nominal thickness.
Backing to be mortar as used
in the facework void between
masonry and the face to
be filled with well rammed
mortar

two coats of pitch epoxy waterproofing
to rear face of precast units

precast prestressed
concrete planks

applied prestress

length of unit	force	eccentricity
>5000 mm	1240 kN	0
<5000 mm	650 kN	0

225 mm × 450 mm long porous
blockwork drainage in pillars
at 1.2 m centres

40 mm diameter PVC weep
pipes at each joint in facing
units at back of verge level

finished ground level

900 mm height of brickwork in
stretcher bond in Class B
engineering bricks

*Fig. 8.28. Full-height
prestressed facing unit, faced with
masonry*

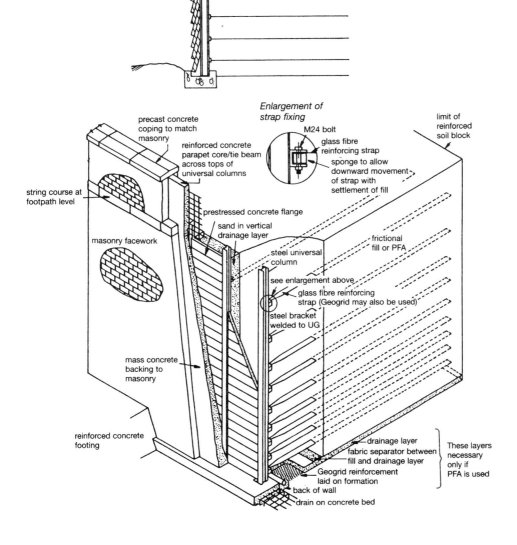

vertical steel joist

infill precast concrete planks

tiling

Enlargement of strap fixing

M24 bolt

glass fibre reinforcing strap

sponge to allow downward movement of strap with settlement of fill

precast concrete coping to match masonry

reinforced concrete parapet core/tie beam across tops of universal columns

limit of reinforced soil block

string course at footpath level

prestressed concrete flange

sand in vertical drainage layer

frictional fill or PFA

masonry facework

steel universal column

see enlargement above

glass fibre reinforcing strap (Geogrid may also be used)

steel bracket welded to UG

mass concrete backing to masonry

reinforced concrete footing

drainage layer

fabric separator between fill and drainage layer

Geogrid reinforcement laid on formation

back of wall

drain on concrete bed

These layers necessary only if PFA is used

Fig. 8.29. Full-height facing formed from post and infill planking

259

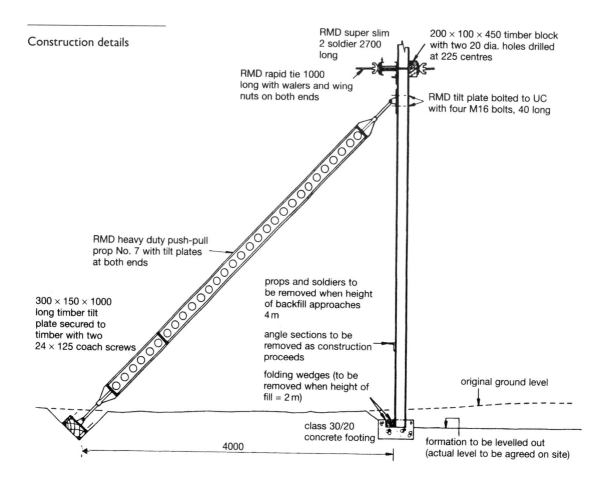

RMD super slim
2 soldier 2700
long

200 × 100 × 450 timber block
with two 20 dia. holes drilled
at 225 centres

RMD rapid tie 1000
long with walers and wing
nuts on both ends

RMD tilt plate bolted to UC
with four M16 bolts, 40 long

RMD heavy duty push-pull
prop No. 7 with tilt plates
at both ends

300 × 150 × 1000
long timber tilt
plate secured to
timber with two
24 × 125 coach screws

props and soldiers to
be removed when height
of backfill approaches
4 m

angle sections to be
removed as construction
proceeds

folding wedges (to be
removed when height of
fill = 2 m)

original ground level

class 30/20
concrete footing

4000

formation to be levelled out
(actual level to be agreed on site)

Fig. 8.30. Detail of propping

8.4.2 Elemental facing units

Brick retaining wall using fabric of geogrid reinforcement, Fig. 8.31.

Concrete blockwork facing used with steel grid reinforcement, Figs 8.32, 8.33 and 8.34.

Glass-fibre reinforced cement facing used with galvanized steel strip or glass-fibre reinforced plastic reinforcement, Figs 8.35, 8.36 and 8.37.

Method of erection
(*a*) Cast 100 mm binding.
(*b*) Erect bottom level of units and half units spacing with damp-proof course felt to a thickness of 6 mm on mortar bed; bolt together.
(*c*) Arrange 2 m long lengths of PVC pipe to pass through 40 mm holes in the flange of the facing unit and stand vertically.
(*d*) Insert first 15 mm diameter mild steel bar into PVC pipe; stagger alternate lengths.
(*e*) Cast porous concrete to a height of 150 mm either side of units.
(*f*) Fix next level of units, spacing with felt; fix reinforcement straps.

(g) Backfill in 300 mm layers compacting to specification.
(h) Extend PVC pipes as necessary using 2 m extension pieces; extend reinforcement with 2 m lengths as necessary.
(i) At end of day's work, grout up PVC tubes to a point 100 mm short of top of reinforcement.

Reinforced concrete beam facing used with steel grid reinforcement, Figs 8.38–8.42 (after California Highway Authority).

Timber facing used with steel grid reinforcement, Figs 8.43–8.46 (after California Department of Transportation).

8.5 Bridge abutments and bank seats
8.5.1 Abutment details of footbridge
Abutment details of footbridge, Figs 8.47 and 8.48.

8.5.2 Abutment details for highway bridge
Abutment details for highway bridge, Figs 8.49 and 8.50.

8.6 Reinforcements and reinforcement connections
8.6.1 Strip reinforcement
(a) Fixed connection associated with the concertina method of construction or the telescope method, Fig. 8.51.
(b) Moving or sliding connection associated with the sliding or York method of construction, Figs 8.52, 8.53 and 8.54.

8.6.2 Grid reinforcement
(a) Fixed connection associated with the telescope method of construction, Figs 8.40 and 8.55.
(b) Moving connection associated with sliding method of construction, Figs 8.56 and 8.57.

8.6.3 Anchor reinforcement
Moving connection associated with the sliding method of construction, Figs 8.58 and 8.59.

8.6.4 Combined strip/anchor reinforcement
Fixed or moving connections associated with the telescope method or the sliding method of construction, Fig. 8.60.

8.6.5 Fabric reinforcement
Fabric wrapped around a rolling block in the concertina method of construction, Fig. 8.61.

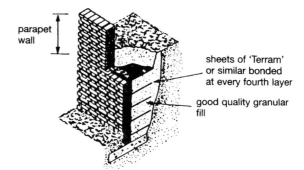

'English Bond' 225 mm
brickwork

parapet
wall

sheets of 'Terram'
or similar bonded
at every fourth layer

good quality granular
fill

*Fig. 8.31. Fabric reinforced
brick retaining wall*

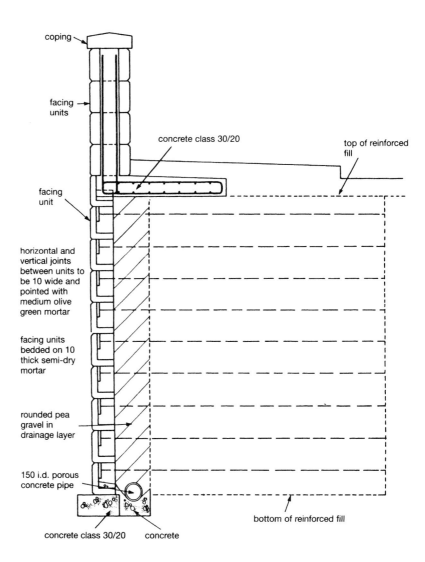

coping

facing
units

concrete class 30/20

top of reinforced
fill

facing
unit

horizontal and
vertical joints
between units to
be 10 wide and
pointed with
medium olive
green mortar

facing units
bedded on 10
thick semi-dry
mortar

rounded pea
gravel in
drainage layer

150 i.d. porous
concrete pipe

bottom of reinforced fill

concrete class 30/20   concrete

*Fig. 8.32. Concrete blockwork
faced structure*

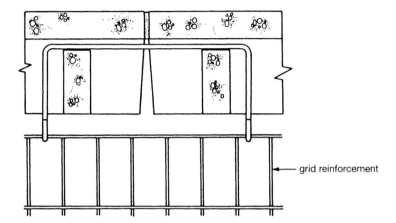

grid reinforcement

Fig. 8.33. Plan

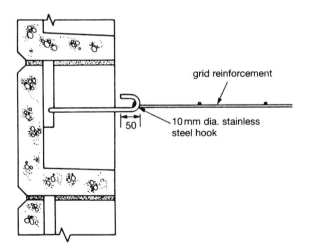

grid reinforcement

10 mm dia. stainless
steel hook

50

Fig. 8.34. Section

Construction details

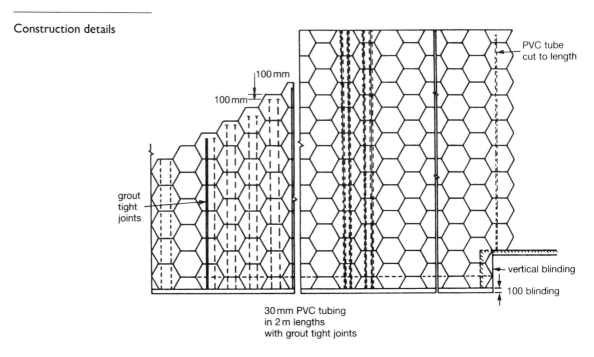

100 mm

100 mm

grout
tight
joints

PVC tube
cut to length

vertical blinding

100 blinding

30 mm PVC tubing
in 2 m lengths
with grout tight joints

Fig. 8.35. Detailed elevation

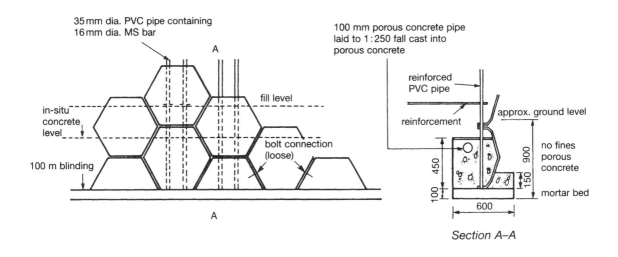

35 mm dia. PVC pipe containing
16 mm dia. MS bar

A

in-situ
concrete
level

fill level

bolt connection
(loose)

100 m blinding

A

100 mm porous concrete pipe
laid to 1 : 250 fall cast into
porous concrete

reinforced
PVC pipe

reinforcement

approx. ground level

no fines
porous
concrete

mortar bed

450

900

100

150

600

Section A–A

Fig. 8.36. Elevation

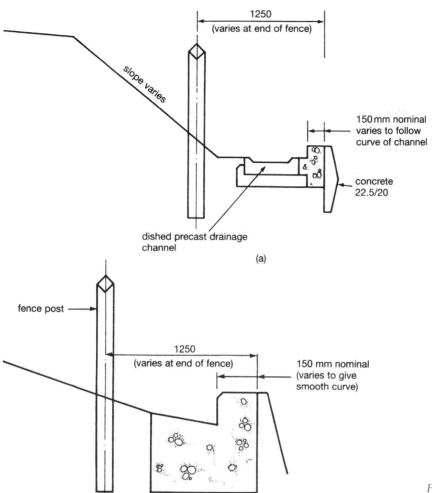

1250
(varies at end of fence)

slope varies

150 mm nominal
varies to follow
curve of channel

concrete
22.5/20

dished precast drainage
channel

(a)

fence post

1250
(varies at end of fence)

150 mm nominal
(varies to give
smooth curve)

*Fig. 8.37. Drainage detail:
(a) type A; (b) type B*

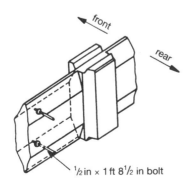

front

rear

$\frac{1}{2}$ in × 1 ft 8$\frac{1}{2}$ in bolt

Fig. 8.38. Isometric

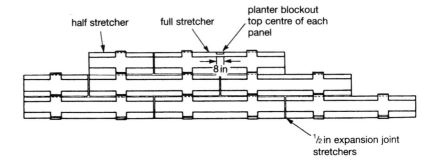

Fig. 8.39. Typical stacking arrangements

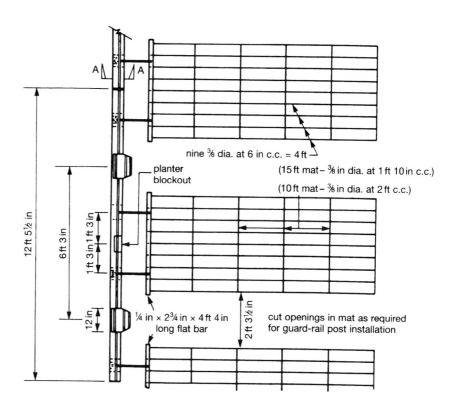

Fig. 8.40. Mechanically reinforced embankment

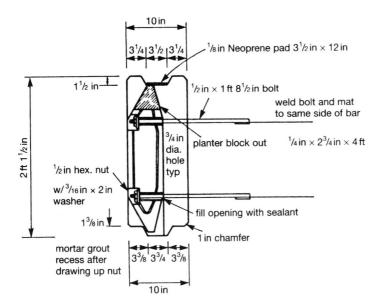

Fig. 8.41. Section A–A

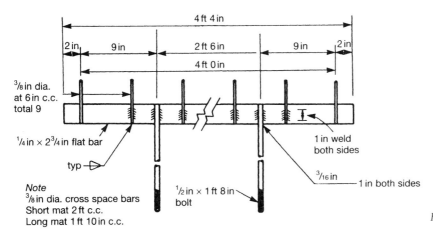

Fig. 8.42. Plan: bolt connector

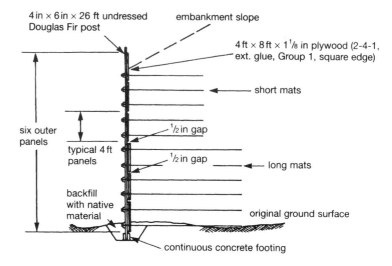

Note
Leave ¹/₂ in gap between each upper and lower panel and above concrete footing for all plywood panels

4 in × 6 in × 26 ft undressed Douglas Fir post

embankment slope

4 ft × 8 ft × 1¹/₈ in plywood (2-4-1, ext. glue, Group 1, square edge)

short mats

six outer panels

¹/₂ in gap

typical 4 ft panels

¹/₂ in gap

long mats

backfill with native material

original ground surface

continuous concrete footing

Fig. 8.43. Timber facing used with steel grid reinforcement

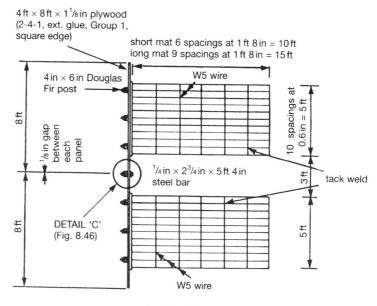

4 ft × 8 ft × 1¹/₈ in plywood (2-4-1, ext. glue, Group 1, square edge)

short mat 6 spacings at 1 ft 8 in = 10 ft
long mat 9 spacings at 1 ft 8 in = 15 ft

4 in × 6 in Douglas Fir post

W5 wire

8 ft

¹/₈ in gap between each panel

10 spacings at 0.6 in = 5 ft

3 ft

tack weld

¹/₄ in × 2³/₄ in × 5 ft 4 in steel bar

DETAIL 'C' (Fig. 8.46)

8 ft

5 ft

W5 wire

Fig. 8.44. Timber facing used with steel grid reinforcement

Section B–B

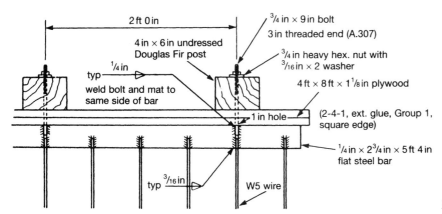

2 ft 0 in

4 in × 6 in undressed
Douglas Fir post

$^3/_4$ in × 9 in bolt
3 in threaded end (A.307)

$^3/_4$ in heavy hex. nut with
$^3/_{16}$ in × 2 washer

typ — $^1/_4$ in

weld bolt and mat to
same side of bar

4 ft × 8 ft × 1$^1/_8$ in plywood

1 in hole

(2-4-1, ext. glue, Group 1,
square edge)

$^1/_4$ in × 2$^3/_4$ in × 5 ft 4 in
flat steel bar

typ $^3/_{16}$ in

W5 wire

Fig. 8.45. Connection details

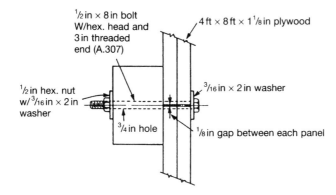

$^1/_2$ in × 8 in bolt
W/hex. head and
3 in threaded
end (A.307)

4 ft × 8 ft × 1$^1/_8$ in plywood

$^1/_2$ in hex. nut
w/$^3/_{16}$ in × 2 in
washer

$^3/_{16}$ in × 2 in washer

$^3/_4$ in hole

$^1/_8$ in gap between each panel

Fig. 8.46. Detail C

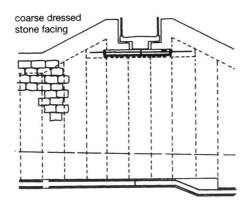

coarse dressed
stone facing

*Fig. 8.47. Abutment details of
footbridge*

269

Construction details

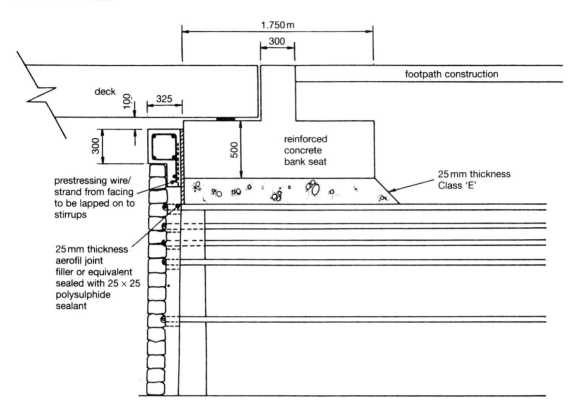

1.750 m

300

footpath construction

deck

100

325

300

500

reinforced
concrete
bank seat

25 mm thickness
Class 'E'

prestressing wire/
strand from facing
to be lapped on to
stirrups

25 mm thickness
aerofil joint
filler or equivalent
sealed with 25 × 25
polysulphide
sealant

*Fig. 8.48. Abutment details of
footbridge*

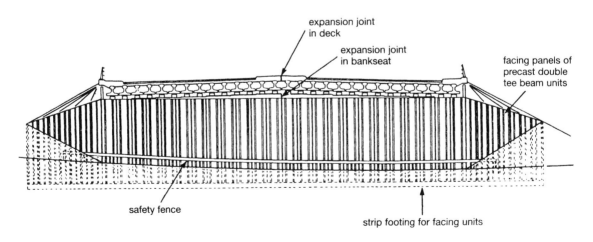

expansion joint
in deck

expansion joint
in bankseat

facing panels of
precast double
tee beam units

safety fence

strip footing for facing units

*Fig. 8.49. Abutment details for
highway bridge*

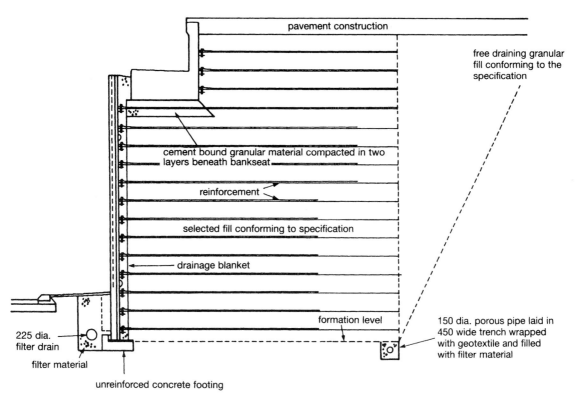

pavement construction

free draining granular
fill conforming to the
specification

cement bound granular material compacted in two
layers beneath bankseat

reinforcement

selected fill conforming to specification

drainage blanket

formation level

225 dia.
filter drain

filter material

unreinforced concrete footing

150 dia. porous pipe laid in
450 wide trench wrapped
with geotextile and filled
with filter material

*Fig. 8.50. Abutment details for
highway bridge*

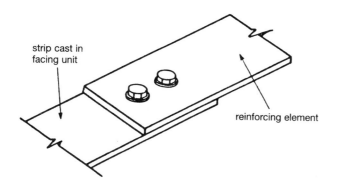

strip cast in
facing unit

reinforcing element

*Fig. 8.51. Fixed connection
associated with concertina method
of construction*

Construction details

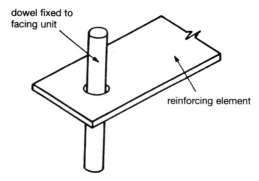

dowel fixed to
facing unit

reinforcing element

Fig. 8.52. Dowel connection

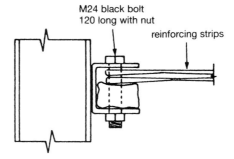

M24 black bolt
120 long with nut

reinforcing strips

*Fig. 8.53. Fixing associated
with facing*

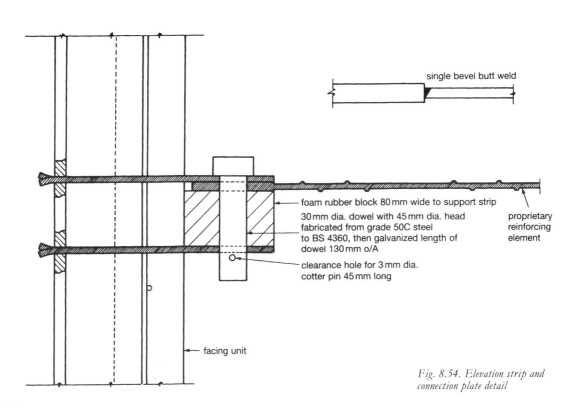

single bevel butt weld

foam rubber block 80 mm wide to support strip

30 mm dia. dowel with 45 mm dia. head
fabricated from grade 50C steel
to BS 4360, then galvanized length of
dowel 130 mm o/A

clearance hole for 3 mm dia.
cotter pin 45 mm long

proprietary
reinforcing
element

facing unit

*Fig. 8.54. Elevation strip and
connection plate detail*

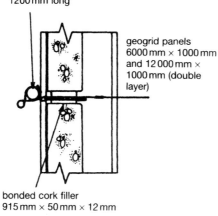

25 mm dia. reinforcing bar, 1200 mm long

geogrid panels 6000 mm × 1000 mm and 12 000 mm × 1000 mm (double layer)

bonded cork filler 915 mm × 50 mm × 12 mm

Fig. 8.55. Fixed connection associated with telescope method of construction

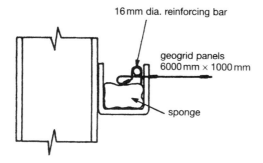

16 mm dia. reinforcing bar

geogrid panels 6000 mm × 1000 mm

sponge

Fig. 8.56. Moving connection associated with sliding method of construction

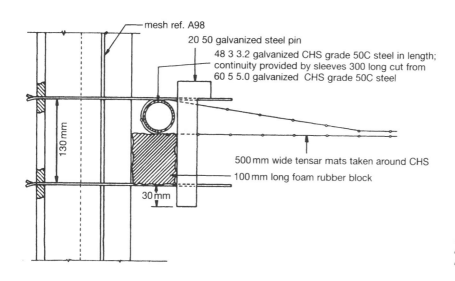

mesh ref. A98

20 50 galvanized steel pin

48 3 3.2 galvanized CHS grade 50C steel in length; continuity provided by sleeves 300 long cut from 60 5 5.0 galvanized CHS grade 50C steel

130 mm

500 mm wide tensar mats taken around CHS

100 mm long foam rubber block

30 mm

Fig. 8.57. Moving connection associated with sliding method of construction

273

Construction details

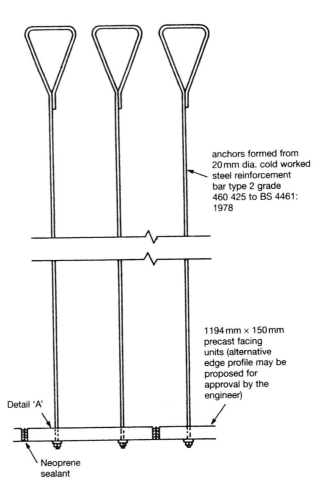

anchors formed from 20 mm dia. cold worked steel reinforcement bar type 2 grade 460 425 to BS 4461: 1978

1194 mm × 150 mm precast facing units (alternative edge profile may be proposed for approval by the engineer)

Detail 'A'

Neoprene sealant

Fig. 8.58. Moving connection associated with sliding method of construction

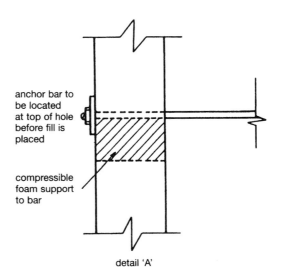

anchor bar to be located at top of hole before fill is placed

compressible foam support to bar

detail 'A'

Fig. 8.59. Detail A

274

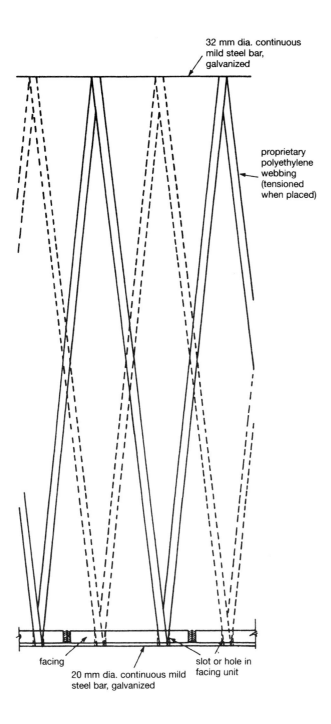

32 mm dia. continuous mild steel bar, galvanized

proprietary polyethylene webbing (tensioned when placed)

facing

20 mm dia. continuous mild steel bar, galvanized

slot or hole in facing unit

Fig. 8.60. Slot or hole in facing unit

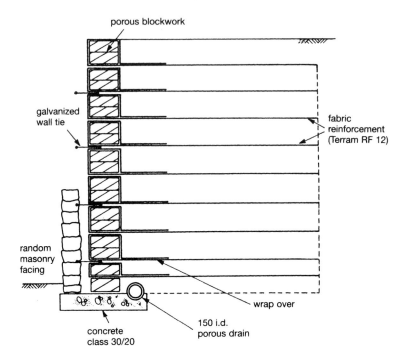

porous blockwork

galvanized
wall tie

random
masonry
facing

fabric
reinforcement
(Terram RF 12)

wrap over

150 i.d.
porous drain

concrete
class 30/20

*Fig. 8.61. Fabric wrapped
around rolling block in concertina
method of construction*

9

Costs and economics

9.1 Economic advantage

The primary advantage gained from the use of reinforced soil may be the improved idealization which the concept permits; thus structural forms which normally would have been difficult or impossible become feasible and economic. An example of the economic benefits which are possible is illustrated in Fig. 9.1, where the total cost of the solution involving the use of earth reinforcement techniques could be half of the conventional solution illustrated, which requires the use of piled foundations.

Reinforced soil structures will not always prove economic when compared with other structural forms, but in many conventional situations where circumstances are favourable, economies may be obtained. The comparative cost of soil structures to conventional structures, found by one group of workers in the UK, is shown in Fig. 9.2, while Fig. 9.3 illustrates the range of economies that have been obtained when soil structures have been substituted for other forms of construction. In Fig. 9.3:

$$\text{Economy} = \frac{(\text{Conventional cost})-(\text{Reinforced soil cost})}{\text{Reinforced soil cost}}$$
$$\times 100 \qquad\qquad (1)$$

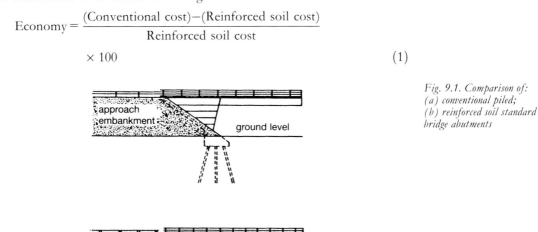

Fig. 9.1. Comparison of: (a) conventional piled; (b) reinforced soil standard bridge abutments

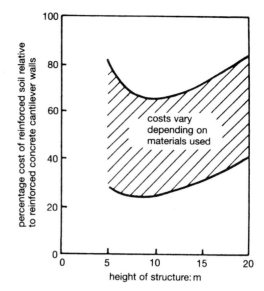

Fig. 9.2. Cost data for
reinforced soil

Figures 9.2 and 9.3 indicate the imprecise nature of cost comparisons.

In accordance with market forces the total cost of soil structures will reflect an element of 'what the market will bear'. With increased competition, costs may be reduced.

9.2 Estimating costs

The derivation of cost estimates for future projects is a significant element in a designer's work. A simple and frequently reliable estimate system is to relate all costs back to a base date and to use a cost index updated to cover inflation and construction industry costs. Table 9.1 and Fig. 9.4 give the base costs and cost index (CI) which may be used for estimating purposes for walls and abutments. Alternatively, Fig. 9.4 may be used to reduce previous scheme prices to base

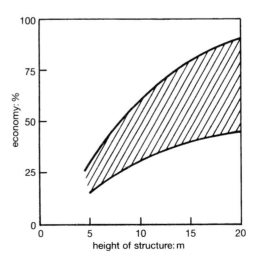

Fig. 9.3. Economics of
reinforced soil

Table 9.1. Base costs for estimating costs of walls and abutments

	1956 base cost of face ($£/m^2$)	Notes
Piled base	20	Base cost relates to North England construction; costs elsewhere may differ
Reinforced concrete retaining wall	20–25	Base cost is for plain concrete structures
Reinforced soil retaining wall	15–20	Piled base assumes three rows of piles to a depth 10–20 m
Reinforced concrete abutment	30–35	Use of reinforced soil abutment may entail 2–4 m increase in deck span
Reinforced soil abutment	20–25	Abutment costs do not include wing walls or excavation costs
		The cost of the reinforced soil structure includes the cost of fill, whereas the reinforced concrete retaining wall excludes the cost of fill
		The fill requirements for a vertically faced soil structure may be less than the volume required for a conventional structure, Fig. 9.5

cost prices for evaluation and comparison. Figure 9.4 may be expressed as:

$$CI = (0.35L + 0.12P + 0.1C + 0.07SG + 0.1T + 0.07F + 0.195S) \times 170 \qquad (2)$$

where L is labour; P is plant; C is cement; SG is sand and gravel; T is timber (soft wood); F is derv fuel; S is steel.

Estimating cost per square metre of facing:

$$= £\left(\frac{\text{Base cost} \times CI}{100}\right) \qquad (3)$$

9.3 Total cost

The total cost of a reinforced soil structure is made up of the following cost elements:

(a) soil fill, C_S
(b) reinforcing connection elements, C_R
(c) facing elements (if required), C_F
(d) labour for transport and construction, C_L
(e) transport materials, C_T
(f) construction (including all ancillary items such as drainage, copings and fencings), C_C
(g) material testing, C_{MT}
(h) profit, P.

279

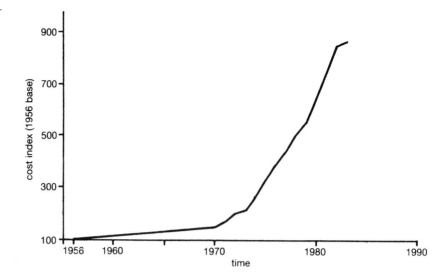

Fig. 9.4. *Cost index based on Baxter Indices related to bridge works*

Thus, total cost,

$$\text{TC} = £(C_S + C_R + C_F + C_L + C_T + C_C + C_{MT} + P) \tag{4}$$

For contractual purposes equation (4) may be reduced to the first three elements, but with the labour, transportation and the ancillary costs included:

$$\text{TC} = £(C'_S + C'_R + C'_F) \tag{5}$$

where C'_S represents the cost of the soil fill, including transport, placing compaction and material testing

C'_R represents the cost of the reinforcement, including transport and fixing

C'_F represents the cost of the facing, including transport and erection. Profit is included.

The elements included in equation (4) are interrelated and the minimum total cost of a structure may be produced by a combination of the most compatible elements in any particular situation. For example, if the necessary material testing systems are unavailable, then the use of reinforcing elements exempt from the testing requirements may provide the economic solution even if these reinforcing elements are highly priced. In the same vein, a combination of construction elements permitting the use of an indigenous fill or a waste fill material such as colliery shale or pulverized fuel ash may be attractive economically.

9.4 Distribution of costs

9.4.1 Design consideration

Using the elemental breakdown given in equation (5) it is possible to illustrate that the distribution of the cost elements vary not only with the relative costs of the constituent materials but also with the dimensions of the structure. Assuming that the relative cost of the three elements of equation (5) are:

soil fill, per unit volume (m^3)	1·0
reinforcing elements per unit area of face (m^2)	1·5
facing elements per unit area of facing	10·0

and where the width, B of the soil structure height, H is defined as $B = H$, then the distribution of costs with respect to the height of a vertically faced retaining wall are shown in Fig. 9.5(a). From Fig. 9.5(a) it can be seen that the relative costs of the three basic elements for a 10 m structure are approximately:

soil fill	30%
reinforcing elements	40%
facing elements	30%

If the relative costs of the three basic elements are changed to:

soil fill, per unit volume (m^3)	0·5
reinforcing elements per unit area of face (m^2)	2·0
facing elements per unit area of facing	10·0

then, although the total cost of the structure remains the same, the distribution of the costs if very different, Fig. 9.5(b).

Figures 9.5(a) and (b) also illustrate the influence that scale or size of construction may have on costs; at the lower heights the influence of the cost of the facing on the overall costs becomes dominant. With small structures, the material requirements for the facing may be of the same order as the material from a conventional structure, a point reflected in Fig. 9.3. At low heights particular attention may be required to reduce the costs of the facing element in order to retain the economic benefit of a reinforced soil structure; one method known to be successful is to use masonry or brick facing normally associated with small-scale construction or building techniques.

A second influence on overall cost, which is associated with the scale of the project, is the contractual arrangements under which the structure is built. For some individual structures, and structures under 3 m in height, the labour requirements are low and the use of specialist subcontractors may prove uneconomic, in which case economic construction of the soil structure may be attained only by the main contractor, local contractors, or by a direct labour organization.

9.4.2 Construction considerations

In conventional structures the placing and compaction of fill may not be associated with the construction. Reinforced soils may be treated similarly and with the fill element removed the distribution of construction costs established in one study is shown in Fig. 9.6. The volume of structural fill required for use with a conventional structure may exceed that used in a reinforced soil structure, Fig. 9.7.

9.4.3 Start-up costs and competition

The dominant factor in reinforced soil construction can be the start-up costs associated with the manufacture of facing elements, and although new forms of

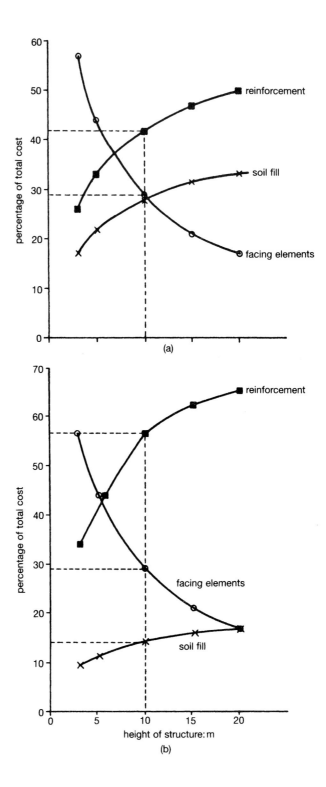

Fig. 9.5 (a) and (b). See text

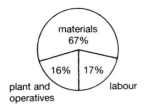

Materials	Facing
	Facing moulds
	Vertical reinforcements
	Horizontal reinforcements
	Drainage
	Others
Labour	Site clearance
	Retaining wall
	Drainage
Plant and operatives	Site clearance
	Retaining wall
	Drainage

Fig. 9.6. Breakdown of construction costs

reinforced soil may be developed by designers, they might not compete on cost against existing proprietary systems. The implication of this is that it can be difficult to introduce competition, and without competition, costs of reinforced soil are likely to rise in respect of market forces.

There are two possible approaches to make in-house designs competitive with proprietary systems in reinforced soil. The first is to make use of the new system compulsory. With this approach, the development costs associated with a new facing are effectively eliminated, as all contractors have to provide the new facing. This is the approach used by the Georgia Department of Transport in the USA. In Georgia, it was found that the *initial cost* of proprietary reinforced soil structures was $28 per square foot. However, the price rose rapidly to $40 per square foot, arguably because of the lack of competition. The introduction of the Georgia Stabilized Earth System (GASE) in 1981 offered an alternative in-house system. The system is based on the use of welded wire grid reinforcement, produced originally by

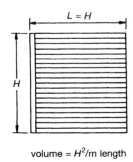

volume = H^2/m length

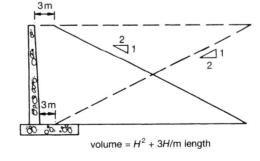

volume = $H^2 + 3H$/m length

Fig. 9.7. Volume of earth fill required in reinforced earth and reinforced concrete retaining walls

CALTRANS in California, together with an incremental facing unit. A key decision in making GASE a success was the decision by the United States Federal Highway Authority to use one of the structures on the proposed Interstate 75 as a demonstration project. This made the use of the GASE system compulsory for all tenders, thereby overcoming the potential problem of the initial set-up cost associated with the manufacture of the formwork for the new facing unit design.

The first application of GASE was for a structure some 18 m high. The benefits of the system became apparent in the second phase of development of Interstate 75 which contained 45 retaining walls. Following the successful demonstration project, the GASE system was chosen for all the walls on the Interstate 75 project by the winning contractor, the reason being the low cost of the GASE system to that offered by the proprietary system. Subsequent to the introduction of the GASE system, the cost of the proprietary system was reduced, thereby confirming that competition was the key element in determining overall cost. Competition from the GASE system has not meant the elimination of the proprietary systems, rather the reduction in costs has resulted in a major increase in the use of reinforced soil, from which all have benefited. Examples exist in Atlanta, Georgia, in which one abutment of a bridge was constructed using the GASE system and the other using one of the proprietary techniques.

The second approach to provide the competition to the proprietary systems is that provided by the York method, whereby it was decided to *adapt* existing materials for use as facing elements and, therefore, to eliminate any start-up costs associated with the facing. The first structure using this approach was built using full-height prestressed concrete planks formed as double tees, Fig. 8.15–8.24. Conventional use of these planks was as flooring units in the construction of medium-sized buildings. As the planks were readily available and used in large numbers, the unit costs were modest, and the cost of the reinforced soil structure was competitive. Subsequent to this, other prestressed flooring units have been used successfully to form the facing of reinforced soil structures. The king post and panel construction method is based on the concept of using readily available existing materials, Fig. 8.29.

9.5 Cost differentials

The cost of fill materials is dependent on local availability and haulage rates. Similarly, the cost of facing materials is a function of locality and custom. Reinforcement materials have different properties and costs vary. Thus, even though the theoretical cost per unit of facing may indicate financial preference for one material, market conditions may give a different trend, Table 9.2. Overriding all considerations is the requirement to obtain the minimum cost (equation (4)).

9.6 Ecology audit

An alternative method of assessing the benefits of earth reinforcing systems, is to use an ecology audit. A major advantage of this approach is that it is

Table 9.2. Relative costs per unit of facing

Reinforcement	Relative theoretical cost per unit of face (after Cole, 1978)	Relative actual cost per unit (after Boden *et al.*, 1979)
Aluminium alloy NS 51-H4, BS 4300/8	101	
Aluminium alloy NS 51-H8	77	
Copper C101, BS 2870	346	
Galvanized mild steel KHR 34/2P, BS 1449: Part 1	100	100
Galvanized high-yield steel KHR 54/35P, BS 1449: Part 1	73	
	73	
Cold rolled stainless steel 316516, BS 1449: Part 2	172	205
Hard rolled stainless steel 316516, BS 1449: Part 2	118	
Glass fibre reinforced plastic	271	360
Polyester fibre	211	56-230
Plastic-coated mild steel Grade 43/25		162
Aluminium-coated embossed mild steel Grade CR4 'Aludip'		56

essentially immune from the commercial distortions which are associated normally with new constructional systems and, therefore, it may produce a more realistic assessment of the true costs.

The increase in energy costs has led to an interest in energy calculations including the energy content of building materials. However, energy is only one of the ecological parameters needed to determine the complete effects (short-term, long-term and side) of engineering works. Of growing importance and interest are the problems created by scarcity of raw materials, the environmental problems created by pollution, both of the atmosphere as well as the land from mining activities, the increase in manpower costs and transportation costs and the cost of maintenance. The choice of structural form used for any scheme influences all of these parameters. Determination of the complete costs to society of a structure may be attempted by studying the ecological parameters represented in the whole cycle necessary for its production, including:

- mining
- raw materials
- process industry
- basic materials
- product/construction industry
- product/structure

- users' maintenance
- waste
- recycling.

In practical terms, the ecological parameters associated with a reinforced soil structure are:

Table 9.3. Energy consumption of construction materials

Material:	
Gravel	0·104 GJ/ton
Sand	0·128 GJ/ton
Ordinary Portland cement (OPC)	8·2 GJ/ton
Blastfurnace cement (HC)	3·0 GJ/ton
Water	0·004 GJ/ton
Mixing of concrete	0·058 GJ/ton
Mild steel reinforcement (bar, grid)	22·8 GJ/ton
Prestressing steel (bar, strand)	28·3 GJ/ton
Plastic (high density polyethylene sheet)	84·0 GJ/ton
Concrete (340 kg HC/m^3)	2·18 GJ/m^3
Concrete (360 kg OPC/m^3)	3·28 GJ/m^3
Concrete tiles and bricks	3·18 GJ/m^3
Consumption of process water in manufacture:	
Concrete	6·30 l/m^3
Steel (reinforcement, prestressing constructional)	55 m^3/ton
Despoiling from production of materials:	
Concrete	0·69 m^2/m^3
Steel	5 m^2/ton
Pollution—SO$_2$ emission:	
Concrete	0·37 kg/m^3
Steel	2·00 kg/ton
Dust emission:	
Concrete	1·29 kg/m^3
Aggregate/fill (sand, gravel)	1·1 kg/m^3
Steel	2·7 kg/ton
Labour—material manufacture and transport:	
Concrete	1 man-h/m^3 manufacture; 1·45 man-h/m^3 transport
Steel	10 man-h/ton manufacture; 0·6 man-h/ton transport

- energy content of the materials forming the structure
- quantity of process water required to manufacture the materials
- despoiling of land necessary to produce the materials
- pollution caused during manufacture and construction
- labour costs for material manufacture, transport, construction and maintenance
- demolition requirements.

The ecological parameters associated with the construction of reinforced soil structures formed using reinforced concrete, steel reinforcement and cohesionless soils are illustrated in Table 9.3 (after Kreijger, 1981).

Figure 9.8 shows ecological parameter values for a 6 m prototype reinforced soil structure compared with an equivalent reinforced concrete cantilever

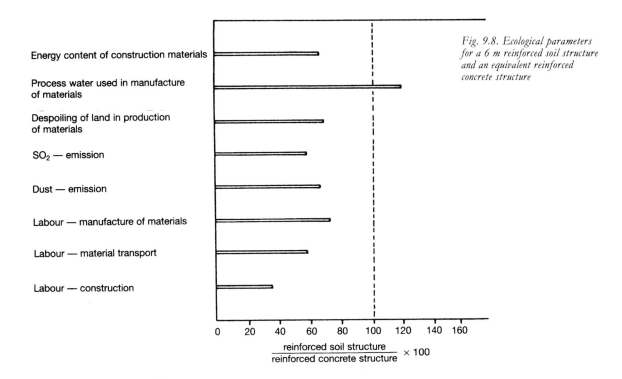

Fig. 9.8. Ecological parameters for a 6 m reinforced soil structure and an equivalent reinforced concrete structure

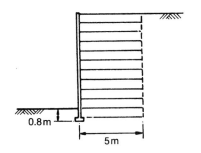
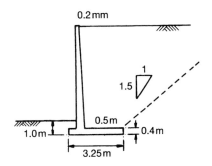

retaining wall. Even though the latter was an optimized design developed using a recognized retaining wall computer program, the reinforced soil structure is significantly more efficient in ecological terms. Arguably, economic parameters have as their ultimate base the ecological parameters, accordingly, Fig. 9.8 is a potent argument that reinforced soil structures are efficient and economic.

9.7 References

ALEXANDER W. O. and APPOO P. M. (1977). Material selection, the total concept. *Design Engng*, Nov., 59–66.

BODEN J. B., IRWIN M. J. and POCOCK R. G. (1978). Construction of experimental walls at TRRL. *Ground Engng*, **11**, No. 7, 28–37.

COLE E. R. L. (1978). Design aspects of reinforced earth construction. *Ground Engng*, Sept., 46–50.

CORNELISSEN H. A. W. (1976). Energie, vervuilings—en schaarsteaspecten bij de fabricage van beton en metselbakstenen in Nederland (energy, pollution and scarceness aspects in the manufacture of concrete and bricks in the Netherlands). *Materialen voor onze Samenleving* (Materials for our society), Publication 22 of Toëkomstbeeld der Techniek (Future Shape of Technology Publications), Nov., 162–73.

DEPARTMENT OF THE ENVIRONMENT (1976). Retwal, Program for the analysis and design of reinforced concrete abutments, piers and retaining walls, HECB/B/8.

KRAUS F., GELDSETZER R. and KOSH U. (1979). Ökologische Kosten von Bauwerken—Lehrstuhl für Baukonstruktion I. Rheinisch–Westfällischen Technischer Hochschule, Aachen.

KREIJGER P. C. (1979). Energy analysis of materials and structures in the building industry. *Appl. Energy*, **5**, No. 2, April, 141–57.

KREIJGER P. C. (1980). Bouwmaterialen en hedendaagse problemen in de samenleving (Matériaux de construction et problèmes actuels d'environment). *Tijdschrift der Openbare Werken van België*, nr. 1–16.

KREIJGER P. C. (1981). Ecology of a prestressed concrete—versus a steel bridge of equal cost. *IABSE Coll. New Look at Traditional Materials*, Imperial College.

SCHLOSSER F. and VIDAL H. (1969). Reinforced Earth. *Bull. des Liaison des Laboratoires Routiers Ponts et Chaussées*, No. 41, Nov., Paris.

10

Durability

10.1 Introduction

In keeping with all structures, durability of reinforced soil is a prime requirement in order that the structure may properly fulfil its designed role. Depending on the role required of the structure, durability becomes more or less important. For reinforced soil structures, other than those of a temporary nature, resistance to corrosion acquires a greater significance than in more conventional constructions. This is because the basic form of these structures involves the integration of reinforcing media or elements within the soil. Soil does not produce the best environment for many materials and, if construction of an earth reinforced structure is undertaken without proper consideration of the environmental hazards, rapid deterioration of parts can occur.

The problem is compounded because underground corrosion (soil structures essentially have their structural elements within the soil) can be difficult to monitor and so areas of critical corrosion may not be apparent until a failure occurs. In addition, the subsequent treatment of corrosion failures may physically be very difficult and will normally be very costly. Thus, reinforced soil structures must reflect the criteria for quality assurance of construction which requires that durability is a function of design life. It is possible to identify three categories of structure, based on design life, and the relative importance of durability and the rate of corrosion of the materials forming the structure, Table 10.1.

10.2 Corrosion

All common engineering materials and metals degenerate, reverting back to similar ores and compounds from which they were extracted. The designer of an earth reinforced structure is concerned with the form and rate of this reversion. In general, metals degenerate as a result of corrosion, and polymeric materials degrade. Corrosion is an electrochemical process and does not affect non-metallic materials. The rate of corrosion is determined by material composition, the geometry of the object, its relationship to the environment and, most importantly, the nature of the surrounding soil.

The importance of the form of corrosion depends on the function of the element subject to attack. General corrosion usually presents few engineering problems as corrosion allowances can be provided. However, local attack can

Table 10.1

Design life and the importance of:	Durability and rate of corrosion

Permanent structures:
    60–100 years US
    120 years UK
	Major consideration
	(except those embankment structures in which reinforcement is required to provide short-time stability only)

Short-life structures: 1–20 years	Minor consideration
Temporary structures: 1–100 weeks	No problem

effectively destroy a structure, as in the case of the perforation of a pipeline. In the case of reinforced soil structures, failure of the connections between the facing and the reinforcement, or a banded attack across the reinforcing element, are the forms of corrosion which cause the most risk. Soils present a complex environment, but past experience provides a means of assessment from which the level of corrosiveness can be obtained; thus soils can be ranked as very corrosive, requiring extensive precautions, or benign, requiring few precautions.

Determination of the actual rate of corrosion is more difficult. The difficulty is compounded by the fact that the true nature of the fill material and the physical conditions within the reinforced soil structure can be determined only during or after construction and, therefore, are not available to the designer. As reinforced soil structures usually last several decades, the engineering solution normally adopted is to select soils which are known to be non–corrosive and to follow a fairly rigid construction code. Thus, a major safeguard to the general designer is through a restrictive specification.

10.2.1 Electrochemical corrosion

Corrosion is an electrochemical process and refers to metals only, i.e. plastics and glass do not corrode. For corrosion to occur there must be a potential difference between two points that are electrically connected in the presence of an electrolyte. This potential difference may be caused by a difference of salt and oxygen concentration in the soil. The microcouple so produced behaves like a short cell travelled by an electric current that leads to corrosion, Fig. 10.1. At the anode the current leaves the metal and corrosion occurs when metal is transferred in solution in the form of positive ions or cations (metal ions deficient in electrons and hence carrying a positive charge). The cathode reaction relates to the electrons remaining in the parent metal; these must be neutralized to enable the anodic reaction to continue. Several reactions are possible, but in earth reinforced soil structures two cathodic reactions usually predominate: hydrogen evolution and oxygen reduction.

(*a*) Hydrogen evolution results from the discharge of protons (hydrogen ions) at the cathodic sites on the metal surface. This reaction can proceed rapidly at low pH, but is of minor significance at neutral and alkaline pH.

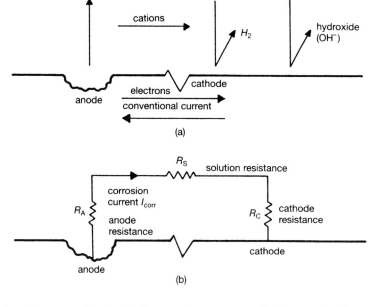

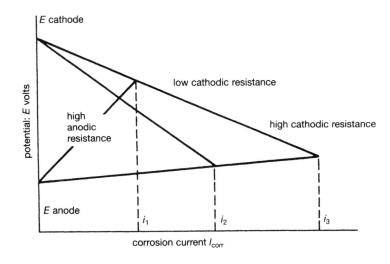

Fig. 10.1. Anodic and cathodic reactions on metal surfaces

(*b*) Oxygen reduction is the most important cathodic reaction in soil. In this case, oxygen combines with electrons and water to produce hydroxide ions. The reaction rate is determined by the rate of defusion of oxygen on the metal surface.

Control reactions which limit the corrosion rate are determined by the level of corrosion current, and the importance of solution resistivity in determining this rate is shown in Fig. 10.2; as the resistivity decreases so corrosion increases.

Further, the electrochemical corrosion provides metal cations and hydroxide, these, together with hydrogen evolution, increase alkalinity. Some

Fig. 10.2. Control reactions which limit the corrosion rate

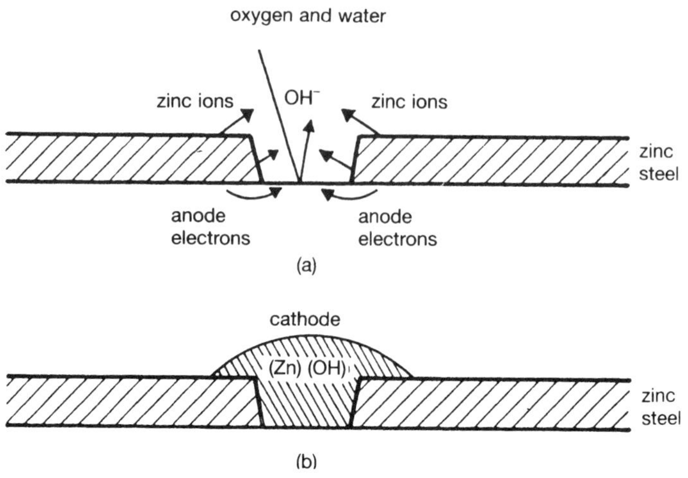

oxygen and water

zinc ions   OH⁻   zinc ions

zinc
steel

anode
electrons

anode
electrons

(a)

cathode

(Zn) (OH)

zinc
steel

(b)

Fig. 10.3. Blockage of damaged areas on galvanizing by the production of zinc corrosion products

metallic cations react with the hydroxide and produce precipitates which in turn can reduce the corrosion rate by stifling both metal dissolution and the oxygen diffusion. Zinc acts in this manner and is the reason why galvanized steel is resistant to corrosion. In the case of galvanized steel, the zinc coating normally corrodes at a slow rate as it is a poor cathode; however, should the zinc be damaged so that the steel substrate is exposed, then the steel becomes a cathode and the zinc the anode. The zinc corrodes and produces corrosion products in the form of hydroxides and carbonates which block the gap in the zinc film, Fig. 10.3.

Aluminium and the stainless steels rely for their resistance to corrosion on the presence of an oxide skin. Should the oxide skin be damaged and not able to reform, the parent metal will corrode very rapidly. As a result, aluminium and stainless steel tend to suffer a pitting attack rather than general corrosion. The presence of chlorides and sulphates both encourage a pitting attack in these metals, Fig. 10.4. Of the two forms of metal attack—general corrosion and pitting—general corrosion is to be preferred, as it is predictable and decreases in strength can be calculated, thereby allowing corrosion allowances to be made. Pitting is unpredictable, with regard to both rate and location; in addition, the attack can be very rapid (King, 1978).

10.2.2 Bacterial corrosion

Metal corrosion can be reduced by the provision of a neutral or alkaline

Fig. 10.4. (a) General corrosion; (b) pitting corrosion (in both cases, the rate of attack depends on the cathodic reaction)

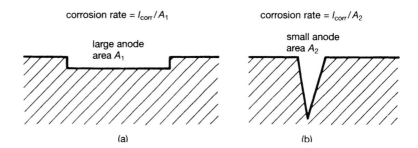

corrosion rate = I_{corr}/A_1

large anode
area A_1

corrosion rate = I_{corr}/A_2

small anode
area A_2

(a)

(b)

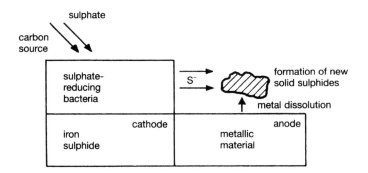

Fig. 10.5. Corrosion processes by the sulphate-reducing bacteria. Production of solid iron sulphides leads to corrosion in the metal adjacent

environment free of oxygen. However, this environment favours the growth of sulphate-reducing bacteria. These bacteria are anaerobic, i.e. they can thrive in the absence of atmospheric oxygen, and grow by obtaining oxygen from sulphate ions, reducing them into sulphide ions in the process. Corrosion by these organisms is both by cathodic stimulation and from the action of sulphides. The aqueous sulphides produced initially migrate through the pore water to the metal and there react to produce corrosive solid sulphides resulting in deep pitting. In this way, this form of corrosion attack may occur at a point distant from the area of growth of the organisms, Fig. 10.5.

Bacteria require an organic food source and sulphate; therefore, to combat bacterial corrosion the designer should select soils which are low in both these elements. As a result, top soil or organic rich soil should be avoided. Certain clay soils, for example London Clay, encourage bacterial corrosion and should not normally be used unless the form of reinforcement is impervious to this type of corrosion (e.g. glass fibre reinforcement or some forms of plastic reinforcement).

In some earth reinforced systems the use of cohesive soils is not permitted within the specification; as a result, the associated bacteriological problem is greatly reduced. In these specifications, criteria relating to bacteriological corrosion is not usually mentioned and the designer may be unaware of this hazard. The possibility of biological corrosion is usually measured by the redox potential of the soil.

10.3 Degradation (polymeric reinforcement)

Modern polymer reinforcements used in reinforced soil are composed of highly durable polymers. However, polymeric materials will eventually degrade as a result of a number of different actions, including ultraviolet light, high energy radiation, oxidization, hydrolysis and chemical reaction. Biological degradation is not considered an issue for polymeric reinforcements formed from high molecular weight polymers and is not discussed further (Koerner *et al.*, 1992).

While each of the actions which can result in polymer degradation are usually assessed individually, they are complicated by temperature, stress and synergism between one another. With respect to temperature, it is established that elevated temperature increases all the listed types of degradation in a predictable manner. With regard to stress, the type and the relative magnitude must be identified. In terms of reinforced soil, stress is associated with tension. The influence of synergism in reinforced soil structures is complicated, and this phenomenon is only now beginning to be explored.

While the actions leading to degradation are complicated and difficult to quantify, the overall impact on polymeric reinforcement is well established. Degradation is associated with chain scission, side chain breaking and cross-linking (Grassie and Scott, 1985). Each of these actions cause the polymer to become aggressively more brittle; thereby, it decreases from its original elongation to gradually lesser values.

10.3.1 Effect of ultraviolet light

The influence of ultraviolet (UV) light on polymeric reinforcements can be eliminated by burying the reinforcement in the soil. However, in some reinforced soil applications the geotextile elements are required to remain exposed to sunlight for extended periods, as in the case of geotextile face slopes and walls. In these cases, the geotextile must be adequately resistant to the effects of UV light exposure. Ultraviolet light causes degradation by reaction with the covalent bonds of organic polymers, causing yellowing and embrittlement. All polymers are susceptible to varying degrees of degradation by this method, with polyester being the least susceptible and polyethylene and polypropylene being the most. In addition to the type of polymer, the structure of the geotextile influences the rate of ultraviolet degradation. Geotextiles with large diameter structural members such as some geogrids or thick and bulky materials exhibit a higher degree of resistance than thin fibrous materials.

Polyester used as a reinforcement has a good resistance to UV light and will retain approximately 80 per cent of its original strength after exposure of continuous sunlight in an aggressive environment. Polyethylene and polypropylene do not have the same UV resistance as polyester and to provide UV resistance, it is normal practice to provide a UV stabilizer into the polymer during manufacture.

Two types of UV stabilizers are used for polyethylene and polypropylene, namely, passive and active stabilizers. Passive stabilizers work by shielding the polymer molecules from UV radiation. The most common passive stabilizer is carbon black, which has been shown to be an effective barrier for UV absorbed by polyethylene. The carbon black type and the dispersion characteristics are crucial to performance. In order to ensure extended UV protection, the carbon black must be channel type with a particle size less than or equal to 20 nanometres (20×10^{-9} m); a minimum concentration of 2 per cent is required and it must be well dispersed. The result of the addition of carbon is to render the polymer black in colour. Carbon black stabilizers are often used in conjunction with active stabilizers (e.g. hindered amines) which absorb the high UV radiation energy and release lower non-destructive energy. In converting the high energy UV radiation into low energy, the active stabilizer is consumed and hence the UV resistance life of the stabilized polymer depends on the quantity of stabilizer originally added during the manufacturing process.

10.3.2 Effect of oxidation on polyolefins; polyethylene (PE) and polypropylene (PP)

Degradation due to oxidation occurs as a result of heat (thermo-oxidation) and exposure to ultraviolet light (photo-oxidation). Photo-oxidation is considered above. Oxidation is not considered a problem with polyester, but can have an effect on polyethylene and polypropylene. The application of oxygen and heat cause a breakdown and cross-linking of the molecular chains, resulting in embrittlement of the polymer.

Controlling the oxidation of polyethylenes is a well-developed science supported by long-term experience and a range of applications in the telecommunications cable insulation field. Antioxidants are added to the polymer to prevent oxidation during processing and use. Antioxidant packages calculated to provide over 250 years of life have been designed for specific polypropylene geotextiles (Wisse *et al.*, 1990). In addition to the use of antioxidants, changing the molecular structure through orientation inhibits degradation. In the case of high density polyethylene (HDPE) geogrids, the degree of orientation required by the manufacturing process has been shown to provide significant resistance to oxidation.

10.3.3 Effect of hydrolysis on polyester; polyethylene-terephthalate (PET)

Hydrolysis occurs when water molecules react with polymer molecules, resulting in chain scission, reduced molecular weight and strength loss. Of the polymeric reinforcements used for permanent reinforcement, only polyester is susceptible to hydrolysis.

The chemical reaction which takes place to form polyester is shown in Fig. 10.6. This is a condensation reaction in which terephthalic acid and ethylene glycol forms polyethylene-terephthalate (polyester), with water being produced as part of the reaction process. Hydrolysis is the reverse of this action, in which water reacts with polyester molecules to form short-length chains with acid and hydroxyl end groups. Hydrolysis is a slow reaction influenced by humidity, polyester structure, temperature, external catalysts and externally applied loads.

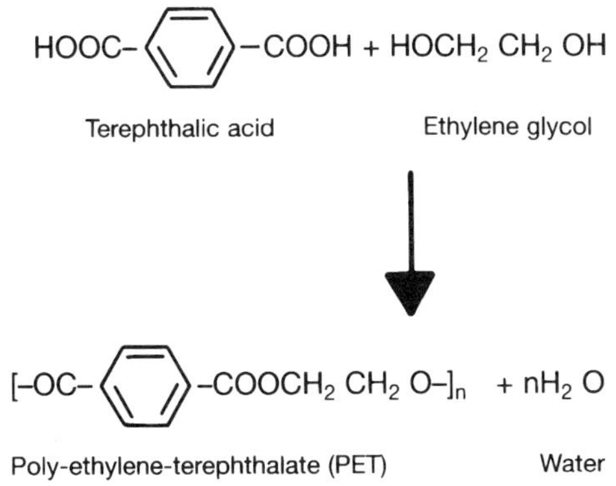

Fig. 10.6. Chemical reaction to form polyester (PET)

$$HOOC\text{--}\langle\bigcirc\rangle\text{--}COOH + HOCH_2\,CH_2\,OH$$

Terephthalic acid   Ethylene glycol

$$[\text{--}OC\text{--}\langle\bigcirc\rangle\text{--}COOCH_2\,CH_2\,O\text{--}]_n\ + nH_2O$$

Poly-ethylene-terephthalate (PET)   Water

295

10.3.4 Humidity

Water must be present for hydrolysis to proceed. In reinforced soil applications, it is assumed conservatively that 100 per cent relative humidity can exist. Therefore, hydrolysis needs to be considered.

10.3.5 Polyester structure

The molecular weight of the polyester affects the rate of hydrolysis, Fig. 10.7. This figure shows that the advances in polyester manufacture since the 1940s have enabled heavier molecular weight polyesters to be produced, with a consequent increase in resistance to hydrolysis. High molecular weight polyesters (average molecular, $M_n > 30\,000$) should be used for technically demanding applications such as reinforced soils.

The type of processing performed on polyester also affects the rate of hydrolysis, and partly drawn polyester film exhibits worse behaviour than highly drawn polyester fibres. The effect of drawing the polymer during processing causes orientation (strengthening) to the molecular structure, with the result that the molecular chains are aligning much closer together. This makes it more difficult for water molecules to penetrate the molecular structure of the polymer. It can be concluded that hydrolysis is not significant if well-engineered materials are used as reinforcement.

10.3.6 Temperature

As with all chemical reactions, raised temperatures affect the rate of hydrolysis; an increase of 300–400 per cent occurs with an increase in temperature from 20° to 30°C.

External catalysts
Chemical agents can act as catalysts in the hydrolysis reaction. In an acid environment (pH < 2), hydrogen ions increase the reaction rate. In an alkaline environment, the OH⁻ ions can also influence the reaction, and when polyester

Fig. 10.7. Effect of: (a) relative humidity on relative rate of hydrolysis of PET (after Risseeuw and Schmidt, 1990); (b) average molecular weight on relative rate of hydrolysis

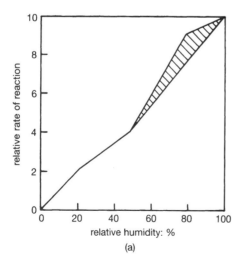

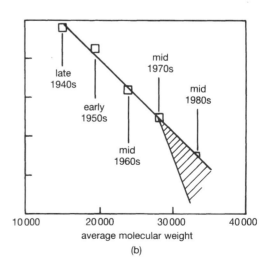

(a) (b)

fibres are directly exposed over long periods of time at pH > 11 the presence of OH⁻ ions can have a detrimental and destructive effect. Therefore, direct exposure of polyester fibres to environments such as curing concrete or calcium hydroxide in an anaerobic environment initiates hydrolysis and reflects poor practice.

To protect polyesters from highly alkaline environments, a robust coating of polyethylene or PVC is used. Both of these coatings ensure that, although water vapour can migrate through the casing, the PE or PVC acts as a barrier to the migration of inorganic ions. Thus, the environment inside the casing remains neutral. If the barrier is punctured during installation, protection can be lost.

10.3.7 Externally applied loads

Tensile loads applied to polymer elements straighten the molecular chains, thus reducing intermolecular distances. In the case of polyester when the molecular chains are straightened, it is more difficult for water molecules to penetrate the structure and to cause molecular scissions. As a result, the rate of hydrolysis is lower for polyester fibres in the stressed condition than when unloaded.

10.4 Physical damage

The durability of a structure is not only affected by electrochemical or biological corrosion, but also by physical damage or wear. In particular, some materials used to construct earth reinforced structures are susceptible to physical damage due to rough handling. For example, glass fibre reinforcement can be damaged by tracked vehicles, as can the protective coatings of metal reinforcement. The placing of reinforcements in soil which is then compacted by mechanical means can result in physical damage. If the level of damage is not known, site damage tests are usually conducted. The purpose of the site damage test is to subject the reinforcement to the environmental and construction conditions associated with the structure, and to measure the influence on the characteristic strength and stiffness of the reinforcement. Reduced material properties are then used in design. A site damage test for any form of reinforcement used in soil structures is described in the UK *Code of practice for reinforced soil*, BS 8006: 1995. Extreme cold is not seen as a durability problem and normal temperatures can be accommodated by most materials. Only steel and reinforced or prestressed concrete appear suitable for conditions subject to fire hazard.

10.4.1 Polymeric materials

The placing and compaction of soil directly against the polymer reinforcement may reduce its tensile properties. The amount of damage inflicted on the reinforcement depends on the actual construction of the geotextile, the size and type of soil being placed and the compaction effort. The effect of installation damage on geotextile reinforcements is to reduce the tensile strength but the modulus (stiffness) is not affected, Fig. 10.8. Fig. 10.8(a) shows the loss of strength of woven polyester exposed directly to the compaction of 30·7 mm diameter crushed limestone. The amount of installation damage is dependent on the nature of the geotextile, the type of soil used and the amount of compacted effort, Fig. 10.8(c). Fig. 10.8(c) shows the influence of geotextile construction and stone size on the amount of installation damage (Lawson, 1990).

The effects of installation damage to the tensile strength of polymeric reinforcements is considered in design by the use of partial factors applied to the tensile strength of the as–manufactured material, Fig. 10.8(b). The partial factor is determined by recovering the geotextiles from test sites and comparing the tensile properties with those of the pre-installed material.

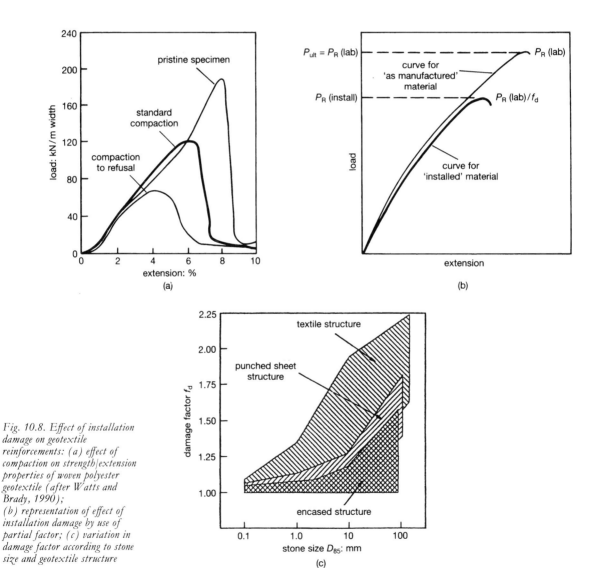

Fig. 10.8. Effect of installation damage on geotextile reinforcements: (a) effect of compaction on strength/extension properties of woven polyester geotextile (after Watts and Brady, 1990); (b) representation of effect of installation damage by use of partial factor; (c) variation in damage factor according to stone size and geotextile structure

10.5 Material compatibility

All metallic components used in earth reinforced structures, i.e. reinforcing elements, connections and metal facings, should be electrolytically compatible. Where this is not possible, effective electrical insulation must be provided.

10.6 Miscellaneous factors

Other soil constituents can affect soil corrosiveness. The more important ones, which have been identified as having a deleterious effect and which can cause serious corrosion, are cinders, carbon particles, coke and coal. Alternatively, chalk and limestone fill may leach out and form deposits on the reinforcement which effectively reduces corrosion.

10.7 Construction factors influencing the corrosion of reinforcement and facings

The major construction factors influencing the corrosion of the reinforcing elements in an earth reinforced structure are the type and nature of fill used and the construction process, particularly with respect to the compaction achieved. In the case of a reticulated or anchored structure, the construction process has little influence; however, the *in situ* soil can be critical.

The type of soil used as fill in earth reinforced structures varies from one application to another. Different soils are more aggressive than others. Table 10.2 provides a general ranking of the soil which illustrates its aggressiveness, while Table 10.3 is an assessment of soil aggressiveness towards metals in particular—arranged in a form suitable for the designer of soil structures.

10.7.1 Cohesionless fill

With earth reinforced structures formed from cohesionless soil, the following are the main considerations influencing the durability of the reinforcement and the connections to the facing:

(a) homogeneity of the fill
(b) degree of unevenness of compaction
(c) water content and drainage.

Homogeneity of the fill is necessary to avoid localized corrosion of the reinforcement or facing connections and can be critical. Provided the reinforcement is contained within a particular soil type, different fill materials can be used. The acceptable and unacceptable mixtures of soils with regard to corrosion are illustrated in Fig. 10.9.

Unevenness of compaction can result in the establishment of discrete areas of corrosion in which the more compacted areas are more liable to corrode, as they are less aerated. As it is not possible to compact uniformly all earth reinforced structures, this point must be borne in mind during design.

Drainage and water content are vital factors which influence the corrosion process. Dry fill is not corrosive, hence structures built in very dry regions will survive even if constructed within plain mild steel reinforcement. The onset of

Table 10.2 (below and facing). Estimation of soil aggressiveness

Parameter	Ranking
Kind of soil:	
Chalk, chalk marl, sand marl or sand	−2
Loam, loam marl, loamy sand or clayey sand	0
Clay, clay marl or humus	2
Peat, mud or bog soil	4
Soil conditions:	
Water present at structure level	1
Disturbed soil	2
Dissimilar soil around structure	3
Water not present	0
Undisturbed soil	0
Homogeneous soil around structure	0
Soil resistivity (ohm-cm):	
Above 10 000	0
5000 to 10 000	1
2300 to 5000	2
1000 to 2300	3
Below 1000	4
Water content:	
Above 20%	1
Below 20%	0
pH value:	
Above 6	0
Below 6	1
Total acidity (mval/kg):	
Below 2·5	0
2·5 to 5·0	1
Above 5·0	2
Redox potential (mV, pH = 7):	
Above 400 (430 for clay)	−2
200 to 400	0
0 to 200	2
Below 0	4
Total alkalinity (mval/kg to pH 4·8):	
Above 1000	−2
200 to 1000	−1
Below 200	0
Hydrogen sulphite/sulphate-reducing bacteria:	
Not present	0
Trace (below 5 ppm sulphide)	2
Present (above 5 ppm sulphide)	4
Coal, coke or cinders:	
Present	4
Not present	0

Table 10.2—continued

Parameter	Ranking
Chloride (ppm):	
Above 100	1
Below 100	0
Sulphate (ppm):	
Above 1000	3
500 to 1000	2
200 to 500	1
Below 200	0
Sum of rank numbers	
Negative: Practically non-aggressive	
0 to 4: Weakly aggressive	
5 to 10: Aggressive	
Above 10: Strongly aggressive	

unacceptable corrosion occurs at 5% water content and increases with increasing water content (Table 10.8, section 10.8.3). Water run-off from embankments or surfaces above the structures can bring corrosive salts into the structure and, in many cases, this may be channelled towards the face and the connections. Because of the hydroscopic nature of most corrosive

Table 10.3. Assessment of soil aggresiveness towards buried metals

Classification/soil property	Aggressive	Selected aggressive soil (average values)	Non-aggressive	Selected non-aggressive soil (average values)
Resistivity (ohm-cm)	< 2000	1156	> 2000	30 400
Redox potential at pH = 7	< 0·400		> 0·400	
Normal hydrogen electrode (volts)		0·263		0·520
	< 0·430 if clay		> 0·430 if clay	
Borderline cases resolved by moisture content (per cent)	> 20	28·5	< 20	12·1

Note: The classification involves a measure of soil resistivity which indicates the possibility of oxidation (electrochemical corrosion). The determination of redox potential provides a means of assessing whether a particular soil is conductive to the activity of sulphate reducing bacteria (biological corrosion).

products, salts leached into the structure will tend to concentrate, resulting in an acceleration of corrosion rates. The selection of suitable structural details to minimize these risks is important.

10.7.2 Cohesive fill

Many of the problems associated with cohesionless fill can be avoided if the structure in question is constructed using cohesive material, provided shrinkage, swelling or frost cracks are avoided. The principal problem associated with cohesive soil may be that of microbiological attack. The risk of this increases with increased sulphur or organic content of the fill. Fig. 10.10 illustrates which area of the structure is at risk from microbiological corrosion.

Care must be taken with galvanized reinforcement used in cohesive soils as some studies have shown that zinc is sensitive to illite. In this case, the corrosion products are absorbed by the clay and hence cannot provide a protective layer on the metal. Another important corrosion factor can be the presence of certain sulphate-reducing vibrions, a form of bacteria. With these, corrosion can be initiated by cathodic depolarization on the periphery of bacterial colonies which form on the surface of the reinforcing elements. Long (1978) considers that the growth of these sulphate-reducing vibrions is favoured by the following conditions:

(*a*) neutral pH value
(*b*) presence of sulphates in aqueous solutions
(*c*) organic matter which is partially oxidized by exposure to the atmosphere of the mud contained in the fill.

10.7.3 Soil nailing

One of the reasons given for the relatively slow acceptance of soil nailing in the UK is concern over the durability of the nails. In the UK, the Transport and

Fig. 10.9. Acceptable and unacceptable mixtures of soils with regard to corrosion (after King, 1978)

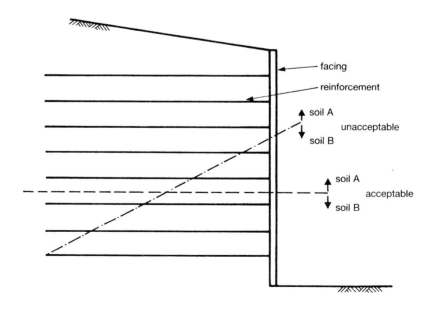

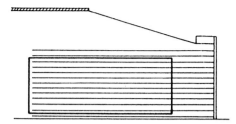

Fig. 10.10. Area of risk of microbiological corrosion in structure of impermeable cohesive fill outlined by black line (after King, 1978)

Road Research Laboratory (TRRL) has recently classified the ground associated with soil nailing into four categories, ranging from non-aggressive to highly aggressive. The French Clouterre recommendations on corrosion allowance/protections adopts a similar approach. The UK categories for soil aggressiveness are based on the conventional soil properties (pH, soluble salts, water content, resistivity and redox potential). The French Clouterre recommendations are based on the same parameters, except redox potential is not considered. In Hong Kong, the Geotechnical Control Office (GCO) has identified the ground conditions under which detailed investigations of ground aggressiveness is required, Table 10.4.

In France, reliance is placed on the use of sacrificial thicknesses on the nails to provide long-term durability. In the UK, for all but the most aggressive soil conditions, a sacrificial approach is adopted for permanent structures defined as having a life of up to 120 years, Table 10.5. In addition, when nails are grouted, the benefits associated with the grout (i.e. high pH) can be used to reduce the ground aggressivity rating to the next lower category. A justification for permitting a more relaxed approach to the corrosion of soil nails than to tiebacks is that the nails are formed from low tensile steel grades which are less susceptible to pitting corrosion and hydrogen embrittlement. However, many UK designers favour the German approach: that of designing to prevent

Table 10.4. Conditions under which detailed investigation of ground aggressiveness is required (after GCO, 1989)

Property	Value
Soil resistivity*	< 50 ohm m
Soil redox potential* corrected to pH = 7	< 0·40 volts < 0·43 volts for clay soils
pH of soil or groundwater	< 5
Chloride ion content of soil†	> 0·2 g/litre
Total sulphate content in soil‡	> 0·2% by weight
Sulphate ion content in soil†	> 1·0 g/litre
Sulphate ion content in groundwater	> 0·3 g/litre

* Based on *in-situ* tests.
† Based on 2 : 1 water/soil extract.
‡ Concentration of sulphates expressed as SO_3.

Table 10.5. Sacrificial thickness in mm applicable to soil nails

Years	Corrosive level	1·5+	5+	10+	30+	40+	50+	70+	100+	120+
France	Slightly corrosive	0		2			4			
	Medium corrosive	0		4			8			
	Very corrosive	2		8						
United Kingdom*	Out of water	0·25			0·35	0·95		1·15		
	In water (fresh)	0·25			0·4	1·3		1·55		
France and United Kingdom	Highly corrosive	Special protection measures needed								
Driven nails† United Kingdom	Slight to medium corrosive, out of water	0·25	1	2·5	3·5	4	5		6	

* Nails galvanized at $1000\,\text{g/m}^2$.
† Nails galvanized at $700\,\text{g/m}^2$.

corrosion and of using protective systems (single or double) rather than just relying on sacrificial thicknesses of steel. Throughout Europe, the use of non-metallic (non-corrosive) nails is expected to grow.

10.8 Measurement of corrosion factors

The main factors in soil corrosiveness are resistivity, redox potential, water

Table 10.6. Test methods for the electrochemical properties of fill

Factor	Test method
pH	BS 1377: Part 3: 1990 Test 9
Chloride content	BS 1377: Part 3: 1990 Test 7·2
Water soluble sulphate	BS 1377: Part 3: 1990 Test 5
Resistivity (saturated sample)	BS 1377: Part 3: 1990 Test 10·4
Resistivity (*in situ*)	BS 1377: Part 9: 1990 Test 5
Organic content	BS 1377: Part 3: 1990 Test 3
Redox potential	BS 1377: Part 9 1990 Test 10
Microbrial activity index	BS 8006: 1995 Annex A
Sulphide content	(See *Encyclopedia of Industrial Chemical Analysis*)

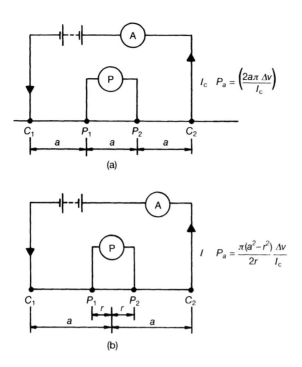

$$I_c \quad P_a = \left(\frac{2a\pi\,\Delta v}{I_c}\right)$$

(a)

$$I \quad P_a = \frac{\pi(a^2 - r^2)}{2r}\frac{\Delta v}{I_c}$$

(b)

Fig. 10.11. (a) Wenner configuration; (b) Schlumberger configuration

content, pH, chloride content, soluble sulphate content, organic content, microbial activity and sulphide content. Some factors are relevant in every application, others depend on specific conditions. Tests for pH, chloride content, water soluble sulphate (SO_3) and resistivity are usually conducted in every application. Tests for organic content are conducted where more than 15% of the particles pass the 63 μm sieve; in this case, a measure of the redox potential or microbial activity is included. The measurement of sulphide content is undertaken if the origin of the fill is likely to contain sulphides. A list of the test methods used in the UK *Code of practice for reinforced soil*, BS 8006: 1995 is shown in Table 10.6.

10.8.1 Resistivity

Resistivity is usually the prime corrosion factor in earth reinforced structures, because in most cases some form of metal element will be present within the structure. This metal element is subject to electrochemical corrosion. The lower the resistance to the flow of electric current through the structure, and particularly through the soil, the higher the corrosion rate. Resistivity is measured electrically by passing a fixed current through the soil and recording the voltage drop across a known length of soil.

Several commercial systems are available to measure resistivity, arguably the most accurate form of device is the Wenner four-pin probe, Fig. 10.11(a). To test for resistivity (BS CP 1013), the four co-linear electrodes are placed in the ground and current passed between the two outer electrodes. A low-frequency alternating current rather than a direct current is used, as this minimizes various inaccuracies caused by the electrodes in the soil. The voltage that

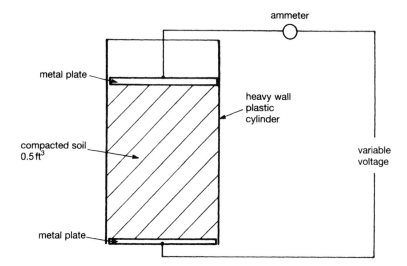

*Fig. 10.12. Laboratory testing
of soils for resistivity*

appears between the two inner electrodes as a result of the electrical field is measured and the apparent resistivity of the ground is given by:

$$P_a = \left(\frac{2a\pi \, \Delta v}{I_c} \right) = 2\pi a R \tag{1}$$

where P_a is the resistivity (ohm-cm)
  a is the electrode spacing (m)
  I_c is the current flow between the outer electrodes (amperes)
  Δv measured potential difference (volts)
  R resistance (ohms).

The following precautions should be observed (Department of Transport, BE 3/78):

(a) The row of electrodes should be orientated in a different direction at each location within the area of the proposed excavation, and testing should not be carried out when the soil is frozen.

(b) The spacing of adjacent electrodes in each row should be not less than 1·5 m and this spacing should be increased by increments of approximately 2 m, for successive tests. The maximum spacing should be equal to the depth to which material is to be excavated.

If the outer electrode separation is gradually increased about a fixed central point, a determination of change in soil sequence with respect to the depth can be made. This is known as the 'Schlumberger configuration'. At close separation, the apparent resistivity will approximate to the resistivity of the upper soil; at wide separation, it is predominantly that of the lower soil, Fig. 10.11(b). Resistivity from the Schlumberger configuration is given by:

$$P_a = \frac{\pi(a^2 - r^2)}{2r} \frac{\Delta v}{I_c} \tag{2}$$

Table 10.7. Representative values of soil resistivity

Soil	Resistivity at saturation (ohm-cm)
Fluvial shale	505
Black shale	7645
Artificial sea sand	2100
Clay sand	3900
Fine sand	4340
Fluvial sand:	21 400
10 ppm Cl^-	12 780
20 ppm Cl^-	10 500
100 ppm Cl^-	5370
300 ppm Cl^-	3270
50 ppm SO_4^-	8100
150 ppm SO_4^-	3630
500 ppm SO_4^-	1450
Altered shales	1420
Sand	3860
Fine sand	23 000
Sand–sand–gravel mixture	7000

Laboratory testing for resistivity is also possible, but because of the size of the sample used, the reliance of the results should be seen as a function of the representativeness of the sample. A typical laboratory test apparatus is shown in Fig. 10.12. Representative values of the resistivity of different soils are shown in Table 10.7.

10.8.2 Redox potential

The redox potential provides evidence of two possibilities: either the presence of widely differing soils, or the presence of highly reducing soils characteristic of the presence of corrosive bacteria.

Earth reinforced structures should be constructed from homogeneous fill. The use of layered fills does not in itself create a corrosion horizon (Fig. 10.9), provided each layer of reinforcement is fully contained within the soil. Thus, redox potentials are usually only of relevance to earth reinforced structures as being indicators of the presence and activity of corrosive sulphate-reducing bacteria.

The redox potential is calculated by measuring the voltage difference between a platinum surface and a reverse electrode, both in intimate contact with the soil, Fig. 10.13. As each type of reverse electrode provides a different base line, the voltage has first to be calculated on a standard potential scale; this is based on the hydrogen reverse electrode. A correction is made to accommodate the pH of the soil in order that the redox potential can be expressed in terms of a neutral condition, i.e. pH 7.

Care has to be taken when undertaking redox potential tests; among those demanded by some specifications (Department of Transport, BE 3/78) are:

(*a*) The redox potential of the fill should be determined at the site of the cutting or of the proposed borrow pit, by measuring the potential of a platinum electrode with respect to a saturated calomel reference electrode. The calomel reference probe should be filled with a saturated solution of potassium chloride. The platinum probe electrodes on the nosepiece should be cleaned using scouring powder and a gentle abrasive action. The nosepiece should be washed once in 70% alcohol and twice in distilled water.

(*b*) Method
   (*i*) The test should be taken in a hole not less than 1 m square in plan at the required depth.
   (*ii*) A sample should be taken from the base of the excavation to determine the pH value of the soil.
   (*iii*) The probes should be located 300 mm apart and pushed into contact with the soil at the base of the excavation.
   (*iv*) The electrical circuit should be completed by connecting one of the electrodes of the platinum probe and the reference probe to the positive and negative terminals of the millivoltmeter respectively. If no reading is obtained the leads should be reversed, with the subsequent readings being treated as negative values.

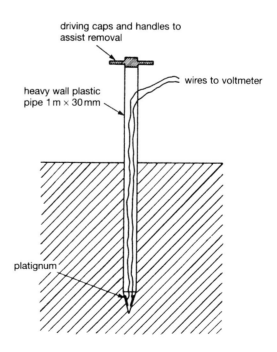

driving caps and handles to
assist removal

wires to voltmeter

heavy wall plastic
pipe 1 m × 30 mm

platignum

Fig. 10.13. Redox probe driven into soil. Two platinum electrodes are used and the average taken for calculation

(v)   Each platinum electrode should be included in the circuit separately, with readings being recorded immediately after the platinum probe has been rotated through one revolution under firm hand pressure. The two electrode readings should be discarded when they differ by more than 10 mV and the electrodes cleaned.

(vi) The mean of two acceptable readings should be recorded as the potential of the platinum probe.

(vii) The redox potential (E_r) for a particular test may be obtained from the expression:

$$E_r = E_p + 250 + 60 \, (\text{pH} - 7) \tag{3}$$

where E_r is the redox potential (mV)
        E_p is the potential of the platinum probe (mV)
        pH   is the value of the acidity of an aqueous solution of the fill.

(c) Where a test is made in fill, rather than in an *in-situ* condition, the fill should be allowed to reconsolidate. This may be achieved by removing soil with the minimum disturbance, recompacting to optimum conditions and storing for five days.

(d) Redox tests should not be taken during periods of frost or drought.

The alternative to taking redox measurements is to analyze the soil in the laboratory for bacteria corrosion susceptibility. A high proportion of organisms coupled with the presence of sulphate-reducing bacteria would normally indicate an aggressive soil unsuitable for use in earth reinforced structures.

10.8.3 Water content

Water content is relevant to corrosion in that corrosion occurs only in the presence of water. Thirty per cent moisture content is often recorded as the threshold at which all the metal surface becomes active. The relative effect of water content on corrosion is seen in Table 10.8.

10.8.4 pH

The pH of the soil (the acidity or alkalinity) is relevant depending on the reinforcing materials in use. A different range of pH is applicable to different reinforcing materials. In some specifications the pH range is limited to that

Table 10.8. The relative effects of water content on corrosion

Water content (%)	Effect
< 5	Little or none
5	Onset
> 5 < 30	Corrosion increases with water content
> 30 < 60	Maximum general corrosion
> 60	Corrosion reduces
< 30 > 60	Risk of pitting attack prevalent

which is tolerant to zinc, an acknowledgement that galvanized steel is one of the most commonly used reinforcing materials. Aluminium and stainless steel together with non-metallic reinforcements have different and usually greater tolerance to acidity or alkalinity.

Although pH of the soil is the criterion normally quoted, this is not strictly the correct indicator. The important factor is the total acidity or alkalinity actually present in the soil. Some soils exhibit neutral pH values but have high reservoirs of acid; thus, although the corrosion rate is low, it will persist, and the corrosion products will not blank off the reinforcement surface and reduce the rate of corrosion over a period of time.

10.9 Durability of existing reinforcing materials

10.9.1 Mild steel and galvanized steel

The use of plain or galvanized steel in soils dates back a number of years. The practice of using steel in soil has provided some knowledge of its long-term behaviour which can be used to estimate the loss of thickness and the resistance of steel reinforcement to corrosion.

(a) *Buried pipework in plain steel.* In 1957, the Department of Water Works and L'Electricité de France examined 24 old buried pipes ranging up to 60 years in age. The results of these studies showed that approximately one-third were still in good or very good condition after an average of 50 years, and that one-tenth had deteriorated to a bad condition. It was concluded that pitting attack over a short time-scale was the prime concern.

(b) *Galvanized steel culverts.* The results of the study of 111 culverts buried in various soils in the US have been recorded by Haviland *et al.*, (1968). The thickness of the steel used in the culverts varied between 1·5 and 2·6 mm and had been in service from 2 to 35 years. All were reported to be in good condition. Due to the method of use, in which one side of the culvert is backed by soil and the other side free to the atmosphere, it can be concluded that this study is most relevant to the use of metal facings in earth reinforced structures. Significantly, it was observed that the corrosion was greatest on the side remote from the soil.

(c) *Sheet piles.* Sheet piles have been in service for a great number of years and are known to behave well, except in certain tropical climates where bacterial action accelerates corrosion attack. In a series of studies undertaken in France by the Département Ponts et Chaussées (Darbin *et al.*, 1978), it was noted that the rate of corrosion decreased with time, even in marine conditions. Over a period of 40 years the average loss of thickness of piles was attributed to 0·06 mm per year in cold water and 0·1 mm per year in warm climates. The greatest loss observed was during the first five years of service when a rate of the order of 0·3 mm per year was recorded. Pitting was observed, with the pits reaching a depth of three to four times the average loss of thickness. These tests corroborated field research undertaken in Germany which produced a figure of total loss of thickness of 1 mm per 100 years.

(d) *National Bureau of Standards* (Romanoff, 1959). The most comprehensive data available on underground corrosion are the results of the field testing

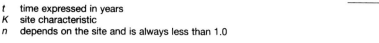

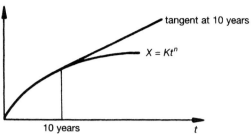

Fig. 10.14. Development of corrosion

of metal culverts in plain and galvanized steel undertaken by the US National Bureau of Standards (NBS), in a programme originating in 1910 and conducted until 1955. More than 333 types of ferrous and other metal, with or without protection, were buried in 128 separate locations. Plain and galvanized steel were tested in 47 sites covering a variety of soils including muds, peats and clays.

The studies undertaken by NBS confirmed that the speed of corrosion decreases significantly with time, and Romanoff (1959) indicated that the damping of corrosion was a more significant parameter than the speed of corrosion. The following relationship has been proposed to calculate the average loss of thickness of plain steel, x relative to a function of time, Fig. 10.14.

$$x = Kt^n \qquad (4)$$

where t is time expressed in years
       K is site characteristic
       n is site characteristic with a value less than 1.

The Romanoff studies did not produce a formula for calculating the loss of thickness of zinc-coated steel as a function of time, but it can be assumed that galvanized steel will have a rate of corrosion less than that of plain steel and that plain steel and galvanized steel can be related, as shown in Fig. 10.15. In this it is assumed that the onset of normal steel corrosion is delayed until all the zinc has been used up.

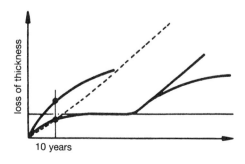

Fig. 10.15. Comparison between galvanized steel and unprotected steel

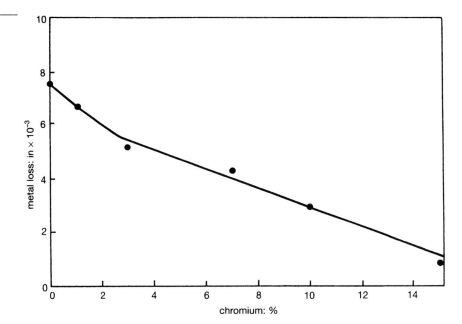

Fig. 10.16. Atmospheric corrosion tests—industrial area—8 months

10.9.2 High-alloy steel

The term stainless steel is applied to a wide range of iron based alloys which rely on their chromium content for their corrosion resistance. Chromium has a high affinity for oxygen and readily forms an oxide that is stable and produces a condition known as passivity. Steel is termed 'stainless' provided it has a chromium content in excess of 11%. In general, corrosion decreases as the chromium content increases, Fig. 10.16. Stainless steels can be divided into three groups:

(a) Martensitic stainless steels: iron and chromium steels with high carbon contents. They have a high internal energy and are not particularly resistant to corrosion.

(b) Ferritic stainless steels: similar to Martensitic steels but with a low carbon content.

(c) Austenitic stainless steels: based on iron, chromium and nickel having 10% Cr and 8% Ni.

The third group is the most corrosive resistant and the most widely used for applications where corrosion performance is paramount.

Corrosion in stainless steel can take various forms, including uniform corrosion, pitting, stress cracking, intergranular, erosion, galvanic, hydrogen cracking and fretting, as well as biological corrosion. Of these forms only pitting/crevice corrosion is a likely hazard in soil as normally the environment will be insufficiently acidic to cause general attack. Stress corrosion will not normally occur at temperatures under 60°C.

Use of stainless steel in earth reinforced structures
Potential candidates for use as reinforcement in a structure are the Martensitic steel Types 410 and 430, and the Austenitic steels Types 304 and 316.

Table 10.9. *Chemical compositions of stainless steels*

AISI type	% C	% Cr	% Ni	% other elements
Martensitic chromium steels				
410	0·15 max	11·5–13·5	—	—
416	0·15 max	12–14	—	Se, Mo or Zr
420	0·35–0·45	12–14	—	—
431	0·2 max	15–17	1·25–2·5	—
440A	0·60–0·75	16–18	—	—
Ferritic non-hardenable steels				
405	0·08 max	11·5–14·5	0·5 max	0·1–0·3 Al
430	0·12 max	14–18	0·5 max	—
442	0·25 max	18–23	0·5 max	—
446	0·20 max	23–27	0·5 max	0·23 N max
Austenitic chromium–nickel steels				
201	0·15 max	16–18	3·5–5·5	5·0–7·5 Mn 0·25 N max
202	0·15 max	17–19	4–6	7·5–10 Mn 0·25 N max
301	0·15 max	16–18	6–8	2 Mn max
302	0·15 max	17–19	8–10	2 Mn max
304	0·08 max	18–20	8–12	1 Si max
304L	0·03 max	18–20	8–12	1 Si max
310	0·25 max	24–26	19–22	1·5 Si max
316	0·10 max	16–18	10–14	2–3 Mo
316L	0·03 max	16–18	10–14	2–3 Mo
317	0·08 max	18–20	11–14	3–4 Mo
321	0·08 max	17–19	8–11	Ti 4 × C (min)
347	0·08 max	17–19	9–13	Nb + Ta 10 × C (min)

However, the straight chromium steels Types 410 and 430 have been shown by the NBS tests to be severely pitted in aggressive clayey soils while the Austenitic steels are virtually unaffected. Data related to work on Types 409 (11% Cr) and 304 (19% Cr and 10% Ni) indicate that Type 304 may be attacked in poorly drained, high-chloride soils, although elsewhere they may be satisfactory.

From Table 10.9 it can be seen that these two are very nearly at opposite ends of the range of stainless steels and have a considerable difference in their resistance to corrosion, particularly with respect to pitting. Lee and Edwards (1977) suggest that Type 430 will suffer significant pitting in clay or sandy soils, while Type 316 will suffer pitting in only very low pH soils. However, Type 316 is usually significantly more expensive than Type 430 and may not be an economical choice. Bearing in mind the degree of selectivity exercised for fills, Type 304 may be a safe alternative. Further, by modification of Type 430, adequate pitting resistance combined with the economic advantage of the steel may be possible. To improve stainless steels to resist pitting, molybdenum has been added and it has been shown that 1% Mo added to 17% Cr steel has the pitting resistance of Type 304. At present this steel is produced as Type 434.

10.9.3 Glass-fibre reinforcing elements

Glass-fibre reinforced plastic (GRP) is a material formed by suitably orientated strands of glass fibre embedded within resin. As both elements of GRP are non-metallic and good electrical insulators, the concepts of electrolytic corrosion do not apply; GRP does not rust. For many years glass fibre products have been used in underground conditions, particularly where soil conditions are aggressive or corrosive chemicals have to be contained. In the main these have been successful due to advances in the critical formulation of resin resistant to chemical attack. It is now considered that composites can be produced with outstanding durability.

Long-term durability
Although GRP is not subject to corrosion, some degradation in strength is manifest when the material is kept in wet conditions for long periods of time. The degradation process is complex, involving weakening both of the resin and the resin/glass bond. At ambient temperatures degradation is sufficiently slow to be insignificant except after long exposure. As the projected lifetime of earth reinforced structures is in excess of 100 years, studies have been undertaken to determine the long-term, load-carrying capacity of GRP. Studies using accelerated testing have been undertaken based on the understanding that the degradation of thermosetting resins by water can be accelerated by submersion in boiling water. Algra and van der Beek (1970) claim an acceleration factor of 250–1000 for accelerated testing of this form.

As glass fibre is not subject to creep, some reliability in acceleration factors is

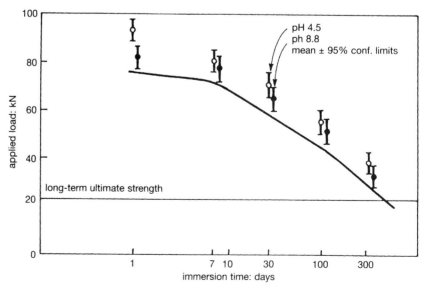

Fig. 10.17. Strength retention of Fibretain straps after immersion in aqueous solutions at a minimum of 95°C

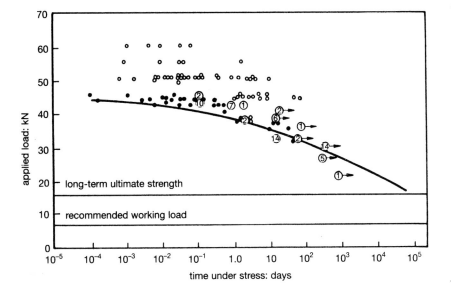

Fig. 10.18. Stress rupture curve
for Fibretain 96/1 end fixings

possible. Fig. 10.17 shows the percentage strength of GRP reinforcement as a
function of exposure time to water buffered to two pH values maintained
above 95°C. The lower boundary of the figure defines the degradation curve of
the aqueous solution examined. These include deionized water, a saturated
solution of potassium dihydrogen orthophosphate (pH 4·5–4·0), a saturated
solution of borax (pH 4·5–8·8), sea water and a 0·1M hydrochloric acid (pH 1).
The curve illustrates the insensitivity of the degradation process to the
presence of absorbed ions which are critical in metallic corrosion.

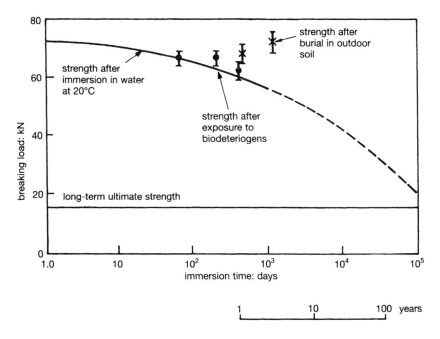

Fig. 10.19. Strength retention of
halfwidth Fibretain 96/1 after
exposure at ambient temperatures
to water, soil and selected
biodeteriogens

Table 10.10. Factors affecting performance of reinforcement materials

Material	Cost	Handling	Fill quality	Corrosion resistance
Galvanized mild steel	Low/medium	Heavy and needs care	Reasonable quality	Good
Aluminium and alloys	Low/medium	Light but tolerates abuse	High quality	Fair
Stainless steel	High	Tolerates abuse	Reasonable quality	Excellent except in anaerobic conditions
GRP	Medium	Needs care	Irrelevant	Should be excellent (but untried)
Fabric	Low	Very light; susceptible to wind	Reasonable quality	Should be excellent (but untried)
Geogrid	Low/medium	Tolerates abuse	Irrelevant	Good

Stress rupture

The weakest part of GRP reinforcement is at the connecting loop. To measure the endurance of this end fixing, stress rupture information obtained from samples tested in water at 20°C have been obtained (Mallinder, 1978). The results of these tests are shown in Fig. 10.18.

Biological degradation

Tests covering a wide range of biological organisms have been applied to GRP reinforcement including fungal cultures, anaerobic and aerobic bacteria and termites. To date, no deterioration has been observed as a result of biological activity, Fig. 10.19.

Physical durability

As GRP is a composite form of glass, attention must be given to its physical durability. Following tests in building sand, 20 mm limestone chips and 80 mm crushed rock, together with cobbles and half bricks with sharp edges, it has been concluded that no deterioration will occur provided the tracked cleats of vehicles are separated from the material by a minimum distance of 50 mm.

10.9.4 Tyres

The extreme durability of car tyres is a prime reason for the difficulty of their disposal. Life expectancy well in excess of 100 years can be assumed with confidence provided they are not subjected to combustion.

10.9.5 Comparative performances

The relative performance of the usual reinforcing materials, relative to fill quality and durability, is shown in Table 10.10.

10.10 References

ALGRA E. A. H. and VAN DER BEEK M. H. B. (1970). Ageing of plastics. *Plastica*, **23**, 45–55.

BRITISH STANDARDS INSTITUTION (1990). *Methods of tests for soils for civil engineering purposes*. BSI, London, BS 1377.

BRITISH STANDARDS INSTITUTION (1995). *Code of practice for strengthened/reinforced soils and other fills*. BSI, London, BS 8006.

DARBIN M., JAILLOUX J.-M. and MOUTRELLE J. (1978). Performance and research on the durability of reinforced earth reinforcing strips. *ASCE Symp. Earth Reinforcement*, Pittsburg.

DEPARTMENT OF TRANSPORT (1978). *Reinforced Earth Retaining Walls and Bridge Abutments for Embankments*. Tech. Memo. BE 3/78.

GEOTECHNICAL CONTROL OFFICE (1989). *Model specification for prestressed ground anchors*. GeoSpec 1, Hong Kong, 167.

GRASSIE N. and SCOTT G. (1985). *Polymer degradation and stabilization*. Cambridge University Press.

HAVILAND J. E., BELLAIR P. J. and MORELL V. D. (1968). *Durability of Corrugated Metal Culverts*. US Highway Research Record, 242.

KING R. A. (1978). Corrosion in Reinforced Earth. *Proc. Symp. Reinforced Earth and other Composite Soil Techniques*, TRRL and Heriot-Watt University, TRRL Supp. Report 457, 276–285.

KING R. A. and NABIZADEH H. (1978). Corrosion in Reinforced Earth Structures. *ASCE Symp. Earth Reinforcement*, Pittsburg.

KOERNER R. M., HSUAN Y. and LORD A. E. (1992). Remaining technical barriers to obtain general acceptance of geosynthetics. *The 1992 Mercer Lectures*, ICE, London, 48.

LAWSON C. R. (1990). Hollaway (ed.). Geosynthetics. *Polymers and polymer composites in construction*. Thomas Telford, London, 205–245.

LEE B.V. and EDWARDS A.M. (1977). The use of alloy steels in resisting corrosion. See King R. A. (1978), 286–291.

LONG N. T. (1978). Some aspects about fill material in Reinforced Earth. See King R. A. (1978), 246–249.

MALLINDER F. P. (1978). The use of FRP vs reinforcing elements in reinforced soil systems. See King R. A. (1978), 262–275.

RISSEEUW P. and SCHMIDT H. M. (1990). Hydrolysis of HT polyester yarns in water at moderate temperatures. *Proc. 4th Int. Conf. on Geotextiles, Geomembranes and Related Products*, The Hague, **2**, Balkema, 691–696.

ROMANOFF M. (1959). *Underground Corrosion*. US National Bureau of Standards, circular 579, April.

WATTS G. R. A. and BRADY K. C. (1990). Site damage trials on geotextiles. *Proc. 4th Int. Conf. on Geotextiles, Geomembranes and Related Products*, The Hague, **2**, Balkema, 603–608.

WISSE J. D. M., BROOS C. J. H. and BOCHS W. H. (1990). Evaluation of the life expectancy of polypropylene geotextiles used in bottom protection structures around the Ooster Schelde Storm Surge Barriers. *Proc. 4th Int. Conf. on Geotextiles, Geomembranes and Related Products*, The Hague, **2**, Balkema, 697–702.

Worked examples

11.1 Introduction

The following worked examples are included in the text to illustrate the relative influence of the various elements of the analytical procedures and also to show that the selection of the material components must be made in a systematic manner.

11.2 Example 1

11.2.1 The problem

The design of a vertical retaining wall supporting a highway, Fig. 11.1. It is assumed that the wall varies in height from 3 m to a maximum full height of 8 m, excluding pavement construction.

Assume: Design in accordance with the UK Department of Transport criteria using the tie-back analysis.

Material properties of fill:

$$\gamma_{\text{fill}} = 19 \text{ kN/m}^3; K_0 = 0.5; c' = 0$$
$$\phi' = 35°; K_a = \tan^2(45° - \phi'/2) = 0.27$$
$$E_{\text{fill}} = 260 - 25 \text{ MN/m}^2$$

Material properties of surfacing:

$$\gamma_{\text{surfacing}} = 24 \text{ kN/m}^3; \nu_{\text{fill}} = 0.25$$

Material properties of reinforcement:

Ratio: Reinforcement (cross-section)/Soil (cross-section) $= 0.6 \times 10^{-3}$

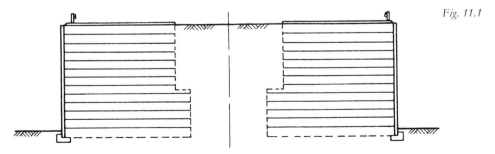

Fig. 11.1

Periphery: Reinforcement/reinforcement vertical spacing, $S_v = 0.24$
$E_r = 200 \text{ GN/m}^2$; $G_r = 10 \text{ GN/m}^2$

Reinforcement slip: $\mu = \tan \phi = 0.4$

Foundation:

$E'_{\text{foundation}} \approx 5\text{–}50 \text{ MN/m}^2$; $\nu_{\text{foundation}} = 0.25$
K_e (pore fluid bulk modulus) $= 2 \text{ GN/m}^2$
$E'_{\text{foundation}} \text{ (total stress)} = 1.2\ E'_{\text{foundation}}$
$\nu_{\text{foundation}} \text{ (total stress)} = 0.5$

Allowable bearing pressure $= 300 \text{ kN/m}^3$

11.2.2 Basic stability condition

Note: This parametric study is not necessary for the analysis, but is included to illustrate the effect of changes in material properties.

The basic stability of the structure may be determined by the finite element method (see Chapters 4 and 6). The results of a mathematical model analysis using the unit cell concept to idealize the reinforced soil structure, based on the material properties given above, produces the following results:

(*a*) Displacement of front face of wall:

$$\approx 5\text{–}10 \text{ mm when } E'_{\text{foundation}} \rightarrow 50 \text{ MN/m}^2$$
$$\approx 20 \text{ mm when } E'_{\text{foundation}} \rightarrow 5 \text{ MN/m}^2$$

(*b*) Reinforcement tension, see Table 11.1. This table indicates that soft foundations produce larger reinforcement tensions in the base of the structure.

(*c*) The finite element results suggest that K_a (Rankine) is not conservative.

(*d*) A reduction in fill stiffness causes higher reinforcement tensions within the limit $K_a \rightarrow K_0$. Consequently, reduction in reinforcement stiffness causes lower reinforcement tensions. Stiff reinforcement will permit a relatively stiffer compaction state to develop, therefore it may be advantageous to use an equivalent quantity of reinforcement having a lower elastic modulus.

(*e*) Reinforcement slip will not occur.

Table 11.1. Reinforcement tension

Depth	Reinforcement tension (kN/m)		Rankine active state of stress (kN/m)
	Stiff foundation	Soft foundation	
3	7	17	11
4	7	21	15
5	5	29	18
6	6	54	22

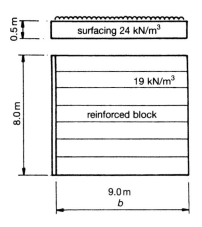
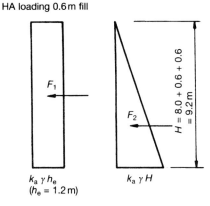

surfacing 24 kN/m³

19 kN/m³

reinforced block

9.0 m
b

HA loading 0.6 m fill

F_1

$k_a \gamma h_e$
(h_e = 1.2 m)

F_2

$k_a \gamma H$

$H = 8.0 + 0.6 + 0.6$
$= 9.2$ m

Fig. 11.2

Conclusion

The use of non-metallic reinforcement, where $E_r < E_{steel}$ will tend to reduce reinforcement tensions relative to those developed within steel reinforcement. Deflections will be increased by a small degree as will the shear stress in the soil.

11.2.3 Check on external stability

Consider Fig. 11.2.

Material properties:

$\gamma_{fill} = 19 \cdot 0$ kN/m³; $\phi' = 35°$; $c' = 0$
$K_a = \tan^2(45° - \phi'/2) = 0 \cdot 27$
$\gamma_{surfacing} = 24$ kN/m³
Therefore, surfacing equivalent to $\left(0 \cdot 5 \times \dfrac{24}{19}\right) = 0 \cdot 63$ m

Parapet loading: Department of Transport Memo BE 3/78 Cl. 2.6.3.1)

$1 \cdot 2 \times 50$ kN over length of $3 \cdot 0$ m distributed through anchor slab.

$(1 \cdot 2 \times 50) \div (3 \cdot 0 \times 0 \cdot 45) = \mathbf{17 \cdot 4}$ **kN/m run** (F_3)

Sliding

$$
\begin{aligned}
\text{Force} \quad F &= F_1 + F_2 + F_3 \\
&= K_a \gamma b_c H + \tfrac{1}{2} K_a \gamma H^2 + 17 \cdot 4 \\
&= (0 \cdot 27 \times 19 \cdot 0 \times 1 \cdot 2 \times 9 \cdot 2) \\
&\quad + (\tfrac{1}{2} \times 0 \cdot 27 \times 19 \times 9 \cdot 2^2) \\
&\quad + 17 \cdot 4 \\
&= 56 \cdot 6 + 217 \cdot 1 + 17 \cdot 4 = \mathbf{291 \cdot 1 \ kN/m}
\end{aligned}
$$

Resisting force, $R_F = \mu_f W$
$W = (9 \cdot 2 \times 9 \cdot 0 \times 19 \cdot 0) = 1573 \cdot 2$ kN/m
Assume ϕ for fill beneath structure $= 30°$
Therefore, $\mu_f = \tan \phi$, $R_F = \tan 30° \times 1573 \cdot 2 = 908 \cdot 3$ kN/m

Factor of safety (F.O.S.) $= \dfrac{R_F}{F} = 908 \cdot 3 / 29 \cdot 1 = \mathbf{3 \cdot 1}$

Overturning

Overturning moment,

$$M_0 = F_1 H/2 + F_2 H/3 + F_3 H$$

$$= \left(56{\cdot}6 \times \frac{9{\cdot}1}{2}\right) + \left(217.1 \times \frac{9{\cdot}1}{3}\right) + (17{\cdot}4 \times 9.1)$$

$$= 257{\cdot}5 + 658{\cdot}5 + 158{\cdot}3 = \mathbf{1074{\cdot}3\ kNm/m}$$

Resisting moment,

$$R_0 = 1373{\cdot}2 \times 9/2 = \mathbf{7079{\cdot}4\ kNm/m}$$
$$\text{F.O.S.} = R_0/M_0 = \mathbf{6{\cdot}6}$$

Bearing pressures
Consider Fig. 11.3.

Assume trapezoidal distribution beneath structure.
(*Note*: lack of information does not permit use of other methods.)

Where

$$I = \frac{bd^2}{12}; \quad Z = \frac{I}{y}; \quad P = W$$

Bearing pressure

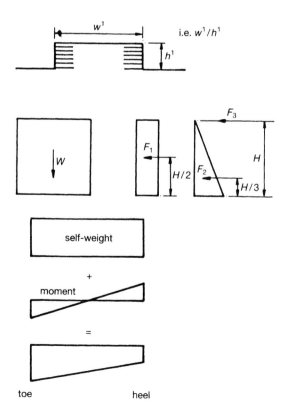

Fig. 11.3

$$= \frac{P}{A} \pm \frac{M}{Z} = \frac{1573\cdot 2}{9} \pm \frac{1074\cdot 2}{9^2} \times 6$$

$$= 174\cdot 8 \pm 79\cdot 6$$

Maximum $= 254\cdot 4 \ \text{kN/m}^2$

Conclusion

(a) Assume maximum bearing pressure beneath base is $254\cdot 4 \ \text{kN/m}^2$ and make all other bearing pressure calculations fit this criterion.

(b) Lower parts of the wall will produce a bigger width/height ratio between walls, i.e. w'/h'.
Therefore, overall stability will be greater (Jones and Edwards, 1981).
Therefore, it will be safe to decrease the breadth of the walls of reduced height, all within the criterion of (a) above.

(c) Overall stability criterion satisfied. (*Note:* A slip circle analysis is required in order to justify this condition, but is outside the scope of this text.)

Stability criteria for reduced-height walls

7 m wall

Minimum width base according to Department of Transport criteria
$= 0\cdot 8 \times 7\cdot 0 = 5\cdot 8 \ \text{m}$, say **5·75 m**

$$\text{Thus } F = F_1 + F_2 + F_3 = K_a \gamma h_e H + \tfrac{1}{2} K_a \gamma H + 17\cdot 4$$
$$= 43\cdot 0 + 125\cdot 7 + 17\cdot 4$$
$$= \textbf{186·0 kN/m}$$

$$W = 7\cdot 0 \times 5\cdot 75 \times 19\cdot 0 = \textbf{764·8 kN/m}$$

Sliding criteria satisfied

Overturning moment

$$(M_0) = (F_1 \times H/2) + (F_2 \times H/3) + (F_3 \times H)$$
$$= 150\cdot 5 + 293\cdot 3 + 17\cdot 4 = \textbf{461·2 kNm/m}$$

Bearing pressure

$$= \frac{P}{I} \pm \frac{M}{Z}$$

$$= \frac{764\cdot 8}{5\cdot 75} \pm \frac{461\cdot 2}{5\cdot 75^2} \times 6 = 133\cdot 0 \pm 83\cdot 7$$

$$= \textbf{216·7 kN/m}^2$$

Conclusion
On the criteria that maximum toe pressure can be $254\cdot 4 \ \text{kN/m}^2$ it is possible to have 7 m wall with base of **5·75 m width**.

5 m wall
Minimum width of base according to Department of Transport criteria
$= 0\cdot 8 \times$ height or 5 m
By inspection of results for 7 m high wall,
use 5 m base on 5 m high wall.

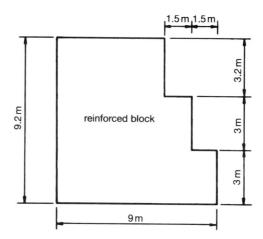

Fig. 11.4

3 m wall
By inspection of results for 7 m high wall,
use 5 m base on 3 m high wall.

11.2.4 Steps in wall

9.2 m section
Consider Fig. 11.4.

7 m section
Consider Fig. 11.5.
This does not conform to minimum criterion for length of reinforcement:
make top section **5 m long**.

5 m section
No stepping possible due to code restrictions.

11.2.5 Structural form

Where the structure is less than 5 m tall the economics of a facing made up of individual units becomes difficult. Therefore, use single-section facing and for convenience use single-section facing for complete wall.

Fig. 11.5

7 m section

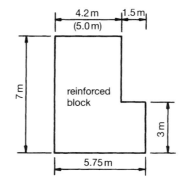

11.2.6 Internal stability

Material properties

$\gamma_{fill} = 19.0 \text{ kN/m}^3; \phi' = 35°; c' = 0$
$K_a = \tan^2(45° - \phi'/2) = 0.27; K_0 = 0.5$

Permissible axial tensile stress in reinforcement element,
$P_{at} = 33.2 \text{ kN/m}^2/\text{m}$. (The reinforcement chosen in this example is non-metallic; accordingly, the permissible axial stress is relatively low.)
Use F.O.S. = 2
Therefore, working stress in reinforcement $= \dfrac{33.2}{2} = \textbf{16.6 kN/m}^2\textbf{/m.}$

Analysis
Consider the ith layer of reinforcing elements:

$$T_i = T_{hi} + T_{wi} + T_{si} + T_{fi} + T_{mi}$$

(Equation (8), Chapter 6)

where T_{hi} is the tensile force generated by fill above ith layer; T_{wi} is the tensile force from surcharges (w_s); T_{si} is the tensile force from external loading (S_i); T_{fi} is the tensile force from horizontal shear force; T_{mi} is the tensile force from bending moment (M_i).

Consider tensile force at 1.0 m, 3.0 m, 5.0 m, 7.0 m, and 9.2 m below top of wall.

$$T_{hi} = V(K_a\gamma h - 2c'\sqrt{K_a}),$$

where V = vertical spacing of reinforcement

$$T_{wi} = K_a w_s V,$$

where $w_s = (0.6 \times 0.63) \times 19.0 = \textbf{23.4 kN/m}$

$T_{si} = 0$ (no external loading)

$$T_{fi} = 2VF_1 Q(1 - Qh_i),$$

where $F_1 = 17.4 \text{ kN/m}, (F_1 = F_3)$

$$Q = \tan(45° - \phi'/2)(d + h/2) = 0.26$$

Therefore, effective to a depth of only **4 m**

$$T_{mi} = \frac{6K_a V M_i}{L_i^2},$$

where $M_i = 17.4 \times h_i$ kNm/m;
$\qquad L_i = 6.0, 7.5,$ or 9.0 m, depending on h_i

See Table 11.2.

Local stability check (pullout)

$$T_i = P/2[\mu L_i(\gamma h_i + w_s) + c'_r L_i]$$

where P_i is the horizonal width of top and bottom faces of reinforcement on ith layer m/m, i.e. **2 m/m** in case of grid reinforcement.

Table 11.2. Tensile force in reinforcements at different depths and vertical spacings

b_i (m)	Spacing $V = 0.25$ m				Spacing $V = 0.3$ m				Spacing $V = 0.4$ m				Spacing $V = 0.5$ m			
	T_{hi}	T_{wi}	T_{fi}	T_{mi}	T_{hi}	T_{wi}	T_{fi}	T_{mi}	T_{hi}	T_{wi}	T_{fi}	T_{mi}	T_{hi}	T_{wi}	T_{fi}	T_{mi}
1													2·56	3·16	4·68	1·17
3													7·69	3·16	1·39	1·17
5																
7									10·5	2·54	0·0	1·0				
9·2	11·80	1·58	0·0	0·80	10·07	1·9	0·0	0·96								

	Depth 7–9·2 m, Spacing 0·25 m	Depth 5–7 m, Spacing 0·3 m	Depth 3–5 m, Spacing 0·4 m	Depth 0–3 m, Spacing 0·5 m
Consider	$[11\cdot8 + 1\cdot58 + 0\cdot80] \times \dfrac{1\cdot2}{1\cdot0}$	$[10\cdot07 + 1\cdot9 + 0\cdot96] \times 1\cdot2$	$[10\cdot50 + 2\cdot54 + 1\cdot0] \times 1\cdot2$	$[2\cdot56 \times 3\cdot16 + 4\cdot68 + 0\cdot39] \times 1\cdot2$
1·2 m module*	$= 15\cdot8$ kN	$= 15\cdot6$ kN	$= 16\cdot8$ kN	$= 13\cdot0$ kN
	F.O.S. 2·10	F.O.S. 2·12	F.O.S. 1·98	F.O.S. 2·55

* Preferred facing based upon double-tee prestressed concrete unit 1·2 m wide.

μ is the coefficient of friction $= \alpha \tan \phi'$
(α is usually 0·5. In case of grid reinforcement, $\alpha = 1·0$ owing to mechanical anchorage)
$= \mathbf{0·70.}$

$$T_i = \frac{2}{2}[0·7L_i(19·0b_i + 23·4) + 0]$$

$=$ (where depth 0–3 m, $b_i = 3$, $L_i = 6$)
$= \mathbf{337·7\ kN/m}$
(where depth 3–6 m, $b_i = 6$, $L_i = 7·5$)
$= \mathbf{721·3\ kN/m}$
(where depth 6–9·2 m, $b_i = 9·2$, $L_i = 9$)
$= \mathbf{1248·6\ kN/m}$

Conclusion
Local stability requirements satisfied.

Spacing of reinforcement

	Alternative 1	
Depth (m)	Spacing (m)	Reinforcement layers
Top		
0–3	0·5	22–26
3–5	0·4	16–21
5–7	0·3	9–20
7–9·2	0·25	1–8
Bottom		

	Alternative 2	
Depth (m)	Spacing (m)	Reinforcement layers
Top		
0–2	0·5	12–15
2–5	0·8	9–11*
5–7	0·6	5–8*
7–9·2	0·5	1–7*
Bottom		

Alternative 1 requires 26 layers of reinforcement
Alternative 2 requires 15 layers of reinforcement
(layers 1–11 inclusive* use double grid with permissible resistance of 33·2 kN/m).

Connection system and reinforcement cut-off
Consider Fig. 11.6.

The use of two layers of grid together permits a degree of cut-off in reinforcement. Assume the distribution of tension in the reinforcement in accordance with Fig. 11.7.

From observation, T_{max} (tension in reinforcement) occurs at $L/3$, measured from face of structure.

T_x (tension at face less than T_{max} but (BE 3/78) assumes $T_x = T_{max}$).

Assumption: Tension in reinforcement at $A-A = T_{max}/2$

Conclusion
In Fig. 11.6, length of layer 1 = $2L/3$, layer 2 = L.

11.2.7 Check wedge stability

Consider wedge stability at depth **4·5 m** and **9·2 m**.
Fig. 11.8. Failure plane inclination,

$$\beta' = 45°\phi'/2 = \mathbf{27·5°}$$

Consider 4.5 m

Weight of wedge = $19·0 \times \frac{1}{2}(4·5)^2 \tan 27·5° = 100·1$ kN/m
Surcharge = $23·4 \times 4·5 \times \tan 27·5° = \underline{54·8}$ kN/m
Therefore total vertical force, $W = \underline{154·9}$ kN/m

Horizontal shear force, $F = 17·4$ kN/m/m (F_3)
Frictional force, $\mu N = \alpha' \tan \phi' N \quad 0·5 \tan 35° N = 0·35 N$
Resolving vertically,

$N \sin \beta' + P \cos \beta' = W$
$0·46N + (0·89 \times 0·35N) = 154·9$
Therefore, $N = 154·9/(0·46 + 0·31) = \mathbf{201}$ **kN/m**

Resolving horizontally,

$T + P \sin \beta' = F + N \cos \beta'$
$T + (0·46 \times 0·35N) = 17·4 + 0·89N$
$T = 17·4 + (0·89 - 0·16)N = \mathbf{164·1}$ **kN/m**

Fig. 11.6

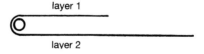

Fig. 11.7

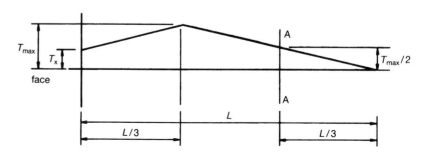

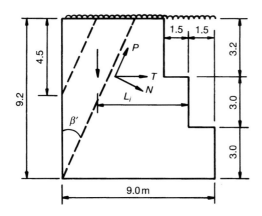

Fig. 11.8

Tensile resistance of layers   12 to 15 = 33·2 kN/m
                                                10 to 11 = 66·4 kN/m

Frictional resistance of reinforcements beyond wedge,

$$R_F = \frac{P}{2}\mu L_i(\gamma b_i + w_s)$$

$$= \frac{2}{2} \times 0.7 L_i(19.0 b_i + 23.4)$$

$$= 13.3 b_i L_i + 16.4 L_i$$

Considering individual layers of reinforcement:

Layer	L_i (m)	b_i (m)	R_F (kN/m)
15	3·8	0·4	20·2 + 62·3 = 82·5
14	4·1	0·9	49·1 + 67·2 = 116·3
13	4·4	1·4	
12	4·7	1·9	Therefore, use full strength
11	5·2	2·9	
10	6·9	3·7	Total $R_F \gg T$

Conclusion

By inspection, R_F greatly exceeds T.

Consider 9.2 m

Weight of wedge $= 19.0 \times \frac{1}{2} \times (9.2)^2 \tan 27.5° = 418.5$ kN/m
        Surcharge $= 23.4 \times 9.2 \times \tan 27.5° = 112.0$ kN/m

Therefore total vertical force, $W = 530.5$ kN/m

Frictional force $= 0.35N$ kN/m

Resolving vertically,

$$N \sin \beta' + P \cos \beta' = W$$

329

$$0.46N + (0.82 \times 0.35N) = 530.5$$
$$N = 530.5/(0.46 + 0.31) = \textbf{689.0 kN/m}$$

Resolving horizontally,

$$T + P \sin \beta' = F + N \cos \beta'$$
$$T + (0.46 \times 0.35N) = 17.4 + 0.89N$$
$$T = 17.4 + (0.89 - 0.16N) = \textbf{520.3 kN/m}$$

Tensile resistance layers 0 to 11 = 66.4 kN/m
12 to 15 = 33.2 kN/m

Frictional resistance of each layer, $R_F = 13.3 h_i L_i + 16.4 L_i$

Consider individual layers of reinforcement:

Layer	L_i (m)	h_i (m)	R_F (kN/m)
15	1·4	0·4	7·4 + 22·9 = 30·3
14	1·7	0·5	20·3 + 27·9 = 48·2
13	2·0	1·4	37·2 + 32·8 = 65·0
12	2·3	1·9	= > 66·4
11	4·8	2·9	= 66·4
1–10			= > 66.4
		Total $R_F \gg T$	

Conclusion
Wedge stability satisfied.

11.2.8 Schedule of reinforcement layout

Facing unit: Use 1·2 m double-tee pre-tensioned concrete beam standing on edge, Fig. 11.9.

Numbering reinforcing layers from *top to bottom*. *Note*: this is different from earlier calculations.

Fig. 11.9

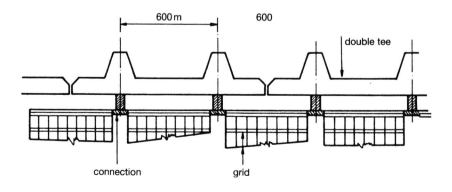

connection   grid

7·0–9·2 m wall

Depth (m)	Reinforcement layer no.	Single/double	Length L_1 and L_2* (m)		Total length† per level (m)
0·5	1	Single	6·0	—	6·0
1·0	2	Single	6·0	—	6·0
1·5	3	Single	6·0	—	6·0
2·2	4	Single	6·0	—	6·0
2·9	5	Double	6·0	4·0	10·0
3·7	6	Double	7·5	5·0	12·5
4·5	7	Double	7·5	5·0	12·5
5·1	8	Double	7·5	5·0	12·5
5·8	9	Double	7·5	5·0	12·5
6·4	10	Double	9·0	6·0	15·0
7·0	11	Double	9·0	6·0	15·0
7·5	12	Double	9·0	6·0	15·0
8·0	13	Double	9·0	6·0	15·0
8·5	14	Double	9·0	6·0	15·0
9·0	15	Double	9·0	6·0	15·0

6·0–7·0 m wall

Depth (m)	Reinforcement layer no.	Single/double	Length L_1 and L_2* (m)		Total length† per level (m)
0·5	1	Single	5·0	—	5·0
1·0	2	Single	5·0	—	5·0
1·5	3	Single	5·0	—	5·0
2·2	4	Single	5·0	—	5·0
2·9	5	Double	5·0	3·5	8·5
3·7	6	Double	5·0	3·5	8·5
4·5	7	Double	5·0	3·5	8·5
5·1	8	Double	5·0	3·5	8·5
5·8	9	Double	5·75	4·0	9·75
6·4	10	Double	5·75	4·0	9·75
7·0	11	Double	5·75	4·0	9·75

0·0–6·0 m wall

Depth (m)	Reinforcement layer no.	Single/double	Length L_1 and L_2* (m)		Total length† per level (m)
0·5	1	Single	5·0	—	5·0
1·0	2	Single	5·0	—	5·0
1·5	3	Single	5·0	—	5·0
2·2	4	Single	5·0	—	5·0
2·9	5	Double	5·0	3·5	8·5
3·7	6	Double	5·0	3·5	8·5
4·5	7	Double	5·0	3·5	8·5
5·1	8	Double	5·0	3·5	8·5
5·8	9	Double	5·0	3·5	8·5

* L_1 refers to bottom layer; L_2 to top layer in double reinforcement.
† Each grid 500 mm wide on a 600 mm module, i.e. two grids per double tee.

11.2.9 Connections

Use slidable system/attachment in accordance with the sliding or York method of constructions (Chapters 7 and 8).

Connecting pin
Use 20×30 mm bar as connecting pin, Fig. 11.10.
  Maximum load in pin from reinforcement $= 33\cdot2$ kN

$$\text{Maximum moment in pin, } M = \frac{WL}{4} = \frac{33\cdot2}{2} \times \frac{75}{2} \times 10^3 \text{ Nmm}$$

$$I_{\text{pin}} = \frac{bd^3}{12}$$

$$\text{Stress in pin} = \frac{M \times y}{I_{\text{pin}}} \text{ (allowable} = 140 \text{ N/mm}^2)$$

$$= \frac{32\cdot2 \times 75 \times 10^3 \times 10 \times 12}{4 \times 20 \times 30^3} = 138 \text{ N/mm}^2$$

Check shear stress, maximum allowable $= 20 \times 30 \times 120 = \textbf{72 kN}$

Brackets
Use 90×5 mm mild steel brackets, Figs 11.10 and 11.11.
  Maximum load per bracket $= 33\cdot2/2 = \textbf{16\cdot6 kN}$
  Maximum allowable force in bracket

$$= (90 - 34\dagger - 1\cdot5^*) \times (5 - 1\cdot5^*) \times 120$$
$$= 54\cdot5 \times 3\cdot5 \times 120 = \textbf{22\cdot9 kN}$$

Pullout force $= 16\cdot6$ kN (embedded 100 mm into concrete unit)

$$\text{Local bond stress} = \frac{16\cdot6 \times 10^3}{2 \times 90 \times 100} = \textbf{0\cdot99 N/mm}^2$$
$$(\text{Allowable} = 1\cdot47)$$

* Corrosion allowance; † Diameter of hole for connecting pin.

Fig. 11.10

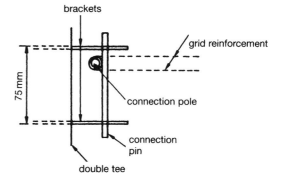

brackets

grid reinforcement

75 mm

connection pole

connection pin

double tee

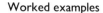

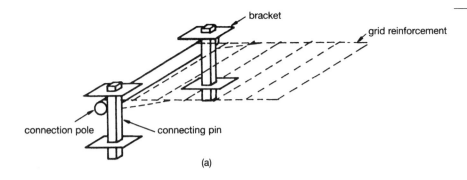

(a)

* corrosion allowance
\# diameter of hole for connecting pin

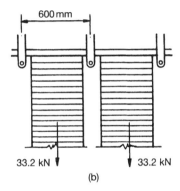

(b)

Fig. 11.11

Connecting pole
Consider connecting pole made from continuous circular hollow section (or similar), Fig. 11.11.

$$M_{\text{pole}} \cong \frac{WL}{12} = \frac{33 \cdot 2 \times 0 \cdot 600}{12} = \mathbf{0 \cdot 83 \times 10^6 \ Nmm}$$

Assume tube made from grade 50C steel $f_p = 232$
Therefore, required section modulus

$$Z = \frac{0 \cdot 83 \times 10^6}{232} = 3 \cdot 5 \times 10^3 \ \text{mm}^3$$

Use tube 48·3 mm diameter and 3·2 mm wall thickness $= \mathbf{4 \cdot 8 \times 10^3 \ mm^3}$

Joints in connecting pole made from 200 mm lengths of 60·3 mm diameter tube, 5·0 mm wall thickness.

Conclusion
Connections are adequate.

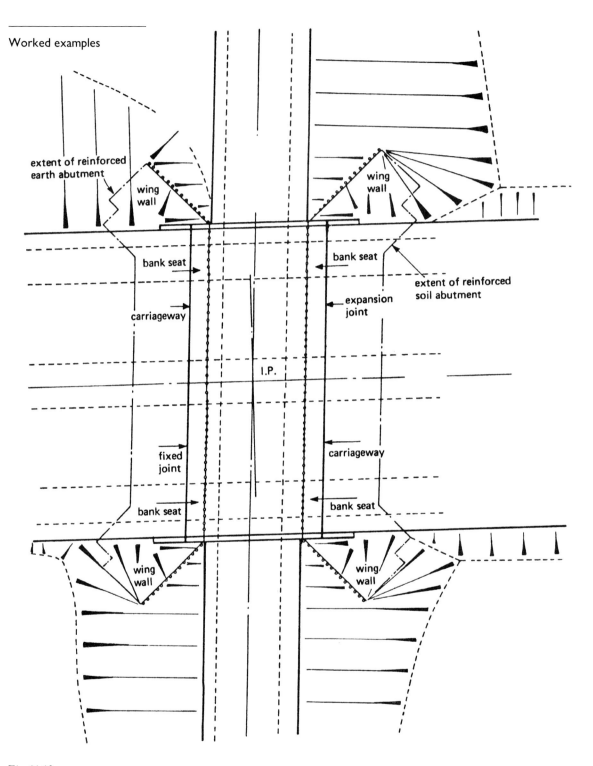

Fig. 11.12

11.3 Example 2

11.3.1 The problem

The design of a reinforced soil abutment supporting a conventional bridge deck capable of carrying British Standards *Highway loading* in accordance with BS 5400: Part 2: 1978, Fig. 11.12.

Assume: Design in accordance with the UK Department of Transport criteria using this tie-back analysis.

Material properties of fill:

$\gamma_{fill} = 20 \text{ kN/m}^3; K_a = 0{\cdot}3; c' = 0; \phi' = 32°$

Foundation:

Allowable bearing pressure 450 kN/m²*; $c' = 0; \phi' = 30°$

Layout of bridge is in accordance with Figs 11.12 and 11.17.

From experience the layout of the abutment is assumed to be similar to Fig. 11.13.

* A high bearing capacity is assumed to simplify the example; when weak foundation conditions exist the techniques illustrated in Chapter 6 may be used.

Fig. 11.13

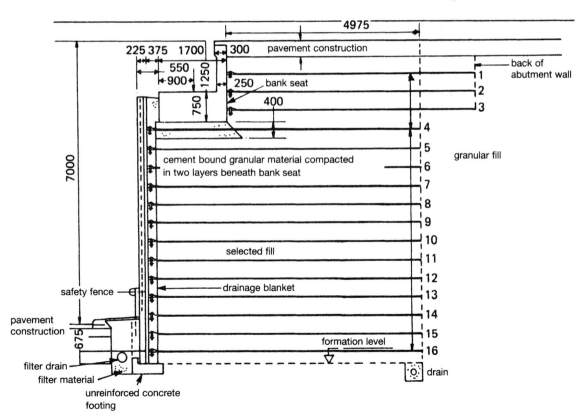

11.3.2 Internal stability

Height of reinforced fill above the layer of elements

$$T_{hi} = [K_a \gamma b_i - 2c' \sqrt{K_a}] V$$
$$0 \cdot 3 \times 20 b_i \times 0 \cdot 45$$
$$= 2 \cdot 7 b_i (3 \cdot 0 b_i \text{ for } V = 0 \cdot 5; 6 b_i \text{ for } V = 1 \cdot 0)$$

Uniform surcharge on top of the wall

HB surcharge $= 1 \cdot 2$ m of fill $= 1 \cdot 2 \times 20 = 24$ kN/m^2

$$T_{wi} = K_a w_s V = 0 \cdot 3 \times 24 \times 0 \cdot 45$$
$$5 \cdot 4 \text{ kN for strap 1;}$$
$$= 3 \cdot 6 \text{ kN for straps 2–4 inc.}$$
$$= 3 \cdot 24 \text{ kN for straps 5–15 inc.}$$

Vertical loading of bank seat

Dead load bearing force
 $= 85 \cdot 25$ kN (from bridge deck calculations)
Live load bearing force
 $= 213 \cdot 22$ kN (from bridge deck calculations)
Dead load force of bank seat, Fig. 11.14
 $= (1 \cdot 25 \times 0 \cdot 25 \times 24) + (1 \cdot 7 \times 0 \cdot 75 \times 24)$
 $= 7 \cdot 5 + 30 \cdot 6$
 $= 38 \cdot 1$ kN

Total dead load $= 38 \cdot 1 + 85 \cdot 25$
$$= \mathbf{123 \cdot 35 \text{ kN}}$$

$$\bar{x} = \frac{(7 \cdot 5 \times 0 \cdot 125) + (30 \cdot 6 \times 0 \cdot 85) + (85 \cdot 25 \times 0 \cdot 8)}{123 \cdot 35}$$

$$= 0 \cdot 77 \text{ m}$$

Therefore, dead load $e = 0 \cdot 08$ m; live load $e = 0 \cdot 05$ m

$b = \mathbf{1 \cdot 7};\ d = \mathbf{1 \cdot 175}$

For $b_i < (2d - b) = (2 \times 1 \cdot 175) - 1 \cdot 7 = 0 \cdot 65$;
$D_i = b_i + 1 \cdot 7$

For $b_i > (2d - b) = 0 \cdot 65$; $D_i = 1 \cdot 175 + \dfrac{(b_i + 1 \cdot 7)}{2}$

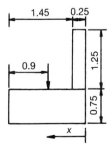

Fig. 11.14

336

$$T_{si} = K_a V \frac{S_i}{D_i}\left[1 + \frac{6e}{b}\right] = 0{\cdot}3 \times 0{\cdot}45 \frac{S_i}{D_i}\left[1 + \frac{6e}{1{\cdot}7}\right]$$

$$= 0{\cdot}135 \frac{S_i}{D_i}[1 + 3{\cdot}53e]$$

For dead load case:

$$T_{si} = 0{\cdot}135 \times 123{\cdot}35 \times \frac{[1 + (3{\cdot}53 \times 0{\cdot}08)]}{D_i} = \frac{21{\cdot}35}{D_i}$$

For live load case:

$$T_{si} = 0{\cdot}135 \times 213{\cdot}22 \times \frac{[1 + (3{\cdot}53 \times 0{\cdot}05)]}{D_i} = \frac{33{\cdot}87}{D_i}$$

See Tables 11.3 and 11.4.

Horizontal forces
Consider Fig. 11.15.

$$\text{Let } Q = \frac{\tan(45° - \phi'/2)}{d + \dfrac{b}{2}}$$

$$\text{Height} = \frac{d + \dfrac{b}{2}}{\tan(45° - \phi'/2)} = \frac{1}{Q}$$

$$\text{Pressure } z = \frac{y(x - H)}{x}$$

$$F_i = \left[y + \frac{y(x - H)}{x}\right]\frac{H}{2}$$

$$\text{Pressure } y = \frac{2F_i}{H} \bigg/ \left[1 + \frac{(x - H)}{x}\right] = \frac{2F_i x}{H(2x - H)}$$

$$TF_i = \frac{2F_i}{H}\frac{x}{(2x - H)} \times \frac{x - b_i}{x} \times V = \frac{2F_i}{H}\frac{(x - b_i)}{2x - H}V$$

$$= \frac{2F_i V}{H}\frac{\left(\dfrac{1}{Q} - b_i\right)}{\dfrac{2}{Q} - H} = \frac{2F_i V}{H}\left(\frac{1 - b_i Q}{2 - HQ}\right)$$

$x < H$, then $H = x = 1/Q$ and the formula reduces to:

$$TF_i = \frac{2F_i V}{1/Q}\left(\frac{1 - b_i Q}{2 - Q/Q}\right) = 2F_i VQ(1 - b_i Q)$$

HB braking forces

Case 1
A 5 m length of the braking area is taken directly by the abutment, Fig. 11.16.

337

Table 11.3. *Tensile force in different layers of reinforcement due to self-weight*

Layer	1	2	3	4	5	6	7	8	9	10	11	12	13	14	15	16
b_i	0·75	1·25	1·75	2·25	2·7	3·15	3·6	4·05	4·5	4·95	5·4	5·85	6·3	6·75	7·2	7·5
T_{hi}	4·5	3·75	5·25	6·75	7·29	8·51	9·72	10·94	12·15	13·37	14·58	15·80	17·01	18·23	19·44	20·66

Table 11.4. *Tensile force in different layers of reinforcement due to external loads*

Layer	1	2	3	4	5	6	7	8	9	10	11	12	13	14	15	16
H_i	—	—	—	0·25	0·7	1·15	1·6	2·05	2·5	2·95	3·4	3·85	4·3	4·75	5·2	5·65
D_i	—	—	—	1·95	2·375	2·6	2·825	3·05	3·275	3·5	3·725	3·95	4·175	4·4	4·625	4·85
T_{si} (dead)	—	—	—	10·95	9·00	8·21	7·56	7·0	6·52	6·1	5·73	5·40	5·11	4·85	4·62	4·40
T_{si} (live)	—	—	—	17·37	14·26	13·03	12·00	11·10	10·34	9·68	9·09	8·57	8·11	7·70	7·32	6·98

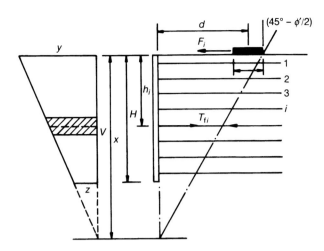

Fig. 11.15

$$\text{Total force} = \frac{5}{9} \times 450 = 250 \text{ kN}$$

(a) If this force is distributed over a 3 m width,

Force/metre $= 250/3 = 83 \cdot 3$ kN/m

(b) If the force is taken as distributed at 45° from a point at the centre of the loaded area, i.e. distributed over 5 m width of abutment,

Force/metre $= 250/5 = 50$ kN/m

(c) A force/metre $= \mathbf{65\ kN/m}$ is adopted for the calculations

$$\text{Force acting on deck} = \frac{4}{9} \times 450 = 200 \text{ kN}$$

Considering the force to be resisted solely by the bearings at the fixed end, Fig. 11.17.

Let the force due to the moment on bearing A be F_A

$$\text{Moment of resistance of group} = \frac{F_A}{t_a} \Sigma x^2 = \text{force acting on deck} \times \text{eccentricity}$$

Therefore, $\dfrac{2F_A}{6 \cdot 5}(6 \cdot 5^2 + 5 \cdot 5^2 + 4 \cdot 5^2 + 3 \cdot 5^2 + 2 \cdot 5 + 1 \cdot 5^2 + 5^2) = 200 \times 3 \cdot 5$

$F_A = 21 \cdot 16$ kN

Therefore, load due to moment $= 21 \cdot 16$ kN

Load due to shear $= 200/14 = 14 \cdot 28$ kN

$$\text{Therefore, maximum bearing load} = 21 \cdot 16 + 14 \cdot 28$$
$$= \mathbf{35 \cdot 44\ kN}$$

$$\text{The loading at bearing B} = \frac{200 \times 3 \cdot 5 \times 3 \cdot 5}{2 \times 107 \times 5} + 14 \cdot 28$$

$$= \mathbf{25 \cdot 68\ kN}$$

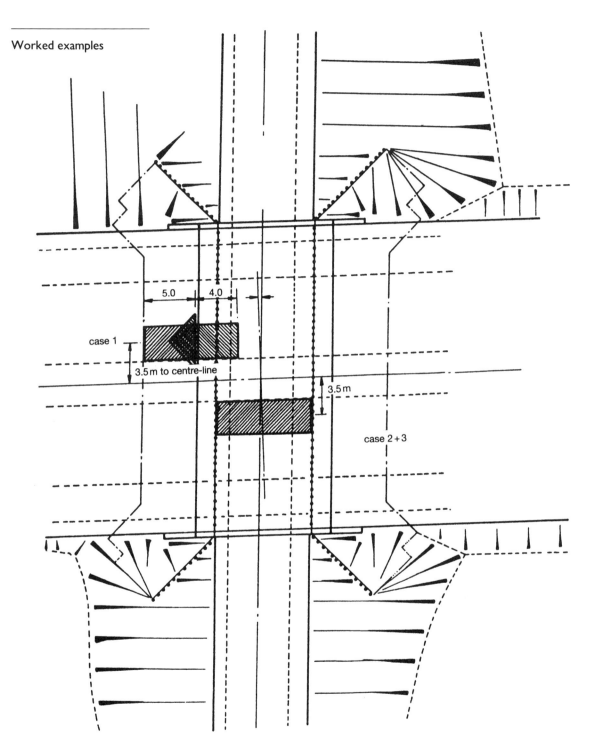

Worked examples

case 1

5.0

4.0

3.5 m to centre-line

3.5 m

case 2 + 3

Fig. 11.16

340

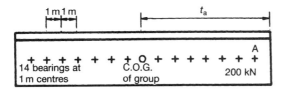

Fig. 11.17

In case 1, Fig. 11.18, it is assumed that of the stresses due to the 35·44 kN load, 50% is taken directly by 1, 2 and 3 reinforcing layers.

Average force in reinforcements 1, 2 and 3 = 5·9 kN

The remaining 17·74 kN is assumed distributed as follows:

$$Q = \tan(45° - 16°)/2·075 = 0·267$$

$1/Q = 3·74$ m. Therefore, straps 4–11 are affected.

$$\begin{aligned} TF_i &= 2F_i V Q (1 - h_i Q) \\ &= 2 \times 17·74 \times 0·45 \times 0·267[1 - (h_i \times 0·267)] \\ &= 4·263(1 - 0·267h_i) \end{aligned}$$

See Table 11.5.
For the 65 kN load, $Q = \tan 29°/7·0 = 0·0792$

Therefore, all reinforcement stressed

$$TF_i = \frac{2F_i V (1 - h_i Q)}{H(2 - HQ)} = \frac{2 \times 65 \times 0·45(-0·0792h_i)}{7·925(2 - 7·925 \times 0·0792)}$$

$$= 5·379(1 - 0·0792h_i)$$

See Table 11.6.

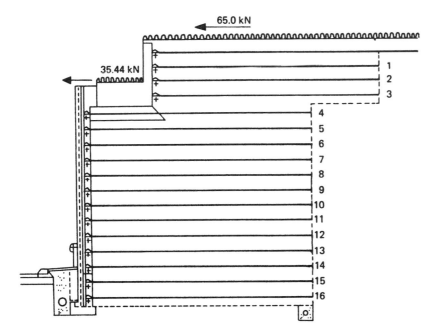

Fig. 11.18

Table 11.5. Tensile force from external load—Case 1, Fig. 11.18

Layer	1	2	3	4	5	6	7	8	9	10	11
b_i	—	—	—	0·25	0·70	0·15	1·60	2·05	2·50	2·95	3·40
TF_i	6·85	5·9	4·95	4·03	3·52	3·00	2·50	1·98	1·47	0·96	0·45

Table 11.6. Tensile force from external load—all reinforcements stressed

Layer	1	2	3	4	5	6	7	8	9	10	11	12	13	14	15	16
b_i	0·75	1·25	1·75	2·25	2·70	3·15	3·6	4·05	4·5	4·95	5·4	5·85	6·3	6·75	7·2	7·65
TF_i	11·24	4·84	4·63	4·42	4·23	4·04	3·85	3·65	3·46	3·27	3·08	2·89	2·69	2·50	2·31	2·12
Additional force	6·85	5·9	4·95	4·03	3·52	3·00	2·50	1·98	1·47	0·96	0·45	—	—	—	—	—
Total force	18·09	10·74	9·58	8·45	7·75	7·04	6·35	5·63	4·93	4·23	3·53	2·89	2·69	2·50	2·31	2·12

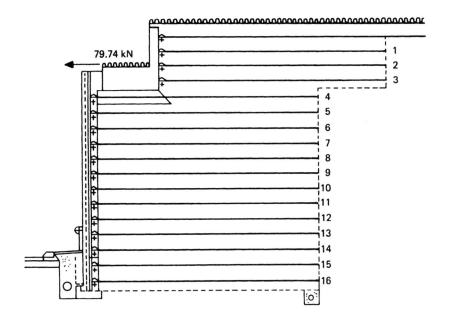

Fig. 11.19

Case 2

The entire braking force is on the deck.

Maximum bearing load $= 35\cdot44 \times 9/4 = 79\cdot74$ kN

In case 2, Fig. 11.19, it is assumed that reinforcements 1, 2 and 3 take no load.

$Q = \tan (45°-16°)/2\cdot075 = 0\cdot267$

$1/Q = 3\cdot74$ m. Therefore, straps 4–11 are affected.

$$\begin{aligned} TF_i &= 2F_i VQ(1 - b_i Q) \\ &= 2 \times 79\cdot74 \times 0\cdot45 \times 0\cdot267 \left[1 - (b_i \times 0\cdot267)\right] \\ &= 19\cdot16(1 - 0.267 b_i) \end{aligned}$$

See Table 11.7.

If all the force were to be taken by reinforcements 1, 2 and 3 say in proportion to their depths, then Table 11.8 applies.

Case 2A

In Case 2A, Fig. 11.20, a failure wedge passing close to the bank seat is considered.

$1/Q = 5\cdot24$ m, therefore, $Q = 0\cdot191$

Table 11.7. Tensile force due to breaking load reinforcements 4–16 loaded

Layer	4	5	6	7	8	9	10	11
b_i	0·25	0·70	1·15	1·60	2·05	2·50	2·95	3·40
TF_i	17·88	15·58	13·28	10·97	8·67	6·37	4·07	1·77

343

Table 11.8. *Tensile force in reinforcements 1–3*

Layer	1	2	3
b_i	0·75	1·25	1·75
TF_i	15·95	26·58	37·21

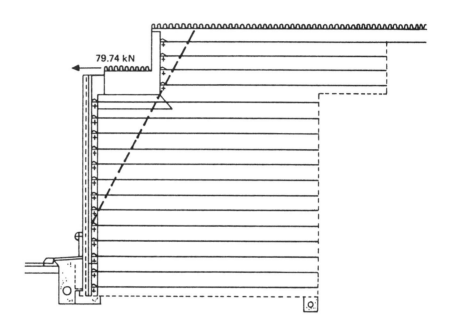

79.74 kN

Fig. 11.20

Table 11.9. *Tensile force in reinforcements, Case 2A*

Layer	1	2	3	4	5	6	7	8	9	10	11
b_i	0·25	0·75	1·25	1·75	2·20	2·65	3·1	3·55	4·00	4·45	4·90
TF_i	14·51	13·05	11·60	10·14	7·95	6·77	5·59	4·41	3·24	2·06	0·88

Table 11.10. *Tensile force in reinforcements, Case 3*

Layer	1	2	3	4	5	6	7	8	9	10	11
TF_i	15·41	13·28	11·14	9·07	7·92	6·75	5·63	4·46	3·31	2·16	1·01

Table 11.11. HB braking forces summary

Layer	1	2	3	4	5	6	7	8	9	10	11	12	13	14	15	16
Case 1	18·09	10·74	9·58	8·45	7·75	7·04	6·35	5·63	4·93	4·23	3·53	2·89	2·69	2·50	2·31	2·12
Case 2	15·95	26·58	37·21*													
Case 2A	14·51	13·03	11·60	10·14	7·95	6·77	5·59	4·41	3·24	2·06	0·88					
Case 3	15·41	13·28	11·14	9·07	7·92	6·75	5·63	4·46	3·31	2·16	1·01					
Envelope	18·09	13·28	11·6	17·88	15·58	13·28	10·97	8·67	6·37	4·23	3·53	2·89	2·69	2·50	2·31	2·12
Case	1	3	2A	2	2	2	2	2	2	1	1	1	1	1	1	1

Additional row under Case 2: 17·88 | 15·58 | 13·28 | 10·97 | 8·67 | 6·37 | 4·07 | 1·77 (layers 4–11)

* Ignore as unrealistic idealization of distribution of braking force, Table 11.8.

$$TF_i = 2F_iVQ(1-h_iQ)$$
$$= 2 \times 79.74 \times 0.45 \times 0.191[1-(h_i \times 0.191)]$$
$$= 13.71(1-0.191h_i)$$

See Table 11.9.

Case 3

As with case 2, this entire braking force is applied to the deck. If 50% of the 79.74 kN force is taken by reinforcements 1, 2 and 3, as in Fig. 11.19, the reinforcement forces are as for Case 1.

$$\text{Load} \times \frac{79.74}{35.44} = 2.25$$

See Table 11.10.

A summary of the HB braking force condition is shown in Table 11.11.

Bending moment caused by external loading acting on the wall (excluding bank seat loads) see Fig. 11.21.

$$T_{mi} = \frac{6K_aVM_i}{L_i^2}$$

Bending moment at level h_i

$$h_i = \frac{K_a\gamma h_i^3}{6} + \frac{K_aw_sh_i^2}{2} + F_ih_i$$

$$I \text{ of section } \frac{1 \times L_i^3}{12}. \ Z \text{ of section } = \frac{L_i^2}{6}$$

Maximum vertical stress on section

$$= \frac{K_a\gamma h_i^3}{L_i^2} + \frac{3K_aw_sh_i^2}{L_i^2} + \frac{6F_iH_i}{L_i^2}$$

Therefore, maximum stress in reinforcement

$$T_{max} = K_aV\left(\frac{K_a\gamma h_i^3}{L_i^2} + \frac{K_aw_sh_i^2}{L_i^2}\right)$$

$$= K_s^2V\frac{h_i^2}{L_i^2}(\gamma h_i + 3w_s) + K_aV6\frac{F_ih_i}{L_i^2}$$

Fig. 11.21

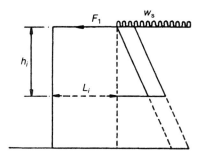

Table 11.12. Calculation of T_{max}

Layer	1	2	3	4	5	6	7	8	9	10	11	12	13	14	15	16
b_i	0·75	1·25	1·75	2·25	2·70	3·15	3·6	4·05	4·5	4·19	5·4	5·85	6·3	6·75	7·2	7·65
$0.0165b_i^3$	0·03	0·07	0·19	0·19	0·32	0·52	0·77	1·10	1·50	2·00	2·60	3·30	4·13	5·07	6·16	7·39
$0.0595b_i^2$	0·13	0·21	0·40	0·30	0·43	0·59	0·77	0·98	1·20	1·46	1·74	2·03	2·36	2·71	3·08	3·48
$1.103b_i$	0·83	1·38	1·93	2·48	2·98	3·47	3·97	4·47	4·96	5·46	5·96	6·45	6·95	7·45	7·94	8·44

Table 11.13. Summary of internal stability analyses

| Layer | T_{h_i} (H of fill) | T_{w_i} (surcharge) HB | Vertical loads on bank seat | | T_{f_i} (HB braking force) | Moment due to external loading | | | | | Permanent load combination 2 + 4 + 7 + 10 | 12 + HB vertical loads QN deck 12 + 5 + 11 | 12 + HB vertical loads QN deck with HB surcharge on abutment and approaches $(\frac{1}{2} \times 5) + 8 + 3 + (\frac{1}{2} \times 11)$ | 13 + 6 HB vehicle braking on deck | 14 + 6 + 9 HB vehicle braking on abutment with HA on deck | Envelope |
| | | | T_{s_i} (dead) | T_{s_i} (live) | | T_{m_i} (active) | T_{m_i} (surcharge) | T_{m_i} (HB braking) | $T.M_i$ (bank seat dead load) | $T.M_i$ (bank seat HB live load) | | | | | | |
1	2	3	4	5	6	7	8	9	10	11	12	13	14	15	16	17
1	4·5	5·4	—	—	18·09	0·03	0·13	0·83	—	—	9·11	17·13	18·65	35·22	37·57	37·57
2	3·75	3·6	—	—	13·28	0·07	0·21	1·38	—	—	8·4	16·42	16·22	29·70	30·88	30·88
3	5·25	3·6	—	—	11·6	0·19	0·40	1·93	—	—	10·02	18·04	18·03	29·64	31·56	31·56
4	6·75	3·6	10·97	17·37	17·88	0·19	0·30	2·48	4·58	8·02	22·47	47·86	39·07	65·74	59·43	65·74
5	7·29	3·24	9·00	14·26	15·58	0·32	0·43	2·98	4·58	8·02	21·19	43·47	36·00	59·05	54·56	59·05
6	8·51	3·24	8·21	13·03	13·28	0·52	0·59	3·47	4·58	8·02	21·82	42·87	36·175	56·15	52·93	56·15
7	9·72	3·24	7·56	12·00	10·97	0·77	0·77	3·97	4·58	8·02	22·63	42·65	36·65	53·62	51·59	53·62
8	10·94	3·24	7·0	11·10	8·67	1·10	0·98	4·47	4·58	8·02	23·62	42·74	37·4	51·41	50·54	51·41
9	12·15	3·24	6·52	10·34	6·37	1·50	1·20	4·96	4·58	8·02	24·75	43·11	38·37	49·46	49·70	49·70
10	13·37	3·24	6·1	9·68	4·23	2·00	1·46	5·46	4·58	8·02	26·05	43·75	39·6	47·96	49·29	49·29
11	14·58	3·24	5·73	9·09	3·53	2·60	1·74	5·96	4·58	8·02	27·49	44·60	41·03	48·13	50·52	50·52
12	15·80	3·24	5·40	8·57	2·89	3·30	2·03	6·95	4·58	8·02	29·08	45·67	42·65	48·56	51·99	51·99
13	17·01	3·24	5·11	8·11	2·69	4·13	2·36	6·95	4·58	8·02	30·83	46·96	44·50	49·65	54·14	54·14
14	18·23	3·24	4·85	7·70	2·50	5·07	2·71	7·45	4·58	8·02	32·73	48·45	46·54	50·95	56·49	56·49
15	19·44	3·24	4·62	7·32	2·31	6·16	3·08	7·94	4·58	8·02	34·80	50·14	48·79	52·45	59·04	59·04
16	20·66	3·24	4·40	6·98	2·12	7·39	3·48	8·44	4·58	8·02	37·03	52·03	51·25	54·15	61·81	61·81

Table 11.14. Bond length requirements for grid reinforcement

Layer	1	2	3	4	5	6	7	8	9	10	11	12	13	14	15	16
b_i	0·75	1·25	1·75	2·25	2·7	3·15	3·6	4·05	4·5	4·95	5·4	5·84	6·3	6·75	8·2	7·6
L_i	0·11	0·065	0·046	0·036	0·03	0·026	0·022	0·02	0·018	0·016	0·015	0·014	0·013	0·012	0·011	0·0
Envelope free (Table 11.11)	37·6	31·0	31·6	65·7	59·0	56·1	53·6	51·4	49·3	49·3	50·5	52·0	54·1	56·5	59·0	6·18
Bond length (m)	4·1	2·0	1·5	2·4	1·8	1·5	1·2	1·0	1·0			less than 1 m				

Table 11.15. Bond length requirements for ribbed strip reinforcement

Layer	1	2	3	4	5	6	7	8	9	10	11	12	13	14	15	16
b_i	0·75	1·25	1·75	2·25	2·7	3·15	3·6	4·05	4·5	4·35	5·4	5·84	6·3	6·75	7·2	7·65
L_i	0·84	0·5	0·36	0·28	0·23	0·2	0·17	0·35	0·14	0·13	0·11	0·1	0·1	0·1	0·1	0·1
Envelope free (Table 11.13)	37·6	31·0	38·6	65·7	59·0	56·1	53·6	51·4	49·7	49·3	50·5	52·0	54·1	56·5	59·0	61·8
Bond length (m)	31·6	15·5	13·8	10·8	13·5	11·2	9·1	7·7	7·0	6·4	5·5	5·2	5·4	5·6	5·9	6·1

(The effectiveness of grid reinforcement with respect to bond is demonstrated by comparing Tables 11.14 and 11.15)

349

In order that the moment at level 16 is correct, the case 2 horizontal force of 79·74 kN is replaced by a force of

$$79{\cdot}74 \times \frac{6{\cdot}4}{7{\cdot}65} = 66{\cdot}71 \ \text{kN}$$

Acting at the carriageway surface:

$$K_a^2 \frac{V h_i^2}{L_i^2} \gamma h_i = \frac{0{\cdot}3^2 \times 0{\cdot}45 \times h_i^3 \times 20}{7^2} = 0{\cdot}0165 h_i^3$$

$$K_a^2 \frac{V h_i^2}{L_i^2} 3 w_s = \frac{0{\cdot}3^2 \times 0{\cdot}45 h_i^2 \times 3 \times 24}{7^2} = 0{\cdot}059 h_i^2$$

$$\frac{6 F_i h_i}{L_i^2} K_a V = \frac{6 \times 66{\cdot}71 \times h_i}{7^2} \times 0{\cdot}3 \times 0{\cdot}45 = 1{\cdot}103 h_i$$

See Table 11.12.

Bending moment caused by vertical bank seat loads

Total dead load = 123·35 kN

Eccentricity about reinforced earth block
= 3·5 − 0·325 − 1·7 + 0·77 = 2·245 m

HB live load = 213·22 kN

Eccentricity about reinforced earth block
= 3·5 − 0·325 − 1·7 + 0·8 = 2·275 m

Reinforcement tension due to dead load moment

$$= \frac{6 K_a V M_i}{L_i^2}$$

$$= \frac{6 \times 0{\cdot}3 \times 0{\cdot}45}{7^2} \times 123{\cdot}35 \times 2{\cdot}24$$

$$= 4{\cdot}58 \ \text{kN}$$

Reinforcement tension due to live load moment

$$= \frac{6 \times 0{\cdot}3 \times 0{\cdot}45}{7^2} \times 213{\cdot}22 \times 2{\cdot}27$$

$$= 8{\cdot}02 \ \text{kN}$$

Note: This is sensitive to the relative position of the bank seat with respect to the centre of gravity of the reinforced block.

See Table 11.13.

Bond requirements

$$\text{Perimeter required } P_i = \frac{T_i}{\dfrac{\mu L_i}{2}(\gamma h_i + w_s) + \dfrac{c_r' L_i}{2}}$$

In granular soils $L_i = \dfrac{2T_i}{\mu P_i(\gamma h_i + w_s)}$

Take $\mu = \alpha' \tan 32° = 0.62\alpha'$

($\alpha = 1$ for grid reinforcement, 0.9 for ribbed strip, and 0.6 for plain strip).

Considering the case of no surcharge and 1 kN/m force T_i, the required length L_i:

(a) For grid reinforcement 1 m width,

$$L_i = \frac{2 \times 1}{0.62 \times 1 \times 2}(20h_i) = \frac{0.0806}{h_i}$$

(b) For ribbed strip reinforcement 80 mm $\times$ 5 mm,

$$L_i = \frac{2 \times 1}{0.62 \times 0.9 \times 0.283}(20h_i) = \frac{0.63}{h_i}$$

See Tables 11.14 and 11.15.

11.3.3 External stability

Consider Fig. 11.22.

Sliding

Active force $= K_a \gamma h^2/2 = 0.3 \times 20 \times 7.925^2/2 = 188.4$ kN
Surcharge force $= K_a w_s h = 24 \times 0.3 \times 7.925 \quad = 57.1$ kN
Braking force $= 65 + 35.44 \qquad\qquad\qquad = 100.5$ kN

$\qquad\qquad\qquad\qquad\qquad\qquad\qquad\qquad \Sigma\ 346.0$ kN

Resistance to sliding $= \{20 \times [(5 \times 7.925) + (2 \times 5.925)] + 123.4\} \tan 30°$
$\qquad\qquad\qquad\quad = 665.6$ kN

F.O.S. against sliding $= \dfrac{665.6}{346.0} = 1.92$

The actual F.O.S. will be greater than the calculated value since, although braking forces have been included, the mass of the force carrying these forces has not been included in calculating the resistance to sliding.

Overturning

Active moment $= 188.4 \times 7.925/3 \qquad\qquad = 497.7$ kNm
Surcharge moment $= 57.1 \times 7.925/2 \qquad\qquad = 226.3$ kNm
Braking moment $= (65 \times 7.925) + (35.44 \times 5.925) = 725.1$ kNm

$\qquad\qquad\qquad\qquad\qquad\qquad\qquad\qquad \Sigma\ 1449.1$ kNm

Resisting moment $= 20(5 \times 7.925 \times 4.5) + (2 \times 5.925 \times 1) + 123.4 \times 1.3$
$\qquad\qquad\qquad\quad = 3964$ kNm

F.O.S. against overturning $= 3964/1449.1$
$\qquad\qquad\qquad\qquad\qquad = 2.74$

Residual moment $= 3964 - 1449.1 = 2514.9$

351

Worked examples

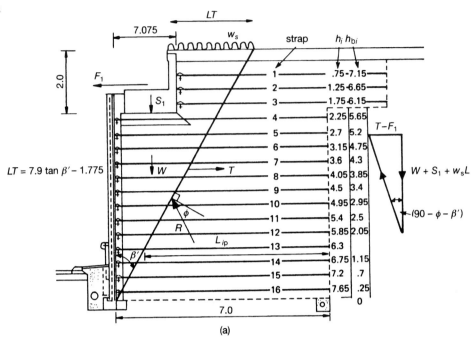

(a)

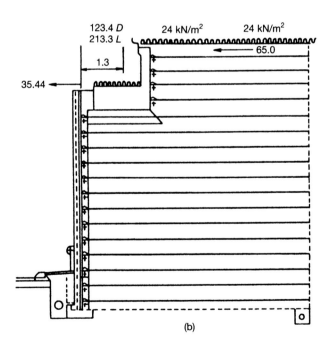

(b)

Fig. 11.22

Resultant $\bar{x} = \dfrac{2514 \cdot 9}{1152 \cdot 8} = 2 \cdot 18$ m

Bearing length $= 2 \cdot 8 \times 3 = 6 \cdot 54$ m

Maximum bearing pressure $= \dfrac{1152 \cdot 8 \times 2}{6 \cdot 54} = 352 \cdot 5$ kN/m^2

Consider additionally the 213·3 kN live load and the 24 kN/m^2 surcharge.

Moment due to 213·3 kN live load $= 213 \cdot 3 \times 1 \cdot 3 = 277 \cdot 3$ kNm
Moment due to 24 kN/m^2
   surcharge $= 5 \cdot 3 \times 24 \times 4 \cdot 35 = 553 \cdot 3$ kNm
Moment due to increased resisting
   moment $= 3964 + 553 \cdot 3 + 277 \cdot 3$
      $= 4795$ kNm
F.O.S. against overturning $= 4795/1449 \cdot 1 = 3 \cdot 3$
Residual moment $= 4795 - 1449$ $= 3346$ kNm

Resultant $\bar{x}$
$$= \frac{3346}{1152 \cdot 8 + 213 \cdot 3 + 127 \cdot 2} = \frac{3346}{1493 \cdot 3}$$
$$= 2 \cdot 24 \text{ m}$$

Bearing length $= 2 \cdot 24 \times 3 = 6 \cdot 72$ m

Maximum bearing pressure $= \dfrac{1493 \cdot 3 \times 2}{6 \cdot 72} = 444$ kN/m^2

If the additional live load is considered and not the surcharge:

Increased resisting moment $= 4241 - 1449 = 2792$ kNm

$$\bar{x} = \frac{2792}{1153 + 213} = \frac{2792}{1366} = 2 \cdot 04 \text{ m}$$

Maximum bearing pressure $= \dfrac{1366 \times 2}{3 \times 2 \cdot 04} = \mathbf{446 \ kN/m^2}$

If the braking moment is not used then

Residual moment $= \mathbf{2792} + \mathbf{725} = 3517$ kNm

$$\bar{x} = \frac{3517}{1366} = 2 \cdot 58 \text{ m}$$

$$= 0 \cdot 92 \text{ m}$$

$$= \frac{1366}{7} \left(1 + \frac{6 \times 0 \cdot 92}{7} \right)$$

$$= \mathbf{349 \ kN/m^2}$$

In the overturning case there will be greater distribution of braking force than assumed in the cases giving maximum bearing pressures of 446 kN/m^2 and it is reasonable to say that it is extremely unlikely that bearing stresses of greater than 400 kN/m^2 will be achieved.

Slip circle
The possibility of this type of failure is not considered in this example.

Wedge analysis
Consider Fig. 11.22(a).

$$T = F_1 + [W + S_1 + w_s \times LT] \tan(90° - \phi' - \beta'), L_{ip} = 7 - h_{bi} \tan \beta'$$

$$W = (7{\cdot}9^2 * \tan \beta' - 2{\cdot}0 \times 2{\cdot}075) \times 20 + 6{\cdot}9 \times 3{\cdot}26 \text{ (facing)}$$

$$W = 624{\cdot}1 \tan \beta' - 60{\cdot}5$$

Anchorage resistance, $R = \sum \dfrac{P_c L_{ip}}{2} (\mu_i \gamma h_i + \mu w_s + c_r')$

With no surcharge or soil cohesion and for grid reinforcement as 1 m slice, $\mu = \tan 32° = 0{\cdot}62$.

$$R = \sum (7 - h_{bi} \tan \beta') \times 0{\cdot}62 \times 20 \times h_i$$

(*a*) Consider the case where $F_1 = 79{\cdot}74$ kN; $S_1 = 336{\cdot}7$ kN

$\quad w_s = 24$ kN/m^2

Try $\beta' = 29°$, then $\quad W = 285{\cdot}4; LT = 2{\cdot}6$ m
$$\qquad T = 79{\cdot}74 + [285{\cdot}4 + 336{\cdot}7 + 24 \times 2{\cdot}6] \tan 29°$$
$$\qquad = 79{\cdot}74 + 684{\cdot}5 \tan 29° = 459{\cdot}16 \text{ kN}$$

Try $\beta' = 32°$, then $\quad W = 329{\cdot}5; LT = 3{\cdot}16$ m
$$\qquad T = 79{\cdot}74 + [329{\cdot}5 + 336{\cdot}7 + 24 \times 3{\cdot}16] \tan 26°$$
$$\qquad = 441{\cdot}7 \text{ kN}$$

Try $\beta' = 26°$, then $\quad W = 243{\cdot}2; LT = 2{\cdot}08$ m
$$\qquad T = 79{\cdot}74 + [243{\cdot}2 + 336{\cdot}7 + 24 \times 2{\cdot}08] \tan 32°$$
$$\qquad = 473{\cdot}3 \text{ kN}$$

Try $\beta' = 23°$, then $\quad W = 204{\cdot}4; LT = 1{\cdot}58$ m
$$\qquad T = 79{\cdot}74 + [204{\cdot}4 + 336{\cdot}7 + 24 \times 1{\cdot}58] \tan 35°$$
$$\qquad = 485{\cdot}15 \text{ kN}$$

Try $\beta' = 20°$, then $\quad W = 166{\cdot}65; LT = 1{\cdot}1$ m
$$\qquad T = 79{\cdot}74 + [166{\cdot}65 + 336{\cdot}7 + 24 \times 1{\cdot}1] \tan 38°$$
$$\qquad = 493{\cdot}6 \text{ kN}$$

(*b*) Consider failure plane through heel of bearing seat

$$\beta' = \tan^{-1}(2{\cdot}075/5{\cdot}9) = 19{\cdot}37°$$

Try $\beta' = 19{\cdot}37°$, then $W = 159{\cdot}0; LT = 1{\cdot}00$ m
$$\qquad T = 79{\cdot}74 + [15{\cdot}90 + 336{\cdot}7 + 24] \tan 38{\cdot}83°$$
$$\qquad = 495 \text{ kN}$$

Length L_{ip} of strap $\quad 1 = 7 - hb_i \tan 19{\cdot}37°$
$$\qquad = 7 - 7{\cdot}15 \tan 19{\cdot}37°$$
$$\qquad = 4{\cdot}49 \text{ m}$$

Anchorage resistance of strap $1 = 4{\cdot}49 \times 0{\cdot}62 \times 20 \times 0{\cdot}75$
$$\qquad = 41{\cdot}76 \text{ kN}$$

This is greater than the advisable failure load. All other straps have greater anchorage resistance.

$$\text{Average strap tension} = \frac{495}{16} = 30 \cdot 9 \text{ kN}$$

(c) Consider now a wedge with its base point at the level of strap 12. The angle of the plane passing through the heel of the bearing seat.

$$\beta' = \tan^{-1}(2 \cdot 075/3 \cdot 85) = 28 \cdot 32°$$

$$W = \left[\frac{5 \cdot 85^2}{2} \times \tan \beta - 2 \cdot 0 + 2 \cdot 075\right] \times 20 + (6 \cdot 9 \times 3 \cdot 26)$$

$$W = 342 \cdot 3 \tan \beta' - 60 \cdot 5 = 123 \cdot 96 \text{ kN}$$

$$LT = 5 \cdot 85 \tan \beta' - 1 \cdot 775 = 1 \cdot 38 \text{ m}$$

Therefore, $T = 79 \cdot 74 + [123 \cdot 96 + 336 \cdot 7 + 24 \times 1 \cdot 38] \tan 28 \cdot 32°$
$$= 361 \cdot 2 \text{ kN}$$

Try $\beta = 29°$, then $W = 129 \cdot 2$; $LT = 147$ m
Therefore, $T = 79 \cdot 74 + [129 \cdot 2 + 336 \cdot 7 + 24 \times 1 \cdot 47] \tan 29°$
$$= 357 \cdot 5 \text{ kN}$$

The worst case is when the wedge passes through the heel of the bearing seat.

$$\text{Average strap tension} = \frac{361 \cdot 2}{12} = 30 \cdot 1 \text{ kN}$$

Conclusion
The worst case is where the wedge passes through the toe of the wall. The average reinforced tensions are less than those calculated previously.

11.3.4 Precast unit fixings
Consider Figs 11.23, 11.24 and 11.25.

Tensile
Maximum allowable force in single beam plate

$$= (90 - 24 - 3) \times (3 - 1 \cdot 5) \times 120 \times 10^{-3}$$
$$= 11 \cdot 34 \text{ kN}$$

Shear
Maximum force $= (50 - 1 \cdot 5) \times (3 - 1 \cdot 5) \times 72 \times 2 \times 10^{-3}$
$$= 10 \cdot 48 \text{ kN}$$

Bearing
Maximum force $= 20 \times (3 - 1 \cdot 5) \times 200 \times 10^{-3}$
$$= 6 \cdot 0 \text{ kN}$$

Maximum force/m width:

$$\text{Shear} = \frac{10 \cdot 48 \times 4}{1 \cdot 2} = 34 \cdot 9 \text{ kN}$$

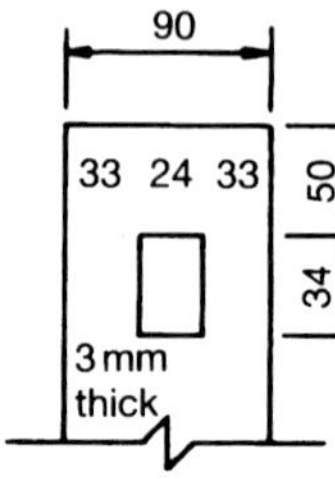

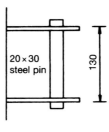

Fig. 11.23

Fig. 11.24

$$\text{Tensile} = \frac{11.34 \times 4}{1.2} = 37.8 \text{ kN}$$

$$\text{Bearing} = \frac{6 \times 4}{1.2} = \textbf{20 kN}$$

Steel pin
Let F kN = max. force permissible

$$\text{Maximum moment} = \frac{F}{2} \times \frac{0.13}{2} = 0.0325F \text{ kNm}$$

$$Z_{\text{bolt}} = \frac{bd^2}{6} = \frac{20 \times 30^2}{6} = 1500 \text{ mm}^3$$

$$\text{Maximum stress} = 120 \text{ N/mm}^2 = \frac{0.0325F \times 10}{1500}$$

$$F = \frac{165 \times 1500}{32.5 \times 10} = 7.62 \text{ kN}$$

Maximum force/m width on pin

$$= \frac{5.54 \times 2}{1.2} = \textbf{12.70 kN}$$

$$\text{Maximum shear in bar} = 20 \times 30 \times 72 \times 10^3$$
$$= 43.2 \text{ kN}$$

Fig. 11.25

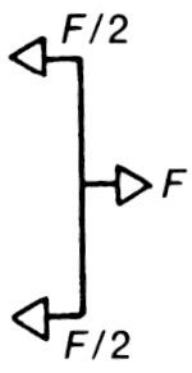

Force to cause pullout from concrete

$$= (1 \times 82 \times 90 \times 2) - (18 \times 30 \times 2) = 13\,680 \text{ kN}$$

$$\text{Force/m for pullout} = \frac{13 \cdot 68 \times 4}{1 \cdot 2} = 45 \cdot 6 \text{ kN}$$

It would be unrealistic to regard the $48 \cdot 3 \times 3 \cdot 2$ 50C CHS as anything other than simply supported. The simply supported length is 600 mm.

Let the maximum uniformly distributed load on the bar be U kN

$$\text{Maximum moment } \frac{U \times 0 \cdot 6}{8} = 0 \cdot 075 U \text{ kNm}$$

$$Z = 4 \cdot 8 \text{ cm}^3$$

$$\text{Maximum permissible stress} = 230 \text{ N/mm}^2$$

$$\text{Therefore, } 230 = \frac{0 \cdot 075 U \times 10^6}{4 \cdot 8 \times 10^3}$$

$$U = \frac{230 \times 4 \cdot 8}{75} = 14 \cdot 72 \text{ kN}$$

Maximum permissible force/m width

$$= \frac{14 \cdot 72}{0 \cdot 6} = \mathbf{24 \cdot 54 \text{ kN}}$$

See Table 11.16.

11.3.5 Prestressed beam

Consider the precast prestressed facing illustrated in Fig. 11.26.

$$\bar{y} = \frac{20\,634\,375}{135 \cdot 750} = 152 \text{ mm}$$

$$I = 335 \times 10^6 + 118 \cdot 7 \times 10^6 = 453 \cdot 7 \times 10^6$$

Prestress assuming 30% losses.

Strand $128 \cdot 8 \times 0 \cdot 7 = 90 \cdot 16$ kN; eccentricity $= 69 \cdot 5$

$$M = Pe = 2 \times 90 \cdot 16 \times 69 \cdot 5 = 12\,532 \text{ kNmm} = 12 \cdot 5 \times 10^6 \text{ Nmm}$$

Table 11.16. Second moment of area of facing beam

Section	A	yb	Ay	y'	Ay'^2	I_{self}
1190×75	89 250	187·5	16 734 375	35·5	112 500 000	41 800 000
$2 \times 150 \times 100$	30 000	75	2 250 000	77	177 900 000	56 300 000
$4 \times 150 \times 55/2$	16 500	100	1 650 000	52	44 600 000	20 600 000
	135 750		20 634 375		335×10^6	$118 \cdot 7 \times 10^6$

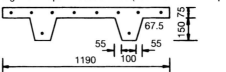

Eight 5 mm prestressed wires (1099 N/mm² initial prestress)

67.5

150.75

55 55

1190 100

Two 12.9 mm 7 wire strand (128.8 initial force per strand)

Fig. 11.26

$$\text{Stresses } \frac{P}{A} + \frac{My}{I} = \frac{2 \times 90 \cdot 16 \times 10^3}{135\,750} + \frac{12 \cdot 5 \times 10^6 \times 152}{453 \cdot 7 \times 10^6}$$

$$= 1 \cdot 32 + 4 \cdot 19 = 5 \cdot 51 \text{ N/mm}$$

$$\frac{P}{A} - \frac{My}{I} = 1 \cdot 32 - \frac{12 \cdot 5 \times 10^6 \times 73}{453 \cdot 7 \times 10^6} = 1 \cdot 32 - 2 \cdot 01 = 0 \cdot 69 \text{ N/mm}$$

Wire 1099 N/mm²; Force $= 1099 \times 8 \times \dfrac{\pi}{4} \times 5^2 \times 0 \cdot 7 = 120 \cdot 84$ kN

Eccentricity $= 33 \cdot 5$.

$$M = Pe = 120 \cdot 84 \times 35 \cdot 5 = 4289 \text{ kNmm} = 4 \cdot 3 \times 10^6 \text{ Nmm}$$

$$\text{Stresses} = \frac{120 \cdot 84 \times 10^3}{135\,750} - \frac{4 \cdot 3 \times 10^6 \times 152}{453 \cdot 7 \times 10^6} = 0 \cdot 89 - 1 \cdot 44 = -0 \cdot 55 \text{ N/mm}^2$$

$$= 0 \cdot 89 + \frac{43 \times 10^6 \times 73}{453 \cdot 7 \times 10^6} = 0 \cdot 89 + 0 \cdot 69 = 1 \cdot 58 \text{ N/mm}^2$$

Stresses due to prestress
Front of unit: $5 \cdot 51 - 0 \cdot 55 = 4 \cdot 96 \text{ N/mm}^2$
Back of unit: $-0 \cdot 69 + 1 \cdot 58 = 0 \cdot 89 \text{ N/mm}^2$

11.3.6 Construction forces

Assuming the facing to be propped at the top and the bottom, the construction forces may be determined as follows, Fig. 11.27.

Active force 3A–4 and 16–16A$= \frac{1}{2}ka\gamma h^2 = 0 \cdot 5 \times 0 \cdot 3 \times 20 \times 0 \cdot 3^2 = 0 \cdot 27$ kN

Active force units from 4 to 16 $= 0 \cdot 5 \times 0 \cdot 3 \times 20 \times 0 \cdot 45^2 = 0 \cdot 61$ kN

$$6 \cdot 7R_B = 0 \cdot 27(0 \cdot 7 + 6 \cdot 6)\left(\frac{1 \cdot 1 + 6 \cdot 05}{2}\right) \times 12 = 420 \text{ kN}$$

Total active force $= (12 \times 0 \cdot 61) + (2 \times 0 \cdot 27) = 7 \cdot 86$ kN

Therefore, $R_B = 7 \cdot 86 - 4 \cdot 20 = 3 \cdot 66$ kN

The maximum moment will occur at the Position of Zero shear which is approximately layer 10.

Moment at layer 10 $= (3 \cdot 5 \times 4 \cdot 2) - (0 \cdot 27 \times 2 \cdot 8) - (6 \times 0 \cdot 61 \times 1 \cdot 275)$
$= 9 \cdot 28 \text{ kNm}$
$(\times 1 \cdot 2^* = \mathbf{11 \cdot 14 \text{ kNm}})$

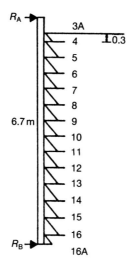

Fig. 11.27

Moment at layer 11 $= (3{\cdot}95 \times 4{\cdot}2) - (0{\cdot}27 \times 3{\cdot}25) - (7 \times 0{\cdot}61 \times 1{\cdot}50)$

$\qquad = 9{\cdot}31 \text{ kNm}$

$\qquad (\times 1{\cdot}2^{*} = \mathbf{11{\cdot}18\ kNm})$

Tension of front face due to moment of 11·18 kNm

$$= \frac{11{\cdot}18 \times 10 \times 152}{453{\cdot}7 \times 10} = 3{\cdot}72 \text{ N/mm}^2$$

Compression at rear face $= \dfrac{11{\cdot}18 \times 10 \times 73}{453{\cdot}7 \times 10} = 1{\cdot}80 \text{ N/mm}^2$

Therefore, final stresses are:

Front face, $4{\cdot}96 - 3{\cdot}72 = 1{\cdot}24 \text{ N/mm}^2$ compression.

Rear face, $0{\cdot}89 + 1{\cdot}80 = 2{\cdot}69 \text{ N/mm}^2$ compression.

* Unit 1·2 m wide.

It would be more realistic to consider the earth pressure at-rest

$\qquad K_0 = (1 - \sin \phi') = 1 - \sin 32° = 0{\cdot}47$

So that: Front face tension $= 5{\cdot}83 \text{ N/mm}^2$;

Rear face compression $= 2{\cdot}82 \text{ N/mm}^2$

Final stresses are: Front face, $4{\cdot}96 - 5{\cdot}83 = -0{\cdot}87 \text{ N/mm}^2$ tension.

Rear face, $0{\cdot}82 + 2{\cdot}82 = 3{\cdot}71 \text{ N/mm}^2$ compression.

The tension is allowable since the loading is short term and the K_0 loadings are overestimates of the moment at the centre of beam since the deflection of the beam at the centre will cause the pressure here to be reduced to active values.

Forces at removal of wedges

Assume that full active pressure is developed from layers 14–17 on removal of wedges, Fig. 11.28.

Fig. 11.28

Active force $= \frac{1}{2} \times 0 \cdot 3 \times 20 \times 1 \cdot 2^2 = 4 \cdot 32$ kN

Let force in reinforcement at layer $16 = F$

Then force in reinforcement at layer $15 = F \times \dfrac{5 \cdot 75}{6 \cdot 2} = 0 \cdot 93F$

$4 \cdot 32 \times 6 \cdot 1 = 0 \cdot 93F \times 5 \cdot 75 + 6 \cdot 2F$

Force at layer 16, $F = 2 \cdot 28$ kN; force at layer 15, $F = 2 \cdot 04$ kN

Frictional resistance force to reinforcement, layer $15 = \dfrac{L_i \mu P_i (h\gamma + w_s)}{2}$

$\quad = \dfrac{5 \times 0 \cdot 62 \times 2 \times 20 \times 0 \cdot 45}{2} = 27 \cdot 9$ kN

Frictional resistance force to reinforcement, at layer

$16 = \dfrac{5 \times 0 \cdot 62 \times 2 \times 20 \times 0 \cdot 9}{2}$ (assuming grid reinforcement)

$= \mathbf{55 \cdot 8 \ kN^2}$

Therefore, the anchorage is adequate.

11.4 Example 3

11.4.1 The problem

The design of a foundation mattress beneath an embankment. It is assumed that the embankment is 15 m high and is founded on a relatively thin layer of soft material underlain by a very stiff clay base, Fig. 11.29.

11.4.2 Assumed behaviour

The use of a rigid foundation mattress is assumed to alter the direction of the normal slip circle failure plane by forcing it to pass vertically through the mattress. This in turn deepens the failure surface and takes it into the stiff

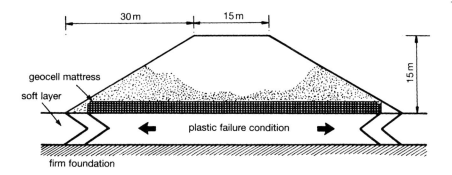

geocell mattress

soft layer

30 m

15 m

15 m

plastic failure condition

firm foundation

Fig. 11.29

underlying base. Before the base material shears, a plastic failure is assumed to occur in the soft layer beneath the embankment. The plastic condition is considered for design.

The foundation mattress is assumed to display the following properties:

(a) sufficient tensile strength to ensure that the full c_u value is mobilized on the base
(b) a rigid mattress to ensure an even distribution of load on to the foundation
(c) roughness at the base of the mattress.

11.4.3 Material properties

Density of the embankment $= 20 \text{ kN/m}^3$
c_u beneath embankment $= 50 \text{ kN/m}^2$
c_u base material $> 200 \text{ kN/m}^2$
ϕ' beneath embankment $= 30°$

11.4.4 Plastic design

The arrangement of the embankment is as in Fig. 11.29. The load distribution in the base may be obtained from the slip-line field diagram for the problem of

Fig. 11.30

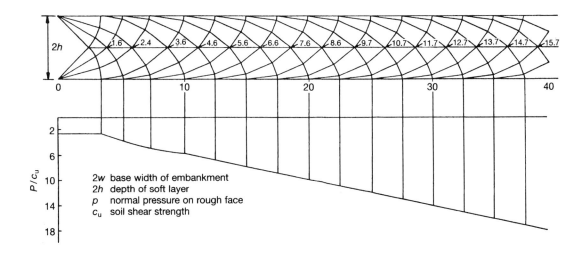

2w base width of embankment
2h depth of soft layer
p normal pressure on rough face
c_u soil shear strength

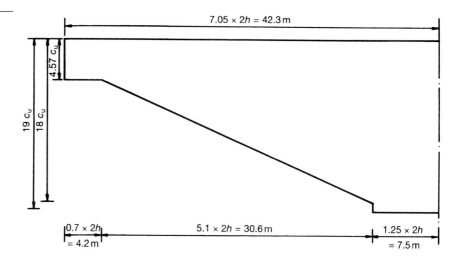

Fig. 11.31

compression between rough parallel platens for various ratios of $2w/2h$, Fig. 11.30, where $2w$ = width of the embankment; $2h$ = depth of soft (plastic) zone.

In Fig. 11.29, $2w/2h = 15$

Using Fig. 11.30 and allowing a value of $2c_u$ to equate to the effect of passive pressure beyond the toe of the embankment, and neglecting the effect of upthrust within $0.45 \times 2h$ of the toe, the pressure distribution on the base, in Fig. 11.31, is obtained.

From Fig. 11.31, the load to failure for half embankment ($c_u = 50 \text{ kN/m}^2$):

$$
\begin{aligned}
4.57 \times 50 \times 4.2 &= & 959.7 \\
0.5(4.57 + 18) \times 50 \times 30.6 &= & 17\,289.0 \\
19 \times 50 \times 7.5 &= & 7125.0 \\
\hline
& & 25\,373.7 \text{ kN}
\end{aligned}
$$

Load from half of embankment:

$$0.5(7.5 + 37.5) \times 15 \times 20 = \mathbf{6750 \text{ kN}}$$

Therefore, the factor of safety against foundation failure

$$= \frac{25\,373.7}{6750} = \mathbf{3.76}$$

11.4.5 Centre of the embankment

For a factor of safety reduced to 1.0, the value of c_u for the foundation would be

$$\frac{50}{3.76} = 13.29 \text{ kN/m}^2$$

From a Mohr Circle construction it can be shown that the horizontal force, T_a to be resisted by the foundation mattress is given by

$$T_a = \frac{c_u}{\sin \phi'} = \frac{13.29}{\sin 30°} = \mathbf{26.6 \text{ kN/m run}}$$

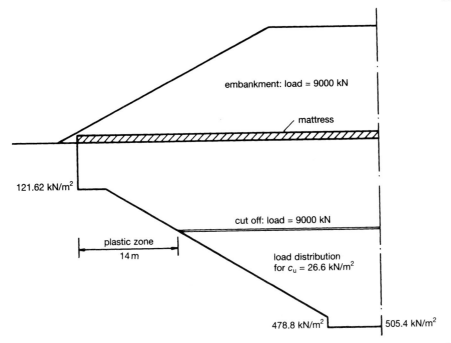

embankment: load = 9000 kN

mattress

121.62 kN/m²

cut off: load = 9000 kN

plastic zone
14 m

load distribution
for c_u = 26.6 kN/m²

478.8 kN/m²   505.4 kN/m²

Fig. 11.32

This is a minimum condition which can be assumed to apply *over the centre of the embankment*.

11.4.6 Edge of embankment

Assume a worst condition where factor of safety drops to **2·0**,

i.e. c_u beneath foundation = 25 kN/m²

The embankment load/foundation support pressure diagram is shown in Fig. 11.32. A cut-off line at foundation support value −9000 kN shows the extent of the plastic zone. Over the plastic zone, extending 14 m from the edge of the mattress the horizontal force T_b to be resisted by the mattress is:

$$T_b = \frac{25 \cdot 0}{\sin 30°} = \textbf{50 kN/m run}$$

11.4.7 Foundation mattress

Assuming the mattress formed from Tensar SR–2 geogrid.

Working load in grid = 23·7 kN/m width (30% of ultimate load)

Then, resistance of 1 m cell

$$= \frac{23·7}{\sqrt{2}} + \textbf{23·7} = \textbf{40·5 kN/m run}$$

and resistance of 0·5 m cell

363

$$= 2\left(\frac{23 \cdot 7}{\sqrt{2}} + 23 \cdot 7\right) = \mathbf{80 \cdot 9 \ kN/m \ run}$$

Therefore, use 0·5 metre cells within 14 m of the edge of the foundation mattress 1·0 metre cells under centre of the embankment, Fig. 11.33.

11.5 Example 4

11.5.1 The problem

The design of a tension membrane supporting an embankment over an area prone to subsidence.

The embankment will be 5 m high and will consist of fill with a bulk unit weight of $\gamma = 18 \ kN/m^3$ and an angle of internal friction $\phi' = 30°$ and $c' = 0$. A surcharge of 10 kN/m^2 is to be included. A subsurface survey has determined that any cavity that occurs beneath the embankment will have a diameter of 3·5 m, and be approximately circular in shape. The design life of the structure will be 120 years and the highway is a principal road. A diagram showing details of the embankment is given in Fig. 11.34. It is assumed that the soil temperature and pH are normal for the UK and will not affect the design.

Design concept
A high strength geosynthetic reinforcing layer will be placed beneath the embankment before fill placement. The reinforcement function is to limit

Fig. 11.33

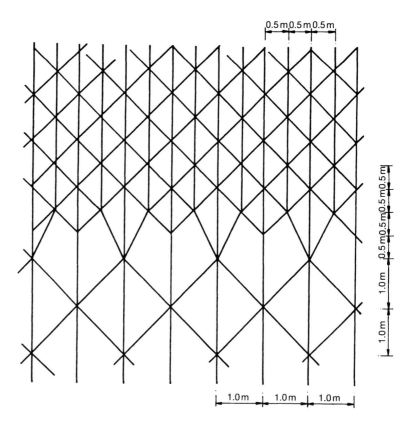

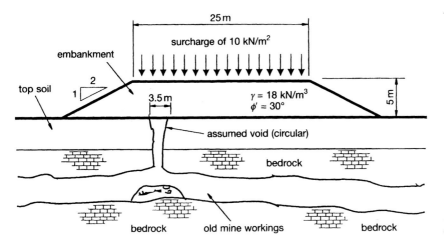

25 m

surcharge of 10 kN/m²

embankment

2
1

top soil

3.5 m

$\gamma = 18$ kN/m³
$\phi' = 30°$

5 m

assumed void (circular)

bedrock

bedrock   old mine workings   bedrock

Fig. 11.34. Details of embankment

surface deformations at the top of the embankment should a void form beneath the embankment at some time during the design life of the structure.

Design method
The analytical procedure assumes the method detailed in Chapter 4, and is compatible with BS 8006: 1995 Code of Practice for strengthened/reinforced soils and other fills (final draft). The following procedure is adopted:

(*a*) determination of the maximum allowable reinforcement strain
(*b*) determination of the reinforcement tensile properties of the reinforcement
(*c*) application of the partial material factors for the reinforcement and the partial factor for the ramifications of failure
(*d*) determination of the creep–rupture characteristics of the reinforcement
(*e*) determination of the initial strain and the creep strain of the reinforcement, and checking that the reinforcement provided meets the strain requirements
(*f*) determination of the bond length at the edges of the embankment and lap length at the joints under the embankment
(*g*) conclusions.

11.5.2 Determination of the maximum allowable reinforcement strain

The deflected shape of the reinforcement spanning the void may be approximated to a parabola where the maximum allowable reinforcement strain for circular voids is

$$\epsilon_{max} = \frac{8\left(\dfrac{d_s}{D_s}\right)^2\left(D + \dfrac{2H}{\tan\theta_d}\right)^6}{3D^6}$$

where ϵ_{max} is the maximum allowable strain in the reinforcement
d_s/D_s is the maximum allowable differential deformation that occurs at the surface of the embankment or pavement
D is the design diameter of the void

H is the height of the embankment

θ_d is the angle of draw of the embankment fill, which is approximately equal to its peak friction.

The maximum surface deformation (d_s/D_s) for a principal road is limited to 1%. (Minor roads are limited to a surface deformation (d_s/D_s) of 2%.)

Thus

$$\epsilon_{max} = \frac{8 \times (0 \cdot 01)^{2\times} \left(3 \cdot 5 + \dfrac{2 \times 5}{\tan 30°} \right)^6}{3 \times (3 \cdot 5)^6}$$

$\epsilon_{max} = 11 \cdot 82 = 1182\%$

$\epsilon_{max} = 1182\%$ is the maximum allowable strain in the reinforcement. The design strain should be less than or equal to ϵ_{max}. The initial design strain shall be taken as $5 \cdot 5\%$.

11.5.3 Determination of the reinforcement tensile properties of the reinforcement

The tensile properties of the deflected reinforcement are calculated from:

$$T_{rs} = 0 \cdot 5\lambda (\gamma_{es}\gamma H + \gamma_q w_s)D\sqrt{\left(1 + \frac{1}{6\epsilon}\right)}$$

where T_{rs} is the tensile load in the reinforcement per metre 'run'

  λ is a coefficient dependent on whether the reinforcement support is to function as a one-way or two-way load shedding system.

(For circular or rectangular voids (spanning two ways) $\lambda = 0 \cdot 67$, while for longitudinal voids (spanning one way) $\lambda = 1 \cdot 0$. In this case, for a circular void $\lambda = 0 \cdot 67$.)

  w_s is the surcharge intensity on top of the embankment

  ϵ is the initial strain in the reinforcement which must be less than or equal to ϵ_{max}

  γ_{es} is the partial load factor for soil unit weight, $(\gamma_{es} = 1 \cdot 3)$, Table 6.3

  γ_q is the partial load factor for external live loads, $(\gamma_q = 1 \cdot 3)$, Table 6.3.

Other values used in calculating T_{rs} are taken from Fig. 11.34.

Therefore,

$$T_{rs} = 0 \cdot 5 \times 0.67 \times (1 \cdot 3 \times 18 \times 5 + 1 \cdot 3 \times 10) \times 3 \cdot 5 \times \sqrt{\left(1 + \frac{1}{6 \times 0 \cdot 055}\right)}$$

$T_{rs} = \mathbf{306 \cdot 6 \ kN/m}$

11.5.4 Application of the partial material factors of safety for the reinforcement and the partial factor for ramifications of failure

Partial material factors are applicable to the reinforcement to take account of the properties of the material itself and the effects of construction and the environment in which it is placed. The partial material factor for reinforcement is made up of a number of components as outlined below.

Partial material factors for reinforcement:

$$\gamma_m = \gamma_{m1} \times \gamma_{m2} \qquad \text{BS 8006 (1995)}$$

where γ_m is the partial material factor

γ_{m1} is a partial material factor related to the intrinsic properties of the material

γ_{m2} is a partial material factor concerned with construction and environmental effects.

$$\gamma_{m1} = \gamma_{m11} \times \gamma_{m12} \qquad \text{BS 8006 (1995)}$$

where γ_{m11} is a partial material factor related to the consistency of manufacture of the reinforcement, and how strength may be affected by this and possible inaccuracy in assessment

γ_{m12} is a partial material factor related to the extrapolation of test data dealing with base strength.

$$\gamma_{m2} = \gamma_{m21} \times \gamma_{m22} \qquad \text{BS 8006 (1995)}$$

where γ_{m21} is a partial material factor related to the susceptibility of the reinforcement to damage during installation in the soil

γ_{m22} is a partial material factor related to the environment in which the reinforcement is installed.

Reinforcement selected for this case is ParaLink, high strength polyester tape. Partial features for Paraweb reinforcement detailed in British Board of Agrément (BBA) Certificate (1995) are:

$\gamma_{m1} = 1\cdot15$, for a 120 year design life   Table 5, BBA (1995)
$\gamma_{m21} = 1\cdot05$, for large particles up to 125 mm   Table 6, BBA (1995)
$\gamma_{m22} = 1\cdot03$, for normal pH values, pH2 $-9\cdot5$ Table 7, BBA (1995)
Therefore, $\gamma_{m2} = 1\cdot05^*1\cdot03$
$\qquad\qquad \gamma_{m2} = 1\cdot08$
Therefore, $\gamma_m = 1\cdot15^*1\cdot08$
$\qquad\qquad \gamma_m = 1\cdot24$

For embankments where failure would result in moderate damage or loss of service, the partial factor of safety (γ_n) against the ramifications of failure is taken to be (Table 6.1)

$$\gamma_n = 1\cdot0$$

The ultimate strength of reinforcement required is given by:

$$T_u = T_{rs} \times \gamma_m \times \gamma_n$$

where T_u is the ultimate tensile strength of the reinforcement
$\qquad T_u = 306\cdot6 \times 1\cdot24 \times 1\cdot0$
$\qquad \mathbf{T_u = 380\cdot2\ kN/m}$

11.5.5 Determination of the creep–rupture characteristics of the reinforcement

The reinforcement should not fail in tension over the design life of the structure. The base strength should be taken as:

$$T_{CR} = \frac{T_u}{CR}$$

where T_{CR} is the peak tensile creep rupture strength at the appropriate temperature

CR is the reduction in ultimate strength for creep rupture.

From the creep–rupture curve for ParaLink with a design life of 120 years, the reduction in ultimate strength for creep rupture is 65%, BBA (1995).

$$T_{CR} = \frac{380 \cdot 2}{0 \cdot 65}$$

$$T_{CR} = \textbf{583·75 kN/m}$$

ParaLink 600 S with an ultimate breaking load of 600 kN/m will meet the requirements for the ultimate tensile strength.

11.5.6 Determination of the initial strain in the reinforcement and the creep strain over the design life of the reinforcement

At the end of the design life of the structure, the strain in the reinforcement should not exceed a prescribed value.

Check the assumed design strain in the reinforcement:
From the short-term stress–strain curve for ParaLink and with a design strain of 5·5%, the reinforcement would carry 50% of the nominal breaking load.
Using ParaLink 600 S with a nominal breaking load of 600 kN/m:

$$T_{rs} = 0 \cdot 5 \times 600$$

$$T_{rs} = 300 \text{ kN/m}$$

This value is close to the initial calculated tension load in the reinforcement, $T_r = 306$ kN/m, calculated in 11.5.3. The initial design strain of 5·5% is **OK**.

Check the initial strain and creep strain:
From the isochronous creep curves for ParaLink, at 50% of the nominal breaking load:

initial strain $\qquad\qquad\qquad\qquad\qquad\qquad \epsilon_i = 5 \cdot 25\%$
strain after 120 years' design life $\qquad\qquad \epsilon_{120} = 6 \cdot 2\%$
The resultant creep strain would be $6 \cdot 2\% - 5 \cdot 25\% = 0 \cdot 95\%$

The initial strain in the reinforcement is lower than the maximum allowable strain $\epsilon_{max} = 1182\%$ calculated in 11.5.2. The creep strain over the design life of the structure is less than the 2% limit given in BS 8006 (1995).

The requirements for ultimate limit state can be met by ParaLink 600 S with a nominal breaking load of 600 kN/m, Fig. 11.35.

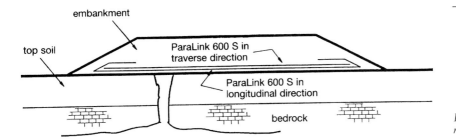

Fig. 11.35. Reinforcement requirements

11.5.7 Determination of the bond length at the edges of the embankment and lap length at joints under the embankment

To generate the tensile load T_{rs} in the reinforcement, adequate bond must exist between the reinforcement and the adjacent soil.

Bond length at edges of embankment

The required bond length of the reinforcement is to be a minimum length as set out below:

$$L_b \geq \gamma b \frac{\gamma_n \gamma_p T_{rs}}{\left(\dfrac{\alpha_1 \tan \phi'_{CV1}}{\gamma_{ms}} + \dfrac{\alpha'_2 \tan \phi'_{CV2}}{\gamma_{ms}} \right)}$$

Where L_b is the required reinforcement bond

γ_n is the partial factor governing the economic ramifications of failure, Table 6.1

γ_p is the partial factor applied to the pullout resistance of the reinforcement, Table 6.3.

T_{rs} is the tensile load in the reinforcement per metre 'run'

b is the average height of the embankment fill above the reinforcement length L_b

γ is the unit weight of the embankment fill

α'_1 is the interaction coefficient relating to soil/reinforcement bond angle to $\tan \phi_{CV1}$ on one side of the reinforcement

α'_2 is the interaction coefficient relating the soil/reinforcement bond angle to $\tan \phi_{CV2}$ on the other side of the reinforcement

γ_{ms} is the partial material factor applied to $\tan \phi'_{CV}$, Table 6.3.

The partial factor of safety against the ramifications of failure is taken to be:

$\gamma_n = 1\cdot 0$ Table 6.1

Soil/reinforcement interaction factors, pullout resistance of reinforcement:

$\gamma_p = 1\cdot 3$ Table 6.2

Soil material factors, to be applied to $\tan \phi'_{CV1}$:

$\gamma_{ms} = 1\cdot 0$ Table 6.3

From pullout tests:

$\alpha'_1 = 0\cdot 8$

$\alpha'_2 = 0\cdot 8$

369

Bond length in the longitudinal direction

For calculation of h, it will first be assumed that L_b is 5 m; this will give $h = 5$ m.

$$L_b = \frac{1 \cdot 0 \times 1 \cdot 3 \times 306 \cdot 6}{18 \times 5 \times \left(\dfrac{0 \cdot 8 \times \tan 30°}{1 \cdot 0} + \dfrac{0 \cdot 8 \times \tan 30°}{1 \cdot 0} \right)}$$

$L_b = 4 \cdot 79$ m

This is close to the assumed bond length of 5 m and there is no need to carry out further checks. **Provide a bond length of 5 m in the longitudinal direction.**

Bond length in the traverse direction

For calculation of h, it will first be assumed that L_b is 10·0 m; this will give $h = 2 \cdot 5$ m.

$$L_b = \frac{1 \cdot 0 \times 1 \cdot 3 \times 306 \cdot 6}{18 \times 2 \cdot 5 \times \left(\dfrac{0 \cdot 8 \times \tan 30°}{1 \cdot 0} + \dfrac{0 \cdot 8 \times \tan 30°}{1 \cdot 0} \right)}$$

$L_b = 9 \cdot 59$ m

This is close to the assumed bond length of 10 m and there is no need to carry out further checks. **Provide a bond length of 10 m in the transverse direction.**

Bond length at a joint under the embankment

All joints formed in the reinforcement are lapped joints. When a joint is being constructed, the full tensile load in the reinforcement has to be generated across the joint.

For calculation of joint length, h will be equal to the full height of the embankment $h = 5$ m.

$$L_b = \frac{1 \cdot 0 \times 1 \cdot 3 \times 306 \cdot 6}{18 \times 5 \times \left(\dfrac{0 \cdot 8 \times \tan 30°}{1 \cdot 0} + \dfrac{0 \cdot 8 \times \tan 30°}{1 \cdot 0} \right)}$$

$L_b = 4 \cdot 8$ m

Provide a joint length of 4·8 m in both the longitudinal and transverse directions, Fig. 11.36.

11.5.8 Conclusions

The final design solution involves using two layers of ParaLink 600 S, one layer being placed in the longitudinal direction and the other in the traverse direction. In the longitudinal direction the ParaLink reinforcement will be placed with the main elements parallel to the centre-line of the embankment and covering the full width of the embankment. A 5 m bond length will be provided at each end. In the traverse direction the ParaLink reinforcement will be placed with the main elements perpendicular to the centre-line of the embankment. A 10 m bond length will be provided at each end. All joints in

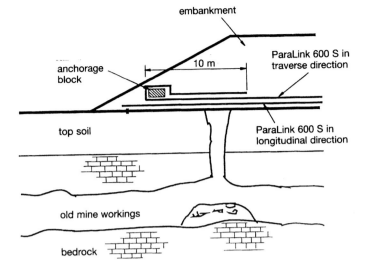

Fig. 11.36. Construction details

the reinforcement shall have a lap length of 5 m in both directions and shall be constructed under the main embankment only, Fig. 11.37. Construction details are shown in Fig. 11.36.

11.6 Example 5

11.6.1 The problem

Checking the design of a bridge abutment for seismic conditions.

Assume: The bridge abutment has been designed in accordance with established procedures. Details of the design problems and the arrangement of the reinforcement are shown in Fig. 11.38. (*Note:* the height of the abutment is 5·1 m, the design height is increased to accommodate the uniform surcharge loading of 85 kN/m^2. Thus, design height of wall $H = 5·1 + (25/20) = 6·35$ m).

From the bridge abutment analysis, the following design criteria have been established:

(*a*) Resultant active force from the retained fill P_L = 113 kN/m
(*b*) Resultant active force due to surcharge P_q = 63 kN/m

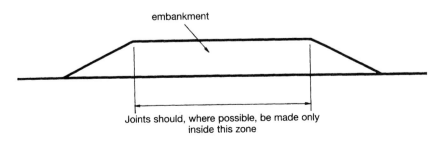

Fig. 11.37. Position of joints in reinforcement

embankment

Joints should, where possible, be made only inside this zone

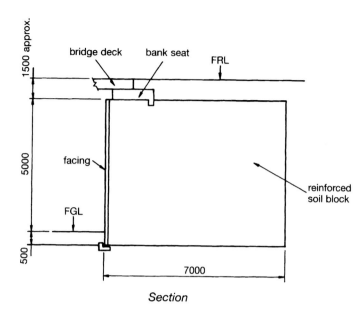

Fig. 11.38

Section

(*c*) Height of structure (equivalent) H = 6·35 m

(*d*) Peak ground acceleration coefficient α_m = 0·416

(*e*) Weight of active wedge of retain soil W = 210 kN/m
     (Where $\phi' = 35°$, $c' = 0$)

(*f*) The angle of the active wedge from the toe of the wall
     upwards, $\beta = (45° - \phi'/2)$ = 17·5°

(*g*) During an earthquake, no horizontal loading due to
     braking or temperature occurs

(*h*) Bank seat and approach loading are included in the
     seismic condition

(*i*) Bank seat dead loading, BL   = 85 kN/m

(*j*) Friction coefficient μ between abutment and
     function N = 0·65

(*k*) Reinforcement, Tensar SR 110

(*l*) Characteristic strength of reinforcement   = 63 kN/m

(*m*) Partial factor of safety for reinforcement γ_m = 1·2

(*n*) Overall factor of safety for design   = 1·35

(*o*) Safety design strength of reinforcement   = 38·89 kN/m

11.6.2 Methodology

The methodology used is based on the procedures adopted in the US Federal
Highway Administration Design Manual for Reinforced Soil Structures (US
Department of Transport, 1989).

11.6.3 Seismic loads

Dynamic horizontal force, $P_{AE} = 0.375\,\alpha_m\gamma H^2 = 0.375 \times 0.416 \times 20 \times 6.35^2 = 126$ kN/m (equation (87), Ch. 6)

Horizontal inertial force, $P_{IR} = 0.5\,\alpha_m\gamma H^2 \tan\beta = 0.5 \times 0.416 \times 20 \times 6.35^2 \times \tan(45° - \phi'/2) = 53$ kN/m (equation (88), Ch. 6)

Total horizontal force resulting from seismic conditions

$$= P_b + P_q + P_{AE} + 0.6P_{IR} \text{ (equation (89), Ch. 6)}$$
$$= 113 + 63 + 126 + (0.6 \times 53) = \mathbf{334\ kN/m}$$

Check overturning

Factor of safety against overturning $= \dfrac{\Sigma \text{ Restoring moments}}{\Sigma \text{ Horizontal moments}}$

$$= \frac{w \times H \times L \times L/2 + (BL \times 1)}{P_q \times \dfrac{H}{2} + \dfrac{H}{3} + P_b \times \dfrac{H}{3} + 0.6 \times P_{AE} \times 6 + 0.6 \times P_{IR} \times 3}$$

$$= \frac{20 \times 6.35 \times 6 \times 3 + (85 \times 1)}{63 \times \dfrac{6.35}{2} + 113 \times \dfrac{6.35}{2} + 0.6 \times 6 \times 126 + 0.6 \times 3 \times 53}$$

$$= \frac{2371}{988} = 2.4$$

Therefore, dynamic factor of safety $= 0.75 \times 2.4 = \mathbf{1.8\ OK}$

Check sliding

Factor of safety against sliding $= \dfrac{\Sigma \text{ Restoring forces}}{\Sigma \text{ Overturning forces}}$

$$= \frac{(W \times H \times L + \text{Surcharge}) \times F + BL}{P_q + P_b + P_{AE} + 0.6\,P_{IR}}$$

$$= \frac{(6.35 \times 6 \times 20 + 6 \times 20) \times 0.65 + 85}{334}$$

$$= 2.0$$

Therefore, dynamic factor of safety $= 0.75 \times 2.0 = \mathbf{1.5\ OK}$

Internal stability

Horizontal inertial force, P_{IR} $= 53$ kN/m
Design weight of active wedge $= 210$ kN/m
Seismic loading conditions $= 53 + 210 = 263$ kN/m

Accept an increase in characteristic strength of reinforcement of 1.5 (Section 6.5.1.1)

Available force, assuming the initial calculation to be stable
$= 1.5 \times 210 > 315$ kN/m

Therefore, factor of safety against rupture $= \dfrac{315}{263} = > \mathbf{1.2\ OK}$

(Pullout resistance of geogrids in structures > 3 m in height can be deemed adequate.)

11.7 References

BRITISH BOARD OF AGRÉMENT (1995). *Exxon Paralink Geocomposite Products* (Draft), 13.

BRITISH STANDARDS INSTITUTION (1990). *Steel, concrete and composite bridges. Part 2. Highway loading.* BSI, London, BS 5400.

BRITISH STANDARDS INSTITUTION (1995). *Code of Practice for strengthened/reinforced soils and other fills.* BSI, London, BS 8006.

DEPARTMENT OF TRANSPORT (1978). *Reinforced earth retaining walls and bridge abutments for embankments.* Technical Memo. (Bridges) BE 3/78.

JONES C. J. F. P. and EDWARDS L. W. (1980). Reinforced earth structures on soft foundations. *Géotechnique*, **30**, No. 2, June, 207–211.

US DEPARTMENT OF TRANSPORT (1989). *Reinforced Soil Structures. Vol. 1, Design and Construction*, Report No. FHWA-RD-89-043, 287.

Index